The release of genetically engineered microorganisms (GEMs) into the environment is possibly the single most controversial aspect of applied biology. Areas of research directly relevant to the use of GEMs in the environment are fully discussed and some aspects of the use of specially selected un-engineered bacteria and fungi are covered. A series of chapters then deals with points important to the release of all types of bacteria, especially aspects of gene transfer and survival. The final chapters discuss specific case-histories of some of the known releases of GEMs. The book concludes with the likely concerns and reactions of legislature of various countries to the issues raised by this branch of biotechnology.

By setting the consequences of the release of GEMs into the historical context of the release of unaltered organisms, the international contributors and editors of this title have also created a state-of-the-art account of work in the forefront of their field.

PLANT AND MICROBIAL BIOTECHNOLOGY RESEARCH SERIES: 2
Series Editor: James Lynch

Release of genetically
engineered and other micro-organisms

PLANT AND MICROBIAL BIOTECHNOLOGY RESEARCH SERIES
Series Editor: James Lynch

Release of genetically engineered and other micro-organisms

edited by

J. C. FRY AND M. J. DAY
University of Wales College of Cardiff

CAMBRIDGE
UNIVERSITY PRESS

Published by the Press Syndicate of the University of Cambridge
The Pitt Building, Trumpington Street, Cambridge CB2 1RP
40 West 20th Street, New York, NY 10011-4211, USA
10 Stamford Road, Oakleigh, Victoria 3166, Australia

© Cambridge University Press 1992

First published 1992

Printed and bound in Great Britain by
Butler & Tanner Ltd, Frome and London

A catalogue record for this book is available from the British Library

Library of Congress cataloguing in publication data

Release of genetically engineered and other
micro-organisms / edited by J.C. Fry and M.J. Day.
p. cm. – (Plant and microbial biotechnology research series)
Includes index.
ISBN 0 521 41756 2 (hardback)
1. Recombinant microorganisms – Enviromental aspects.
2. Recombinant microorganisms – Ecology.   I. Fry, John C.   II. Day, Martin J.   III. Series
QR100.E57 1992
576′.15 – dc20 92–5899 CIP

ISBN 0 521 41756 2 hardback

We dedicate this book to the memory of Dr Steve Cuskey,
who died in February 1990.

Contents

Contributors

Noëlle Amarger. *Laboratoire de Microbiologie des Sols, Institut National de la Recherche Agronomique, B.V. 1540, 21034 Dijon-cedex, France*

Mark J. Bale. *Department of Botany, University of Bristol, Woodlands Road, Bristol BS8 1UG, UK*

Gerard F. Barry. *Monsanto Co., St. Louis, Missouri 63198, USA*

Gilles Billen. *Groupe de Microbiologie des Milieux Aquatiques, Campus Plaine, CP 221, Boulevard du Triomphe, B-1050 Bruxelles, Belgium*

D. H. L. Bishop. *Institute of Virology and Environmental Microbiology, Mansfied Road, Oxford OX1 3SR, UK*

Eva J. Brandt. *Monsanto Co., St. Louis, Missouri 63198, USA*

Gérard Catroux. *Laboratoire de Microbiologie des Sols, Institute National de la Recherche Agronomique, B.V. 1540, 21034 Dijon-cedex, France*

J. S. Cory. *Institute of Virology and Environmental Microbiology, Mansfield Road, Oxford OX1 3SR, UK*

Stephen M. Cuskey. *The United States Environmental Protection Agency, Environmental Research Laboratory, Sabine Island, Gulf Breeze, Florida 32561, USA*

Martin J. Day. *School of Pure and Applied Biology, University of Wales College of Cardiff, PO Box 915, Cardiff CF1 3TL, UK*

David J. Drahos. *Director Research and Development, BP Technologies Inc., 2155 West Park Court, Suite D, Stone Mountain, Georgia 30087, USA*

John C. Fry. *School of Pure and Applied Biology, University of Wales College of Cardiff, PO Box 915, Cardiff CF1 3TL, UK*

D. T. Gooden. *Clemson University, Clemson, South Carolina 29634, USA*

Bruce C. Hemming. *Monsanto Co., St. Louis, Missouri 63198, USA*

Mike Hinton. *Department of Veterinary Medicine, University of Bristol, Langford House, Langford, Bristol, BS18 7DU, UK*

Daniel A. Kluepfel. *Clemson University, Clemson, South Carolina 29634, USA*

Tyler A. Kokjohn. *Program for Molecular Biology and Department of Biochemistry and Biophysics, Loyola University of Chicago, Maywood, Illinois 60153, USA*

Nicholas C. McClure. *School of Pure and Applied Biology, University of Wales College of Cardiff, PO Box 915, Cardiff CF1 3TL, UK*

Martin P. Meadows. *Microbiology and Crop Protection Department, Horticulture Research International, Worthing Road, Littlehampton BN17 6LP, UK*

Robert V. Miller. *Program for Molecular Biology and Department of Biochemistry and Biophysics, Loyola University of Chicago, Maywood, Illinois 60153, USA*

R. D. Possee. *Institute of Virology and Environmental Microbiology, Mansfield Road, Oxford OX1 3SR, UK*

Pierre Servais. *Groupe de Microbiologie des Milieux Aquatiques, Campus Plaine, CP 221, Boulevard du Triomphe, B-1050 Bruxelles, Belgium*

Horace D. Skipper. *Clemson University, Clemson, South Carolina 29634, USA*

Gregory J. Stewart. *Department of Biology, University of South Florida, Tampa, Florida 33620, USA*

José Vives-Rego. *Departmento de Microbiologia, Universitat de Barcelona, Av. Diagonal, 645, Barcelona, Spain*

Andrew J. Weightman. *School of Pure and Applied Biology, University of Wales College of Cardiff, PO Box 915, Cardiff CF1 3TL, UK*

Foreword

THE PRIMARY CONCEPT of this Series of books is to produce volumes covering the integration of plant and microbial sciences in modern biotechnological science. Illustrations abound, for example the development of plant molecular biology has been heavily dependent on the use of microbial vectors and the growth of plant cells in culture has largely drawn on microbial fermentation technology. In both of these cases the understanding of microbial processes is now benefiting from the enormous investments made in plant biotechnology. It is interesting to note that many educational institutions are also beginning to see things this way and integrating departments previously separated by artificial boundaries.

Having set the scope of the Series, the next objective was to produce books on subjects which had not already been covered in the existing literature and, it was hoped, to set some new trends.

There is a widespread and general interest in the opportunities presented to us by our relatively new ability to manipulate organisms genetically, which has been developed to a point where its exploitation is now feasible; the next step in this process is the release of the recombinant organism. The driving force behind this book arose from the desire to analyse the aspects of the process of risk assessment which are, and will be, associated with the release of a novel organism. There have been many releases of novel non-engineered organisms, some of which, to our knowledge (e.g. rabbits into Australia) have given cause for concern. An examination of these examples provides a warning and a basis upon which to design a risk assessment strategy.

Because of the limitations in our knowledge of how organisms, both microscopic and macroscopic, interrelate in natural situations, the effects of release are likely to be difficult to predict. This uncertainty shows clearly the need for more basic science to provide the platform from which to make sensible decisions. Thus, the chapters in this book identify and examine some of these situations and indicate areas of weakness, worthy of further investigation.

Martin Day and John Fry have been very active in establishing important concepts in conjugal gene transfer in aquatic habitats, but now, in bringing this volume together, they have been able to consider the benefits and potential problems of release of both engineered and non-engineered microorganisms into a wide range of habitats. I hope that policy makers and legislators along with students and researchers will read this volume.

13.12.91   *Jim Lynch*

Preface

AN UNDERSTANDING of evolutionary processes, population biology, community interactions and their genetic processes of each are prerequisites for the scientific assessment of the effect of the introduction of a novel organism or species into an environment. Despite this cautionary approach Man has been introducing non-indigenous species into different parts of the world for centuries, usually without much appreciation of the environmental consequences.

New technologies, from fire to pesticides, have always brought environmental consequences. The promises and tangible benefits of all technologies are always to be balanced by the threat they impose. The acceptance and recognition of this threat comes from an analysis of the technology and the establishment of its limitations. Provided that the balance between the benefits and the risks is acceptable then the technology is adopted.

Chapters 1–5 discuss survival, gene exchange mechanisms and the design and incorporation of safety features into gene sequences. The discussion focuses on their ecological relevance, as it is upon these studies that genetically engineered microorganisms (GEMs) will be assessed and decisions made to release or limit their environmental use. Subsequent Chapters, 6–11, each deal with an analysis of field or contained experiments performed with natural and genetically modified organisms. The final chapter summarizes information within the areas of release and assessment and indicates which areas of research need underpinning.

As is expected, the major problems in the assessment process revolve around fundamental areas of research, i.e. they are not directly applied research. It is very difficult to predict the consequences of a release from an untried technology into an environment that is understood poorly at most levels of its organization. Neither is it conceivable for science to examine every possible hazard (real and conjectured). Therefore the assessment decision must be made from a sound scientific base in a way that recognizes the concerns of all parties. As funding bodies, in many countries, have directed funds at applied and biotechnologically related research in recent years, fundamental scientific investigations in many areas of microbial ecology have suffered.

Hopefully, both the contents and the comments in this book will indicate the urgency for the need for such basic information and stimulate others to enter this arena, promote the 'green' theme and subsequently provide the scientific basis for this justifiable use and widespread acceptance of the genetic engineering technology

March 1991

Martin J. Day
John C. Fry

1

Rhizobia as soil inoculants in agriculture

G. Catroux and N. Amarger

Introduction

LEGUME INOCULATION is a common agricultural practice used to supply *Rhizobium* or *Bradyrhizobium*, the microsymbionts of the legume–rhizobia symbiosis, to legume seeds or the soil at the time of sowing.

Inoculation has been used worldwide for almost a century, with the purpose of improving the productivity of leguminous crops through the increase of nitrogen (N_2) fixation. Different crops are used, including forages such as alfalfa and clovers and grain legumes such as beans, pea and soybean. Millions of hectares are inoculated each year. For example, soybean is estimated to be inoculated on 12 million hectares out of the 42 million hectares grown in the major producing countries (Y. Bonhomme, personal communication). This means that, over many years, this agricultural practice has resulted in an enormous deliberate release of bacteria into the soil.

Rhizobial inoculants are used for two purposes:

1. To establish a good legume–rhizobia symbiosis in soils devoid of rhizobia or containing only small rhizobial populations due to adverse conditions. This strategy is relevant to situations where absence or a low number of rhizobia is the limiting factor for establishing symbiosis. These situations are often encountered when a new, non-indigenous leguminous crop is introduced to an area, for example, soybean in western Europe. It is also the situation in most acidic soils, when *Rhizobium meliloti* is used (Obaton 1971), and in neutral and basic soils, when *Bradyrhizobium sp. (lupinus)* is used (Amarger 1980).

2. To improve, or to try to improve, a symbiosis with the use of selected or improved strains in soils containing established or indigenous populations of rhizobia. The assumption that these soil populations give suboptimal symbiosis supports this strategy.

Thus, legume inoculation occurs in a very large range of climatic, soil and agricultural situations around the world.

Rhizobial inoculants are produced using a variety of technologies, from simple to sophisticated processes. As a consequence, inoculants can be pure cultures or nearly equal mixtures of rhizobia and unidentified bacterial or fungal contaminants.

A wide variety of rhizobia, from different genera and species and from strains of different geographical origins, is used for legume inoculation. Legume inoculation therefore represents a unique large-scale history of microbial release into soils, and a release not only of identified strains of rhizobia, but also of many other unknown microorganisms.

We will review some of the problems encountered with the process of legume inoculation. These problems, and those related to the colonization and survival of inoculated rhizobia, will be discussed in connection with questions about the assessment of risk of bacteria released in agricultural soils.

1

Inoculants and methods of inoculation

Inoculants

Rhizobial inoculants are cultures of rhizobia, packaged in a convenient way, that can be used easily by farmers. They can be manufactured in powder, granular or liquid forms (for reviews see Date and Roughley 1977; Smith 1987; Roughley 1988).

Inoculants are generally produced in several steps:

1. Production of pure cultures of rhizobia in flasks or fermentors.
2. Addition of small amounts of the culture to a carrier. The carrier can be unsterilized, pasteurized or fully sterilized and may be supplemented with nutrients.
3. Growth of the bacteria, followed or preceded by packaging and storage.

The strain (or strains) used must be adapted to local conditions – soil, climate and plant genotype – because rhizobia are known to vary in infectivity and effectiveness and successful nodulation and nitrogen fixation are influenced by ecological characteristics (Israel *et al.* 1988; Thompson 1988). They can be good soil colonizers but poor nitrogen fixers. The resulting symbiosis may be of a lower effectiveness than with good nitrogen fixing strains and the full potential yield of the plant may not be expressed.

Inoculants can be made of single or several strains of rhizobia. Whether to use single or multistrain inoculants is still a subject of controversy. The theoretical advantage of a multistrain inoculant is to produce a universal inoculant adapted to different environmental conditions and hosts. The disadvantages are related to the identification of strains during quality control of the inoculants, and also to their effectiveness. There is a strong relationship between the number of bacteria inoculated and nodulation and plant yield (see p. 3) and a mixture of several strains selected to cover a wide variety of agricultural conditions will never apply the same number of bacteria adapted to the specific conditions at the site of inoculation as a single, well selected strain.

Studies with *R. meliloti* and alfalfa (Bordeleau and Antoun 1977) have shown that inoculation of a legume with several strains gives either higher or lower yields than inoculation with single strains, depending on the number and nature of the strains used. A multistrain inoculant of *B. japonicum* was not recommended for some Canadian soybean lines because single strains tested separately had fixed more N_2 than the same strains used as a mixture (Rennie and Dubetz 1984).

Carriers

The carriers used for powdered and granular inoculants vary widely, depending on the compounds available and on the country of manufacture. They include organic compounds like peat, plant derivatives and composts, and inorganic compounds like clays, calcium sulphate, phosphates, etc. To date, finely milled peat is the most used carrier in the world (despite some problems of variability between batches from the same deposit) because it allows the bacteria to survive well. All these carriers are, by their nature, non-sterile and will bring a natural microflora to the inoculant if they are not sterilized. The presence of such unidentified contaminants can present some potential agricultural and environmental risks.

Survival

It is known that rhizobial survival is lower in non-sterile than in sterile peat (Date and Roughley 1977). Thus, the shelf-life of contaminated inoculants is probably lower than that of pure culture inoculants. A lower effectiveness of inoculation may result with contaminated products delivering a low number of bacteria.

The quality control of contaminated inoculants is more difficult (Jansen Van Rensburg and Strijdom 1974). The plate count method becomes useless and plant-infection methods (Hiltbold *et al.* 1980) or serological methods (Olsen and Rice 1989) must be used, making the process more expensive and time-consuming.

Finally, contaminants may present a risk of introduction and dissemination of human, animal or plant pathogens. So far, there has been no evidence of health or environmental problems arising from these contaminants (Beringer and Bale 1988), although non-sterile peat-based inoculants are the most widely used. However, this risk must be taken into account, particularly when considering commercial exchanges of inoculants between countries. This is already covered by regulations controlling the movement of food, seeds and many other products that can transport pathogens. Currently, the authorities in France have a registration scheme for rhizobial inoculants: they must be pure cultures and remain without contaminants during storage (Catroux 1987).

Methods of inoculation

Rhizobia can be provided to the legumes either by inoculation of the seeds, which is still the most common method, or by direct inoculation of the soil (for reviews see Brockwell 1977; Roughley 1988).

Inoculation of the seeds can be done just before sowing by dusting or by slurry inoculation. In slurry treatment, powdered inoculants are mixed with a small amount of water and the suspension is mixed with the seeds. The seeds must be dried before sowing and, during drying, some of the inoculant may fall off, which results in rapid death of the rhizobia. Adhesive agents are often used to avoid these losses of inoculant and to improve rhizobia survival (Elegba and Rennie 1984). Oil-based inoculants have also been proposed as a solution to these problems (Kremer and Peterson 1982).

Preinoculation by seed manufacturers several days (or weeks) before sowing can be advantageous for farmers as sowing is done directly with seeds that are 'ready to use'. However, it is important to ensure the survival of the bacteria between preinoculation and sowing; it usually takes 4 to 8 weeks for processing and sale of preinoculated seed. With currently used processes rhizobial cells survival on the seeds is generally poor. Quality control of commercially pelleted legume seed in New Zealand gave variable results, with poor survival after 28 days storage at 19–22°C (Bonish *et al*. 1980). The same trends were observed in Australia on commercial, preinoculated seeds carrying 100 times less bacteria than seeds preinoculated in the laboratory (Brockwell *et al*. 1985). The recovery of viable *B. japonicum* cells, from seeds with different coating materials, was found to be in the range of 3–19 per cent and 1.5–7 per cent after 3 weeks storage at 15 and 22.5°C, respectively (Davidson and Reuszer 1978). This poor survival can give limited nodulation of the plants and farmers still need fully effective commercially produced preinoculated seeds.

Soil inoculation is done using granular or liquid inoculants. Granular inoculants are delivered to the row, at seed level, in amounts varying from 5 to 15 kg/ha. They are made of peat or inorganic materials like clays, clay derivatives or calcium carbonate. Granular inoculants may be preinoculated or inoculated just at the time of sowing. This latter method allows storage of the inoculant in small volumes (200–400 g/ha), and is safer from contamination from the risks from dessication and temperature than when it is preinoculated (10 kg/ha). Inoculation at sowing with a traditional peat inoculant mixed with inorganic granules similar to those used for delivering pesticides gives good results in France for soybean. Liquid inoculation is done with diluted suspensions of cultures (Schiffman and Alper 1968) or with diluted peat-based inoculants (Hely *et al*. 1976). The amount of dilution in water is from 50 to 250 litres per hectare, and may be a severe limitation to the method. The inoculant is delivered to the soil by gravity, spraying or with a peristaltic pump. This liquid inoculation has the advantage of delivering the bacteria directly to the soil, eliminating the possibility of dessication around the seed or the granules. It can also be used for postemergence inoculation (Rogers *et al*. 1982; Boonkerd *et al*. 1985; Ciafardini and Barbieri 1987).

Inoculant viability

Losses of viable rhizobia can be as high as 1 to 2 log during inoculation (Brockwell *et al*. 1987). The number of nodules generally increases with the log of the number of bacteria inoculated (Weaver and Frederick 1972; Philpotts 1979; Smith *et al*. 1981; Brockwell *et al*. 1988) and care must be taken, when choosing the method of inoculation, to minimize losses of viable bacteria. These losses must be considered in connection with the quality standards adopted in the country concerned. There is no general agreement about the minimum number of viable rhizobia to be inoculated per seed for optimal plant growth. For soybean, values as low as $1–2 \times 10^4$ bacteria per seed (Date and Roughley 1977; Sicardi de Mallorca and Labandera 1978; Skipper *et al*. 1980) and as high as 2×10^5 bacteria per seed (Curley and Burton 1975) are accepted. Quality controls of commercial inoculants have shown that high titre inoculants, delivering $10^5–10^6$ bacteria per seed, are best and that those delivering 10^3 are inefficient (Hiltbold *et al*. 1980).

The quality standard adopted for registration of soybean inoculants by the authorities in France is high: the inoculant must have a titre and be used so that the number of viable bacteria per seed is 10^6 at sowing. This high standard minimizes the effect of uncontrolled losses due to the method of inoculation by farmers.

Is the number of rhizobia released by inoculation significant compared to the total soil microflora and soil indigenous rhizobial population? A good-quality inoculant contains more than 10^9 viable bacteria per gram and the usual rate of inoculation is 200–400 g/ha. If we assume a survival of 10–100 per cent during inoculation, the number of released bacteria is between 2×10^{10} and 4×10^{11}/ha. This does not represent much compared to the total soil microflora, which can be estimated as 10^8–10^9 bacteria per gram soil, or 10^{17}–10^{18} bacteria/ha in the top 20 cm of soil, which weighs about 2000 tonnes. If we consider indigenous rhizobial populations of 10^3–10^5 per gram soil (see p. 6) established in the top 20 cm of soil, then 2×10^{12}–2×10^{14} rhizobia are present per hectare. Thus, the inoculation will bring at best about 0.1–1 per cent of the indigenous population of rhizobia. However, this will become more significant if we consider only the seed bed volume of soil.

Assuming a cylindrical seed bed volume of 6 cm diameter and 4 rows per metre, the seed bed soil in one hectare can be estimated at roughly 10 tonnes. This will contain around 10^{11}–10^{13} rhizobia, which means inoculation adds a number of bacteria equivalent to 1–10 per cent of the indigenous population.

Colonization and survival of inoculated rhizobia

The next question, if we want to assess the possible agricultural and environmental risks these bacteria can cause, is 'Will the inoculated bacteria be able to colonize the surrounding soil and became an established part of soil microflora?'

Methodological problems

The study of the behaviour of rhizobia introduced into soils that may be free from, or populated by, indigenous rhizobia needs appropriate technologies to enumerate the bacteria, to sample and isolate representative strains, to identify the strains and for the fine characterization of the bacterial populations. The currently available technologies have limitations and have been fully reviewed and discussed elsewhere (Parker et al. 1977; Stacey 1985)

but we would like also to emphasize several of them.

It seems that only a part of the total rhizobial population is recovered from soil by the most well-used methods, which give a poor recovery after a known amount of bacteria has been added to soil (Kingsley and Ben Bohlool 1981). This may be due to the attachment of rhizobia to soil particles (Ozawa and Yamaguchi 1986). Thus, the enumeration of added or resident rhizobia may be biased because colonization and survival patterns are not typical. The enumeration of rhizobia in soils is commonly done by inoculation of soil dilutions onto the host plant. Obviously, only rhizobia producing nodules are counted, rhizobia in a 'dormant' state, or strains with nodulation restriction for the test plant, will be underestimated.

The use of immunological methods does not fully solve the nodulation problems unless all the serogroups of all the strains present in the soil were known, and all the corresponding antisera were available for enumeration. This means that not all the rhizobia present in a soil will be taken into account, whatever enumeration method is used.

The isolation of strains from soils suffers the same kind of limitations. Rhizobial strains differ in their competitiveness, which can be defined as the ability of a strain to compete against others for infection and nodulation, colonization and survival in soils. Only nodulating strains are isolated and the most competitive for the plant genotype used will be preferentially selected. As the biochemical determinants involved in competitiveness are unknown (Beringer and Bale 1988), the general validity of sampling by isolation in host plants is doubtful.

The characterization of nodulating strains of the rhizobial soil populations remained very limited until recently – serological methods allowed the recognition of only some of the isolates. Some 34 to 81 per cent of the B. japonicum strains isolated from nodules did not react with any of 15 antisera (Howle et al. 1987). On the other hand, serotyping does not provide a reliable identification because the same serogroup can include genetically diverse strains (Sobral et al. 1990). Finally, the use of molecular biological methods such as plasmid profiles, restriction fragment length polymorphism, hybridization with specific probes, protein patterns, etc. seem the only way to identify rhizobial strains genetically and to study their genetic diversity. The use of these methods is relatively recent and data are scarce (see p. 6).

Factors affecting colonization and survival

These factors have been reviewed extensively (Parker, Trinick and Chatel 1977; Lowendorf 1980; Stacey 1985; Beringer and Bale 1988). Root colonization, plant growth and nodulation play a major role in the establishment of *B. japonicum*. Soybean crops increase the number of *B. japonicum* in soils (Weaver *et al.* 1972). Their number increases during the plant's growth to reach 10^6 and more per gram soil in the winter, and then decrease slowly (Hiltbold *et al.* 1985). This increase is attributed to release of bacteria from disintegrating nodules (Kuykendall *et al.* 1982; Brockwell *et al.* 1987). The same trends were observed with *R. meliloti*, which established at $10^3–10^5$ bacteria per gram soil during the first year of an alfalfa crop (Barber 1982).

Movement of rhizobia in soils is limited, not exceeding 20 cm laterally during the first year of an alfalfa crop (Barber 1982). It is also usual, in field trials, to keep uninoculated plots of soybean nonnodulated at distances as short as 30 cm from inoculated plots (G. Catroux, unpublished results). Movement of *B. japonicum* was found to be limited (Madsen and Alexander 1982) except when percolating water was added, which results in more vertical transport (Breitenbeck *et al.* 1988).

Rhizobia are true soil inhabitants, with generally good saprophytic ability. They are able to use a diverse range of energy sources, from sugars to phenolic compounds (Parker *et al.* 1977; Chakrabarti *et al.* 1981; Stowers 1985). Some strains of *B. japonicum* are also chemoautotrophic and are able to grow in soil using H_2 and O_2 (Viteri and Schmidt 1989).

Some theoretical calculations can be made to predict how an introduced bacterial population will survive in soils. The energy flow involved in the maintenance and turnover of rhizobial populations below 10^5 bacteria per gram of soil is probably very low. Assuming that the dry matter of 10^9 bacteria is 1–2 mg or 0.5–1 mg organic carbon, then 10^5 rhizobia are equivalent to 0.05–0.1 mg carbon biomass. Estimates of the need of organic substrates for growth and survival of such populations can be made using data obtained from the total soil microbial biomass. The average turnover of the soil

microbial biomass has been estimated to be 1.5–2.5 years in temperate soils (Chaussod *et al.* 1988). Assuming a yield for organic carbon utilization of around 40 per cent, populations of 10^5 rhizobia would need 0.05–0.16 mg of organic carbon per year.

Another approach is to use an estimate of the maintenance requirements of actively-metabolizing dormant microbial populations (Anderson and Domsch 1985a,b). Using data from these authors, it can be estimated that populations of 10^5 rhizobia will need 0.003–0.014 mg when' dormant and, when actively metabolizing, 2.1–4.2 mg of organic substrate per year. From these estimates, it can be concluded that a rhizobial population of 10^5 bacteria per gram soil has very low annual energy requirements for survival.

Alternatively, the soil microflora has an estimated volume equivalent to 0.4–1 per cent of the available space in the soil, and microbial colonies are probably non-interacting spatially (Lockwood 1981). This concept is corroborated by the small observed increases of slow-growing populations of *B. japonicum* on organic substrates added to soils, when in competition with other micro-organisms (Pena-Cabriales and Alexander 1983, Viteri and Schmidt 1987).

Finally, these data can be interpreted as follows: an introduced rhizobial population of 10^5 bacteria per gram of soil, properly dispersed into the soil it colonizes, can find enough organic nutrients to survive provided that the physicochemical conditions of the soil, e.g. pH, salinity and temperature, are suitable.

Survival of rhizobia in soil is generally good, except in extreme conditions. Survival of indigenous rhizobial populations is seen in the range of $10^3–10^5$ rhizobia per gram of soil even 1 year after inoculation, and without cropping the host plant. For example, inoculated strains of *R. leguminosarum* in a soil already containing the same species, decreased from $10^5–10^6$ to 10^4/g soil after lying fallow for 3–4 years (Jensen and Sorensen 1987). It was shown in the same experiment that better host and non-host plants improved the survival of the introduced strains. Survival of introduced *B. japonicum* in soils is very good and it is possible to recover isolates of serogroups 123, 6, 46 and 31, which correspond to the strains used in the early commercial inoculation programmes in the USA (Weber *et al.* 1989). Soil populations of *B. japonicum* in a cotton–corn–soybean rotation stabilized at 10^4–10^5/g soil if soil conditions were good, i.e. if Ca, P and K were at suitable concentrations, and fell to 10^2/g soil in acidic conditions (Hiltbold *et al.* 1985). Survival of *B. japonicum* introduced into two soils devoid of this species stabilized at populations >10^4 per gram soil (Crozat *et al.* 1982). The same authors have shown, under laboratory conditions, that the equilibrium population was not dependent upon the initial number present. This equilibrium, also observed in the field, may be explained by predation and by nutrient availability in the soil.

In conclusion, native or introduced rhizobial populations are able to survive for years in soils even in absence of the host plant.

Rhizobial populations in soils

Rhizobial populations must be considered from three points of view: number, genetic diversity and stability.

Size of natural rhizobial populations
The numbers of rhizobial cells in soil are affected by several factors (see p. 5), but one of the most important is soil pH. A survey done on 60 soils of different types and varying pH showed that different rhizobial species behaved differently (Amarger 1980). *R. meliloti* was either absent or present in very low numbers (10^1/g soil) when the soil pH was below 6; their numbers increased to reach a maximum of 10^3/g soil in the pH range 7–8. In contrast, *Bradyrhizobium* sp. (*lupinus*) was absent from soils with a pH above 6.5 and present at $\leqslant 10^2$/g soil when the pH was between 4.5 and 5.5. The behaviour of *R. leguminosarum* was less variable, and was nearly the same for the three biovars (bv. *viciae*, bv. *phaseoli* and bv. *trifolii*). Their numbers decreased with extreme soil pH, acidic or basic, and were lower at these pHs than for *R. meliloti* and *Bradyrhizobium* sp. (*lupinus*), which reached higher populations of 10^4 to 10^5 per gram of soil in the pH range 5.5–7.5. Values as high as 10^6 to 10^7 *R. leguminosarum* bv. *trifolii*/g soil have been found occasionally (Mytton and Livesey 1983). On the other hand, *B. japonicum* is known to stabilize at between 10^2 and 10^4/g soil after introduction (Hiltbold *et al.* 1885). Thus, rhizobial populations range up to 10^7 per gram of soil, depending on the species considered and on the soil.

Diversity
Is there a phenotypic and genetic diversity between isolates from different soils and from the same soil?

Phenotypic diversity between isolates of the same rhizobial species isolated from different soils is well known and remains the basis of strain selection for symbiotic and ecological characteristics (Thompson 1988). For example, the variability in nitrogen fixation effectiveness of indigenous strains from different sites was clearly demonstrated for *R. leguminosarum* bv. *trifolii* (Hagedorn 1978; Mytton and Livesey 1983), *R. leguminosarum* bv. *viciae* (Amarger 1974) and *Bradyrhizobium* sp. (*lupinus*) (Lagacherie *et al.* 1976).

The best studies of diversity have been done on *B. japonicum* isolated from soybean nodules of different soils. These studies were done primarily by serotyping and they revealed a great diversity among the strains. For example, strains isolated from soils from different areas of the United States were related to more than 10 different serogroups (Weber *et al.* 1989; Furhmann 1990). The same trend was observed using polyacrylamide gel electrophoresis of proteins (Kamicker and Brill 1986) and this method was also used to study genomic DNA digestion patterns. It distinguished between isolates of the single serogroup 123 (Sadowsky *et al.* 1987; Sobral *et al.* 1990).

A study done on 10 different soils containing *R. leguminosarum* identified 17 different plasmid profiles from 60 *R. leguminosarum* bv. *phaseoli* isolates obtained from bean nodules, and 10 plasmid profiles from 24 *R. leguminosarum* bv. *viciae* isolates from pea nodules (Kucey and Hynes 1989). The study of genetic diversity among *R. leguminosarum* bv. *phaseoli* from various geographical and ecological sources, by the electrophoretic mobility of 15 metabolic enzymes, demonstrated 46 electrophoretic types in the 51 isolates studied (Pinero *et al.* 1988). Thus, there is great phenotypic and genotypic diversity among rhizobial strains isolated from different soils.

Data on strains isolated from the same soil are less common. However, phenotypic variability between rhizobial isolates from a single site was observed for nitrogen fixation effectiveness of some *R. leguminosarum* bv. *trifolii* isolates associated with *Trifolium subterraneum* (Gibson *et al.* 1975; Schofield *et al.* 1987). A study done on isolates of *B. japonicum* from one soil in Maryland has shown that the isolates belonged to six different serogroups (Caldwell and Weber 1970). More recently, isolates of *R. meliloti* and *R. leguminosarum* bv. *viciae*, *trifolii* and *phaseoli* from a single site in England were examined by enzyme electrophoresis and the results suggested that the isolates comprise several genetically distinct lineages (Young 1985). More detailed study done on *R. leguminosarum* bv. *trifolii* using symbiotic (Sym) plasmid analysis and DNA hybridization with an *R. leguminosarum* nodulation and nitrogen fixation gene probe revealed a great variation in plasmid size, and multiple plasmids with symbiotic genes were found in a large proportion of strains (Harrison *et al.* 1988). Using a probe specific for biovar *trifolii*, the same authors have found Sym plasmids suspected of being interbiovar hybrids. This means that plasmid transfer may occur in the field. This conclusion was also reached by Young and Wexler (1988), who worked on *R. leguminosarum* bv. *viciae* after they found that the distribution of plasmid in different chromosomal backgrounds was not randomly distributed.

The same conclusions arose from the study of *R. leguminosarum* bv. *trifolii* isolated from a single field in Australia (Schofield *et al.* 1987). These workers used three different Sym plasmids – specific DNA hybridization probes – to find the same Sym plasmid in unrelated strains, as well as unrelated Sym plasmids in identical host strains. Each of these experiments indicates the existence of genetic exchange within this soil population. Thus, it may be concluded that there is a great genetic diversity in natural rhizobial population and that genetic transfer may also occur within these soil populations.

Stability
The phenotypic and genetic stability of rhizobial strains in soils has not been studied extensively. The symbiotic effectiveness of inoculant strains of *R. meliloti*, *R. leguminosarum* bv. *trifolii*, *B. japonicum* and *R.* species recovered from inoculated soils showed there was no significant change over several years (Jansen Van Rensburg and Strijdom 1985). A more complete study was done on *B. japonicum* 8–13 years after its introduction into soils previously free of this bacterium (Brunel *et al.* 1988). No significant changes were found for serological and physiological characteristics, symbiotic effectiveness or DNA hybridization analysis.

Thus, it seems reasonable to expect good stabil-ity of strains introduced into soils over relatively short periods of time – 10 years. But for longer periods we must consider the possibility that genetic transfers would result in increased variability within populations and introduce strain instability. However, some genetic characteristics were not randomly distributed, suggesting some genetic stability at the population and community level (Young and Wexler 1988).

General discussion

This brief survey of legume inoculation and related problems can be useful as a basis for the discussion of several general questions on microbial release into soils. Is it possible to introduce a microbial inoculum into a soil and for it to establish? How long will the established population remain viable and active? What kind of problems can occur with established populations? What kind of preventive or curative solutions could we imagine?

Establishment in soils of an introduced micro-organism

It is possible to introduce and establish a soil bacterial strain into soil previously free of the same species, whatever its geographical origin. For example, *B. japonicum* is not native to the United States or to northern Mediterranean countries, and yet it has become established in soils through soybean inoculation. The established populations remain at between 10^3 and 10^5 viable rhizobia per gram of soil, the precise number depends on the soil and the time after soybean cropping (Lagacherie 1978). In France, where only one registered strain of *B. japonicum* is used in commercial inoculants (Catroux 1987), soils are colonized by only one strain. This presents an exceptional ecological situation, one strain is probably easier to manage than multistrain colonized soils (see p. 6). In this case, the introduction of a new strain of *B. japonicum* will be acceptable only if the candidate brings significantly higher yields.

It is also possible to introduce and establish a bacterial strain in soils already containing populations of the same species. For example, this was clearly shown for *R. leguminosarum* inoculated on peas (Meade *et al.* 1985) and for *B. japonicum* and soybean in the USA, where nodule isolates are classified into numerous serogroups including those corresponding to strains used in the past as commercial inoculants (Weber *et al.* 1989). These data can be interpreted as successive introductions of strains and successful establishment in soils. It was also demonstrated experimentally in the field that, after 3 years of inoculation with a non-indigenous strain of *B. japonicum*, this strain became established and gave an increase in nodule occupancy with time (Dunigan *et al.* 1984). However, as previously emphasized, introducing a new strain to such colonized soils is interesting only if the new strain has the ability to give higher plant yield than the indigenous strains.

It should be noted that legume inoculation in soils containing indigenous rhizobia is often reported as unsuccessful. This means that the inoculant strain is not found in the nodules, but it does not mean that the strain is not established in the soil. The inoculant strain may be less competitive for nodulation than the indigenous strains and may not form the nodules, although it is present in the soil. Thus, as long as there is no clear evidence that the inoculant bacteria are not established in the soils, it should not be concluded from an inoculation failure that introduction of a new strain is not possible in soils containing bacteria of the same species.

These examples of introduction and establishment in soils of soil bacteria with good saprophytic abilities like rhizobia, indicate that other introductions may be successful, providing the candidate is a true soil inhabitant and encounters suitable soil conditions.

In summary, it seems necessary to distinguish between two situations: soils with, and soils without, the species of the bacteria to be introduced. Introduction into soils devoid of the bacteria must be considered more carefully as it can present a risk of permanent establishment of a strain with poor or medium effectiveness which might give troubles in the future.

Possible problems encountered

We have emphasized the problem of unintentional introduction of organisms into soils when contaminated inoculants are used. From a technological point of view this problem is easy to solve by producing pure culture inoculants. The manufacturing cost of these inoculants is comparatively high but if the rate of inoculation, storage and total safety are considered, it becomes evident that it is better for farmers to pay more for a good inoculant, giving yield increases, than to pay less but to receive no beneficial effect on crop yield.

Another kind of risk must also be taken into account: the introduction of a bacterium with poor effectiveness. Some examples of this agricultural risk arose from the introduction of *B. japonicum* in soils of the United States. A survey of serological distribution of *B. japonicum* isolated from soybean nodules all around the United States has shown that 24 per cent of the isolates belong to the serogroup 123, which is considered to give very variable results for N_2 fixing (Weber *et al.* 1989). Another survey in Delaware soils (USA), has shown that 23 per cent of the soybean nodule isolates belong to serogroups 76 and 94, which were related to serotype strains that produce a phytotoxin causing a chlorosis of susceptible soybean genotypes (Fuhrmann 1990). Although there is no direct evidence about the yield limitation introduced by these strains it must be taken into account, particularly because screened strains of *B. japonicum* have been described as better N_2-fixers than the serotype strain 123 (Cregan *et al.* 1989). A very careful selection of the characters of the inoculant strains is therefore an important prerequisite to any such release that may have long-term or immediate consequences.

One other agricultural risk to be considered is related to the stability of the introduced bacteria. Again, the example of *B. japonicum* is interesting, as no variation in the genetic characteristics and symbiotic properties were found after 8–13 years in soils (Brunel *et al.* 1988). The same trends were found with different species of *Rhizobium* recovered 4–8 years after their introduction into soil. Very few of the isolates identified as the inoculated strains differed significantly in effectiveness compared to the parent strains (Jansen Van Rensburg and Strijdom 1985). Although from this evidence the risk seems to be low for non-genetically engineered strains, their stability in soil must be studied before their introduction on a large scale.

Strategies for preventing possible agricultural problems

Before considering the problem, we need efficient strategies that are able to minimize or to suppress these risks, based on the two situations described previously: soils with or without the bacterial species to be introduced. In the first situation it is possible to have a preventive strategy with the chosen strains. This choice must take into account apparently opposite interests. In the short-term, farmers want to colonize their soil as quickly as possible to avoid the need to inoculate more than once, and choosing a good colonizer with good survival ability would be the solution. In contrast, in the long-term, for the protection of fertility and ecosystems, the best candidate would be a strain unable to colonize and persist in the soil. Currently, for temperate climatic environments, there is no rhizobial strain unable to survive in agricultural soils. However, in the future, it may be possible to engineer strains that survive poorly, through cold- or heat-sensitivity, the accumulation of toxins or the use of 'killing genes' (Cuskey 1988). We used another strategy with the first introduction of *B. japonicum* in French soils because the strain was to be the only *B. japonicum* strain present. As it is possible to screen the strains for their competitiveness for infection and nodulation (Amarger 1981), it was proposed to introduce a strain with the highest N_2-fixing effectiveness and lowest competitiveness. Thus, it was expected, would allow for an easier introduction of a new strain in the future.

The problems become more difficult when the soils contain inidigenous bacteria from the same species as the strain to be introduced. Several approaches were used for rhizobia and legumes through inoculation or alteration of the plant genotype. Returning to the *B. japonicum* and soybean example in the United States, it is generally recognized that inoculation with good N_2-fixing strains does not produce any yield increase, but neither does it allow the strain to nodulate the plant extensively. Average nodule occupancy by inoculated strain USDA 110 was found to be in the range 0–17 per cent in Louisiana (Dunigan *et al.* 1984). Other data, obtained with strains that are claimed to be competitive, have shown variable nodule occupancy by the inoculant strains between 0.3 and 15.7 per cent in Midwest USA soils (Klubeck *et al.* 1988) and between 0 and 46 per cent in Iowa (Berg *et al.* 1988).

As both competitiveness for infection and the resultant nodule occupancy is dependent on the number of rhizobia in competition at the time of infection (Weaver and Frederick 1974; Amarger and Lobreau 1982), we could expect to improve nodule occupancy by increasing the number of inoculated rhizobia. Increases of nodule occupancy

with up to 50 per cent infection were found by using massive inoculation – as many as 10^9 cells applied to the seed or surrounding soil (Kamicker and Brill 1987). Successive inoculations of strain USDA 110 over 3 years resulted in an increase of nodule occupancy from 0 to 17 per cent at the beginning to up to 54 per cent 7 years after the first inoculation (Dunigan *et al.* 1984).

It is also possible to enhance soybean rhizosphere colonization by the use of antimicrobial agents and *B. japonicum* strains resistant to those inhibitors (Maqbul Hossain and Alexander 1984). It is also claimed that selected strains of *B. japonicum* can be more successfully introduced into soils of a certain region if they are adapted to soils and cultivars of the same region (Klubeck *et al.* 1988).

Another possible strategy could be to use a plant genotype to restrict nodulation by a given strain. Soybean genotypes restricting the nodulation by *B. japonicum* strain USDA 123 were found (Cregan and Keyser 1986) and, more recently, soybean genotypes restricting nodulation by serocluster 123 isolates in favour of an inoculant strain have also been found (Cregan *et al.* 1989).

In summary, the release of rhizobia into soils may cause some agronomic problems as the released rhizobia usually become a permanent part of the soil microflora and may later effectively prevent a newer strain with improved performance from nodulating a legume. Although some curative methods could theoretically be used, there is no evidence of their practical and economical feasibility. This means that it would be better to try to prevent this kind of situation as far as possible. This could be done by authorizing only commercial inoculants with strains of rhizobia, selected or engineered, which have been proved to produce statistically significant positive and reliable effects on the crop, and yield not only commercial benefits.

Conclusion

Legume inoculation is an example of a large-scale release of bacteria into soil and has, until now, led to the establishment of many non-indigenous rhizobia, with different genotypes, in many different regions. So far, these introductions have not caused any environmental problems. Improvement of the symbiotic nitrogen fixing ability of the rhizobial strains by genetic engineering is now contemplated. There is no reason to believe that these strains will behave differently from the non-engineered strains that have been used until now. Some agronomic problems were shown with the soybean example, when introduced *B. japonicum* strains with a high percentage of nodule occupancy were suspected of fixing less nitrogen than some other selected strains. However, there is no clear evidence of the existence of any critical problem, although such problems must be considered in the future, when improved strains will be available.

It would be advantageous to submit micro-organisms intended for release into soils to regulatory control. If this control is to be efficient it should take into account the entire process of release and the choice of the strains and of the inoculant should be considered very carefully. The strains to be released should give an agricultural advantage over the indigenous or previously used non-engineered inoculant strains. This advantage should be reliable and demonstrable in the field under the farmer's working conditions. The inoculant must be considered as a carrier for scientific progress and, in this way, high quality inoculants are essential, especially in titre, shelf-life and purity, as in the use of clearly identified and authorized strain. Such regulations exist in several countries and must be considered as a part of a safety strategy for the future so that scientific, technical and economic progress is not hampered.

References

Amarger, N. (1974). Aspect microbiologique de la culture féverole en France. *Comptes Rendus de l'Académie d'Agriculture de France*, **60**, 105–12.

Amarger, N. (1980). Aspect microbiologique de la culture des légumineuses. *Le Sélectionneur Français*, **28**, 61–6.

Amarger, N. (1981). Selection of *Rhizobium* strains on their competitive ability for nodulation. *Soil Biology and Biochemistry*, **13**, 481–6.

Amarger, N., and Lobreau, J.P. (1982). Quantitative study of nodulation competitiveness in *Rhizobium* strains. *Applied and Environmental Microbiology*, **44**, 583–8.

Anderson, T.H., and Domsch, K.H. (1985a). Determination of ecophysiological maintenance carbon requirements of soil microorganisms in a dormant state. *Biology and Fertility of Soils*, **1**, 81–9.

Anderson, T.H., and Domsch, K.H. (1985b). Maintenance carbon requirements of actively metabolizing microbial populations under *in situ* conditions. *Soil Biology and Biochemistry*, **17**, 197–203.

Barber, L.E. (1982). *Rhizobium meliloti* distribution in the soil following alfalfa inoculation. *Plant and Soil*, **64**, 363–8.

Berg, R.K., Loynachan, T.E., Zablotowicz, R.M., and Lieberman, M.T. (1988). Nodule occupancy by introduced *Bradyrhizobium japonicum* in Iowa soils. *Agronomy Journal*, **80**, 876–81.

Beringer, J.E., and Bale, M.J. (1988). The survival and persistence of genetically-engineered micro-organisms. In *The release of genetically-engineered micro-organisms*, (Ed. M. Sussman, C.H. Collins, F.A. Skinner, and D.E. Stewart-Tull), pp. 29–46. Academic Press Ltd, London.

Bonish, P.M., Neville, F.J., and Lowther, W.L. (1980). Laboratory testing for numbers of viable rhizobia on commercially-pelleted legume seed. *New Zealand Journal of Experimental Agriculture*, **8**, 139–43.

Boonkerd, N., Arunsri, C., Rungrattanakasin, W., and Vasuvat, Y. (1985). Effects of post-emergence inoculation on field grown soybeans. *MIRCEN Journal*, **1**, 155–61.

Bordeleau, L.M., and Antoun, H. (1977). Effet de l'inoculation mixte avec des souches de *Rhizobium meliloti* sur le rendement de la luzerne cultivar Saranac. *Canadian Journal of Plant Science*, **57**, 1071–5.

Breitenbeck, G.A., Yang, H., and Dunigan, E.P. (1988). Water-facilited dispersal of inoculant *Bradyrhizobium japonicum* in soils. *Biology and Fertility of Soils*, **7**, 58–62.

Brockwell, J. (1977). Application of legume seed inoculants. In *A treatise on dinitrogen fixation*, (Ed. R.W.F. Hardy and A.H. Gibson), pp. 277–309. John Wiley and Sons, Chichester.

Brockwell, J., Herridge, D.F., Roughley, R.J., Thompson, J.A., Gault, R.R. (1975). Studies on seed pelleting as an aid to legume seed inoculation. 4. Examination of preinoculated seed. *Australian Journal of Experimental Agriculture and Animal Husbandry*, **15**, 780–7.

Brockwell, J., Gault, R.R., Chase, D.L., Turner, G.L., and Bergersen, F.J. (1985). Establishment and expression of soybean symbiosis in a soil previously free of *Rhizobium japonicum*. *Australian Journal of Agricultural Research*, **36**, 397–409

Brockwell, J., Roughley, R.J., and Herridge, D.F. (1987). Population dynamics of *Rhizobium japonicum* strains used to inoculate three successive crops of soybean. *Australian Journal of Agricultural Research*, **38**, 61–74.

Brockwell, J., Herridge, D.F., Morthorpe, L.J., and Roughley, R.J. (1988). Numerical effects of *Rhizobium* population on legume symbiosis. In *Nitrogen fixation by legumes in Mediterranean agriculture*, (Ed. D.P. Beck and L.A. Materon), pp. 179–93. Icarda, Nijhoff, Dordrecht.

Brunel, B., Cleyet-Marel, J.C., Normand, P., and Bardin, R. (1988). Stability of *Bradyrhizobium japonicum* inoculants after introduction into soil. *Applied and Environmental Microbiology*, **54**, 2636–42.

Caldwell, B.E., and Weber, D.F. (1970). Distribution of *Rhizobium japonicum* serogroups in soybean nodules as affected by planting dates. *Agronomy Journal*, **62**, 12–14.

Catroux, G. (1987). Les controles de qualité des inoculums soja en France. *Colloque Soja*, **8–10 Sept.**, 101–14. Onidol, CIS, CETIOM, Toulouse.

Chakrabarti, S., Lee, S.M., and Gibson, H.A. (1981). Diversity in the nutritional requirements of strains of various *Rhizobium* species. *Soil Biology and Biochemistry*, **13**, 349–54.

Chaussod, R., Houot, S., Guiraud, G., and Hetier, J.M. (1988). Size and turnover of the microbial biomass in agricultural soils: laboratory and field measurements. In *Nitrogen efficiency in agricultural soils and the efficient use of fertilizer nitrogen*, EEC meeting, Edinburgh, (Ed. D.S. Jenkinson), pp. 312–16. Elsevier Applied Science, Amsterdam.

Ciafardini, G., and Barbieri, C. (1987). Effects of cover inoculation of soybean on nodulation nitrogen fixation and yield. *Agronomy Journal*, **79**, 645–8.

Cregan, P.B., and Keyser, H.H. (1986). Host restriction of nodulation by *Bradyrhizobium japonicum* strain USDA 123 in soybean. *Crop Science*, **26**, 911–16.

Cregan, P.B., Keyser, H.H., and Sadowsky, M.J. (1989). Soybean genotype restricting nodulation of a previously unrestricted serocluster 123 bradyrhizobia. *Crop Science*, **29**, 307–12.

Crozat, Y., Cleyet-Marel, J.C., Giraud, J.J., and Obaton, M. (1982). Survival rates of *Rhizobium japonicum* populations introduced into different soils. *Soil Biology and Biochemistry*, **14**, 401–5.

Curley, R.L., and Burton, J.C. (1975). Compatibility of *Rhizobium japonicum* with chemical seed protectants. *Agronomy Journal*, **67**, 807–8.

Cuskey, S.M. (1988). In *The release of genetically-engineered microorganisms*, (Ed. M. Sussman, C.H. Collins, F.A. Skinner, and D.E. Stewart-Tull), pp. 233–4. Academic Press, London.

Date, R.A., and Roughley, R.J. (1977). Preparation of legume inoculants. In *A treatise on dinitrogen fixation*, (Ed. R.W.F. Hardy and A.H. Gibson), pp. 243–76. John Wiley and Sons, Chichester.

Davidson, F., and Reuszer, H.W. (1978). Persistence of *Rhizobium japonicum* on the soybean seed coat under controlled temperature and humidity. *Applied and Environmental Microbiology*, **35**, 94–6.

Dunigan, E.P., Bollich, P.K., Hutchinson, R.L., Hicks, P.M., Zaunbrecher, F.C., Scott, S.G., and Mowers, R.P. (1984). Introduction and survival of an inoculant strain of *Rhizobium japonicum* in soil. *Agronomy Journal*, **76**, 463–6.

Elegba, M.S., and Rennie, R.J. (1984). Effect of different inoculant adhesive agents on rhizobial survival, nodulation and nitrogenase (acetylene-reducing) activity of soybeans (*Glycine max* (L.)Merrill). *Canadian Journal of Soil Science*, **64**, 631–6.

Fuhrmann, J. (1990). Symbiotic effectiveness of indigenous soybean bradyrhizobia as related to serological, morphological, rhizobitoxine and hydrogenase phenotypes. *Applied and Environmental Microbiology*, **56**, 224–9.

Gibson, A.H., Curnow, B.C., Bergersen, F.J., Brockwell, J., and Robinson, A.C. (1975). Studies of field populations of *Rhizobium*: effectiveness of strains of *Rhizobium trifolii* associated with *Trifolium subterraneum* L. pastures in south-eastern Australia. *Soil Biology and Biochemistry*, **7**, 95–102.

Hagedorn, C. (1978). Effectiveness of *Rhizobium trifolii* populations associated with *Trifolium subterraneum* L. in southwest Oregon soils. *Soil Science Society of America Journal*, **42**, 447–51.

Harrison, S.P., Gareth Jones, D., Schunmann, P.H.D., Forster, J.W., and Young, J.P.W. (1988). Variation in *Rhizobium leguminosarum* biovar *trifolii* Sym plasmids and the association with effectiveness of nitrogen fixation. *Journal of General Microbiology*, **134**, 2721–30.

Hely, F.W., Hutchings, R.J., and Zorin, M. (1976). Legume inoculation by spraying suspensions of nodule bacteria into soils beneath seed. *The Journal of the Australian Institute of Agricultural Science*, **42**, 241–4.

Hiltbold, A.E., Patterson, R.M., and Reed, R.B. (1985). Soil populations of *Rhizobium japonicum* in a cotton–corn–soybean rotation. *Soil Science and Society of America Journal*, **49**, 343–8.

Hiltbold, A.E., Thurlow, D.L., and Skipper, H.D. (1980). Evaluation of commercial soybean inoculants by various techniques. *Agronomy Journal*, **72**, 675–81.

Howle, P.K.W., Shipe, E.R., and Skipper, H.D. (1987). Soybean specificity for *Bradyrhizobium japonicum* strain 110. *Agronomy Journal*, **79**, 595–8.

Israel, D.W., Wollum II, A.G., and Mathis, J.N. (1988). Relative performance of *Rhizobium* and *Bradyrhizobium* strains under different environmental conditions. In *ISI atlas of science: animal and plant sciences*, Vol. 1, pp. 95–9. Institute of Scientific Information, Philadelphia.

Jansen Van Rensburg, H., and Strijdom, B.W. (1974). Quality control of *Rhizobium* inoculants produced from sterilized and non-sterile peat in South-Africa. *Phytophylactica*, **6**, 307–10.

Jansen Van Rensburg, H., and Strijdom, B.W. (1985). Effectiveness of *Rhizobium* strains used in inoculants after their introduction into soil. *Applied and Environmental Microbiology*, **49**, 127–31.

Jensen, E.S., and Sorensen, L.H. (1987). Survival of *Rhizobium leguminosarum* in soil after addition as inoculant. *FEMS Microbiology Ecology*, **45**, 221–6.

Kamicker, B.J., and Brill, W.J. (1986). Identification of *Bradyrhizobium japonicum* nodules isolates from

Wisconsin soybean farms. *Applied and Environmental Microbiology*, **51**, 487–92.

Kamicker, B.J., and Brill, W.J. (1987). Methods to alter the recovery and nodule location of *Bradyrhizobium japonicum* on field-grown soybeans. *Applied and Environmental Microbiology*, **53**, 1737–42.

Kingsley, M.T., and Ben Bohlool, B. (1981). Release of *Rhizobium* spp. from tropical soils and recovery for immunofluorescence enumeration. *Applied and Environmental Microbiology*, **42**, 241–8.

Klubeck, B.P., Hendrickson, L.L., Zablotowicz, R.M., Skwara, J.E., Varsa, E.C., Smith, S., Islieb, T.G., Maya, J., Valdes, M., Dazzo, F.B., Todd, R.L., and Walgenback, D.D. (1988). Competitiveness of selected *Bradyrhizobium japonicum* strains in midwestern USA soils. *Soil Science Society of America Journal*, **52**, 662–6.

Kremer, J.R., and Peterson, H.L. (1982). Effect of inoculant carrier on survival of *Rhizobium* on inoculated seed. *Soil Science*, **134:2**, 117–25.

Kucey, R.M.N., and Hynes, M.F. (1989). Populations of *Rhizobium leguminosarum* biovars *phaseoli* and *viceae* in fields after bean or pea in rotation with nonlegumes. *Canadian Journal of Microbiology*, **35**, 661–7.

Kuykendall, L.D., Devine, T.E., and Cregan, P.B. (1982). Positive role of nodulation on the establishment of *Rhizobium japonicum* in subsequent crops of soybean. *Current Microbiology*, **7**, 79–81.

Lagacherie, B. (1978). Survie dans le sol des *Rhizobium* apportés par inoculation des graines de soja. *Informations techniques CETIOM*, **62**, 11–15.

Lagacherie, B., Giniès, P., and Amarger, N. (1976). L'inoculation du lupin. Essais préliminaires. *Comptes Rendus de l'Académie d'Agriculture de France*, **62**, 987–94.

Lockwood, J.L. (1981). Exploitation competition. In *The fungal community, its organization and role in the ecosystem*, (Ed. D.T. Wicklow and G. Carrol), pp. 319–49. Marcel Dekker, New York.

Lowendorf, H.S. (1980). Factors affecting survival of *Rhizobium* in soil. In *Advances in microbial ecology*, Vol. 4, (Ed. M. Alexander), pp. 87–124. Plenum Press, New York.

Madsen, E.L., and Alexander, M. (1982). Transport of *Rhizobium* and *Pseudomonas* through soil. *Soil Science Society of America Journal*, **46**, 557–60.

Maqbul Hossain, A.K., and Alexander, M. (1984). Enhancing soybean rhizosphere colonization by *Rhizobium japonicum*. *Applied and Environmental Microbiology*, **48**, 468–72.

Meade, J., Higgins, P., and O'Gara, F. (1985). Studies on the inoculation and competitiveness of a *Rhizobium leguminosarum* strain in soils containing indigenous rhizobia. *Applied and Environmental Microbiology*, **49**, 899–903.

Mytton, L.R., and Livesey, C.J. (1983). Specific and general effectiveness of *Rhizobium trifolii* populations from different agricultural locations. *Plant and Soil*, **73**, 299–305.

Obaton, M. (1971). Inflence de la composition chimique du sol sur l'utilité de l'inoculation des graines de luzerne avec *Rhizobium meliloti*. *Plant and Soil, special volume*: 273–85.

Olsen, P.E., and Rice, W.A. (1989). *Rhizobium* strain identification and quantification in commercial inoculants by immunoblot analysis. *Applied and Environmental Microbiology*, **55**, 520–2.

Ozawa, T., and Yamaguchi, M. (1986). Fractionation and estimation of particle-attached and unattached *Bradyrhizobium japonicum* strains in soils. *Applied and Environmental Microbiology*, **52**, 911–14.

Parker, C.A., Trinick, M.J., and Chatel, D.L. (1977). Rhizobia as soil and rhizosphere inhabitants. In *A treatise on dinitrogen fixation*, (Ed. R.W.F. Hardy and A.H. Gibson), pp. 311–53. John Wiley and Sons, Chichester.

Pena-Cabriales, J.J., and Alexander, M. (1983). Growth of *Rhizobium* in soil amended with organic matter. *Soil Science Society of America Journal*, **47**, 241–5.

Philpotts, H. (1979). *Rhizobium* strains and inoculation rate for direct drilled lupins. *The Journal of the Australian Institute of Agricultural Science*, **45**, 251–3.

Pinero, D., Martinez, E., and Selander, R.K. (1988). Genetic diversity and relationships among isolates of *Rhizobium leguminosarum* biovar *phaseoli*. *Applied and Environmental Microbiology*, **54**, 2825–32.

Rennie, R.J., and Dubetz, S. (1984). Multistrain vs. single strain *Rhizobium japonicum* inoculants for early maturing soybean cultivars: N_2 fixation quantified by ^{15}N isotope dilution. *Agronomy Journal*, **76**, 498–502.

Rogers, D.D., Warren Jr, R.D., and Chamblee, D.S. (1982). Remedial postemergence legume inoculation with *Rhizobium*. *Agronomy Journal*, **74**, 613–19.

Roughley, R.J. (1988). Legume inoculants; their technology and application. In *Nitrogen fixation by legumes in Mediterranean agriculture*, (Ed. D.P. Beck, and L.A. Materon), pp. 259–269. Icarda, Nijhoff, Dordrecht.

Sadowsky, M.J., Tully, R.E., Cregan, P.B., and Keyser, H.H. (1987). Genetic diversity in *Bradyrhizobium japonicum* serogroup 123 and its relation to genotype-specific nodulation of soybean. *Applied and Environmental Microbiology*, **53**, 2624–30.

Schiffman, J., and Alper, Y. (1968). Inoculation of peanuts by application of *Rhizobium* suspension into the planting furrows. *Experimental Agriculture*, **4**, 219–26.

Schofield, P.R., Gibson, A.H. Dudman, W.F., and Watson, J.M. (1987). Evidence for genetic exchange and recombination of *Rhizobium* symbiotic plasmids in a soil population. *Applied and Environmental Microbiology*, **53**, 2942–7.

Sicardi de Mallorca, M., and Labandera, C. (1978). Quality control of commercial inoculants for legumes in Uruguay (1963–1975). *Revista Latino-americana de Microbiologia*, **20**, 153–60.

Skipper, H.D., Palmer, J.H., Giddens, J.E., and Woodruff, J.M. (1980). Evaluation of commercial soybean inoculants from South Carolina and Georgia. *Agronomy Journal*, **72**, 673–4.

Smith, R.S. (1987). Production and quality control of inoculants. In *Symbiotic nitrogen fixation technology*, (Ed. G.H. Elkan), pp. 391–411. Marcel Dekker, Inc., New York.

Smith, R.S., Ellis, M.A., and Smith, R.E. (1981). Effect of *Rhizobium japonicum* inoculant rates on soybean nodulation in a tropical soil. *Agronomy Journal*, **73**, 505–8.

Sobral, B.W.S., Sadowsky, M.J., and Atherly, A.G. (1990). Genome analysis of *Bradyrhizobium japonicum* serocluster 123 field isolates by using field inversion gel electrophoresis. *Applied and Environmental Microbiology*, 56, 1949–53.

Stacey, G. (1985). Other introductions into the environment. The Rhizobium experience. In *Engineered organisms in the environment: scientific issues*. Proceedings of a cross-disciplinary symposium held in Philadelphia, Pennsylvania, 10–13 June 1985. (Ed. H.O. Halvorson, D. Pramer, and M. Rogul), pp. 109–21. American Society for Microbiology, Washington DC.

Stowers, M.D. (1985). Carbon metabolism in *Rhizobium* species. *Annual Review of Microbiology*, **39**, 89–108.

Thompson, J.A. (1988). Selection of *Rhizobium* strains. In *Nitrogen fixation by legumes in Mediterranean agriculture*, (Ed. D.P. Beck, and L.A. Materon),

pp. 207–23. Icarda, Nijhoff, Dordrecht.

Viteri, S.E., and Schmidt, E.L. (1987). Ecology of indigenous soil rhizobia: response of *Bradyrhizobium japonicum* to readily available substrates. *Applied and Environmental Microbiology*, **53**, 1872–5.

Viteri, S.E., and Schmidt, E.L. (1989). Chemoautotrophy as a strategy in the ecology of indigenous soil bradyrhizobia. *Soil Biology and Biochemistry*, **21**, 461–3.

Weaver, R.W., and Frederick, L.R. (1972). Effect of inoculum size on nodulation of *Glycine max* (L.) Merril, variety Ford. *Agronomy Journal*, **64**, 597–9.

Weaver, R.W., and Frederick, R. (1974). Effect of inoculum rate on competitive nodulation of *Glycine max* (L.) Merril, 1: Greenhouse studies. *Agronomy Journal*, **66**, 229–33.

Weaver, R.W., Frederick, L.R., and Dumenil, L.C. (1972). Effect of soybean cropping and soil properties on numbers of *Rhizobium japonicum* in Iowa soils. *Soil Science*, **114**, 137–41.

Weber, D.F., Keyser, H.H., and Uratsu, S.L. (1989). Serological distribution of *Bradyrhizobium japonicum* from U.S. soybean production areas. *Agronomy Journal*, **81**, 786–9.

Young, J.P.W. (1985). *Rhizobium* population genetics: enzyme polymorphism in isolates from peas, clover, beans and lucerne grown at the same site. *Journal of General Microbiology*, **131**, 2399–408.

Young, J.P.W., and Wexler, M. (1988). Sym plasmid and chromosomal genotypes are correlated in field populations of *Rhizobium leguminosarum*. *Journal of General Microbiology*, **134**, 2731–9.

2

Bacteria and agricultural animals – survival and gene transfer

M.J. Bale and M. Hinton

Introduction

THE CURRENT DEBATE concerning the release of genetically engineered micro-organisms (GEMs) has exposed our inability to predict how they survive in the environment (Beringer and Bale 1988; Colwell *et al.* 1988). Concern has also been expressed about the potential for engineered genes to transfer and persist outside the original bacterial host (Curtiss 1988).

In this review we will discuss the potential for bacteria to survive and for genes to transfer in the farm environment. The bacteria that have been most widely investigated in this context are the animal pathogens. Drug resistance is one of the easier phenotypic traits to study and resistance genes are common in both pathogenic and commensal bacterial species (Hinton *et al.* 1986; Wray 1986). The original source of resistance genes was probably soil micro-organisms, although the route by which they transferred from the originating organisms to present-day animal pathogens and gut commensals was probably tortuous (Saunders 1984).

In the UK the Zoonoses Orders of 1973 and 1989 require that the isolation of salmonellas from food animals be reported to the Ministry of Agriculture, Fisheries and Food and that the identity of a proportion of the isolates is confirmed at either the Central Veterinary Laboratory or the Central Public Health Laboratory. A careful study of these salmonellas has allowed the monitoring of the evolution of clones of various salmonella serovars, e.g. *Salmonella typhimurium* (Threlfall and

Rowe 1984) and *Salmonella enteritidis* (Frost *et al.* 1989).

Survival of bacteria in the farm environment

Bacteria are capable of surviving in many niches on the farm, although their numbers can be expected to vary considerably depending on the genus of the bacterium and prevailing ambient conditions (Wray 1975).

Animals pollute their environment with faeces, and much of the feed they eat will have a high bacterial content as it will not have been heat treated. Similarly, the water they drink will not necessarily be potable and may contain many bacteria.

The time an individual bacterium survives may be relatively unimportant if the environment is continually being recontaminated, as it is with faeces. With the exception of certain potential pathogens, such as anthrax, the salmonellas and commensal organisms like the coliforms and enterococci, there is relatively little detail on the survival of animal-associated bacteria in the farm environment.

Animal feed

Animal feed is contaminated naturally with bacteria of many genera, although of the potential pathogens it is the salmonellas that have received the greatest attention and are most commonly present in feed ingredients of animal and vegetable origin (Wray and Sojka 1977; Williams 1981a,b). The coliforms and enterococci are also of interest either as indicators of faecal pollution or as potential sources of antiobiotic resistance genes.

Bacteria of these genera are capable of surviving in feed for many months when stored dry in ambient conditions (Williams and Benson 1978). The salmonellas and *Escherichia coli* are destroyed if feed is heated above 80°C, as might occur if the feed is pelleted using steam conditioning, although it is unwise to assume that this process will sterilize the feed completely (Stott *et al.* 1975; Cox *et al.* 1983).

In general, the water activity of animal feed is too low for bacterial multiplication to occur. However, feed may become hydrated either accidentally during manufacture or storage or by the drinking water of the animals, which itself may be contaminated with bacteria (Gauger and Greaves 1946). Under these circumstances bacteria, e.g. salmonellas, can multiply and gene transfer may occur (M.J. Bale *et al.* unpublished results).

Animal faeces and slurry

Salmonellas are the principal bacterial pathogens to be spread via manure and slurry, a mixture of faeces, urine and water. Salmonellas may actually grow in solid composted manure if the temperature is in the range of 20–40°C, the moisture content exceeds 20 per cent and the carbon to nitrogen ratio is in excess of 15 : 1. On the other hand, salmonella numbers decline in stored slurry (Jones 1976). Survival times for *Salmonella dublin* were up to 18–33 weeks (Findlay 1972). The rate at which this occurs is dependent on temperature and the concentration of solid matter, being least when the temperature is < 10°C and the solids concentration is 5 per cent or more (Jones 1976). The decline in salmonella numbers is due to the production of toxic acidic compounds by the 'natural' microbiota of the slurry and can be hastened by the addition of disinfectants such as lime, sodium hydroxide or formic acid.

Salmonellas may survive on the pasture for many months, although they persist for longer in the soil than on the herbage. Nevertheless, grazing animals rarely develop an active infection and consequently the spreading of slurry on pasture does not constitute a serious risk to animal health (Jones 1980).

Survival in litter and dust

The litter and dust associated with the animals' accommodation are important sites of infection, especially in intensively reared broiler chickens and hens. Harry (1964) showed that the high counts of *E. coli* found in dust from broiler houses survived for up to 32 weeks at low relative humidities, but declined rapidly at relative humidities of near 100 per cent. Salmonella survive for long periods in built-up litter in broiler houses, which usually consists of wood shavings, spilt food, feathers and faeces. *Salmonella thompson* has been found to survive for up to 20 weeks in fresh litter (Tucker 1967), and longer if the pens were left unoccupied. When old litter was used the same serovar survived for 4–5 weeks. This reduced survival in litter was also observed by Turnbull and Snoeyenbos (1973), who ascribed it to the production of ammonia, and a corresponding high pH, by the other microflora. The death rate was highest at high water activities where the competing microflora was most active. More recently, the survival of laboratory and wild type *E. coli* with and without an engineered plasmid has been examined in a variety of farm environments (Marshall *et al.* 1988). They showed that, after aerosol deposition, all strains survived best on wood shavings in a barn and were detectable after 10–20 days.

Survival on surfaces

There is a considerable body of work on the survival of bacteria on surfaces. The general consensus is that survival is best at low water activities (relative humidities) and on surfaces offering a large number of possible microniches (Bale *et al.* 1990) such as wood and fabrics. At high water activities (> 0.95) some bacterial species will survive well and may grow (McEldowney and Fletcher 1988) and conjugate (M.J. Bale *et al.* unpublished results). It is difficult to envisage a situation where bacteria will be exposed and survive on a surface without a 'carrier' material such as faeces or saliva. In many ways survival experiments done on clean surfaces may not be relevant to survival in the farm environment.

Colonization of animal intestines

As mentioned in the introduction, the survival of bacteria in the farm environment is influenced by the presence of animals. For the many bacterial species able to colonize the gastrointestinal tract of animals, the ability to survive outside the animals is sufficient to ensure that they will be ingested by animals and subsequently excreted in higher numbers. The ecology of the intestinal tract is complex and involves much competition for available colonization sites and nutrients (Freter 1986; Fuller 1989), however, the composition of the microbiota shows large variations with time. Particular clonal types (serotypes, biotypes) of *E. coli* and phage types of *Salmonella* are known to persist in populations of animals for long periods (Threlfall and Rowe 1984; Bennett and Linton 1986).

Other bacterial species may well be incapable of colonizing the intestine either as commensals or pathogens and could therefore be considered unimportant. This may be a false assumption, however, as many of the bacteria forming the natural microbiota in slurry and deep litter are also dependent on animals for their long-term survival. It is probably unwise, therefore, to assume that bacteria of any species will disappear rapidly from the farm environment.

Bacterial gene transfer in animals

There is ample evidence, both direct and indirect, that plasmid transfer occurs in animals. The direct evidence arises from short-term experiments involving the dosing of animals with cultures of plasmid-containing donor and genetically marked recipient bacteria. Selective media are then used to isolate the transconjugants from faeces and gut contents.

The indirect evidence of widespread plasmid transfer in animal populations comes from epidemiological studies, usually of antibiotic-resistance or virulence plasmids. A commonly studied antibiotic-resistance marker is that for trimethoprim, which is one of the few 'novel' antimicrobial compounds to be introduced during the last 40 years (Amyes 1986).

Direct evidence of plasmid transfer

A detailed study of the potential for plasmid transfer *in vivo* was carried out by Smith (1970). In a series of experiments, 1-day-old chicks were given water containing donor strains at approximately 10^8/ml for 3 days. On the fourth day they were fed 10^9 recipient bacteria. Swabs of cloacal faeces were taken immediately, and then after 3 and 6 days. Plasmid transfer occurred between *E. coli* donors and *S. typhimurium* recipients. Six out of fifteen different donor strains, isolated originally from pigs, chickens and humans, transferred resistance markers. These donors were those with the highest transfer frequencies *in vitro* ($> 10^2$ per recipient) and which colonized the intestinal tract satisfactorily. When an isolate that was an efficient donor was used with a variety of *Salmonella* and *E. coli* recipient strains there was no significant difference between the incidence of transfer of the antiobiotic resistance markers studied.

Analysis of the intestinal contents indicated that the numbers of donors and recipient were low in the crop, gizzard and small intestine, whilst the majority of bacteria, including the transconjugants, were present in the caecum and, to a lesser extent, in the cloaca. In the same study separate groups of chicks were inoculated with *E. coli* donors and *S. typhimurium* recipients and the two groups were mixed. Both strains could be identified in birds from both initial groups after 2 days. Antibiotic-resistant recipients (transconjugants) were also isolated from both initial groups, indicating the rapid spread of bacteria within the flock.

The frequency of transfer of antibiotic resistance genes was generally increased when the relevant antibiotics were included in the feed (Smith 1970). When oxytetracycline and sulphadimidine were included at high levels there was a reduction in the isolation of recipients from the faeces. However, this tended to lead to transfer of a wider range of antibiotic-resistance genes, including those selected for by the antibiotics included in the feed. This was not the case for chloramphenicol, which was ascribed to the fact that this antibiotic was degraded before it reached the caecum, where transfer was thought to occur. Jones *et al.* (1984) found that the administration of chlortetracycline in the feed, along with *E. coli* containing plasmid R100, to pigs increased the incidence and persistence of the strain in the faeces. Conjugation *in vivo* only occurred in those pigs fed chlortetracycline and the transconjugants eventually spread and colonized chicks housed with the pigs.

The factors affecting conjugation, such as adequate intestinal colonization and antibiotic-selective pressures, have also been shown to be important in other studies. Transfer was more readily observed in the intestine of young chicks with an immature gut microbiota, which were easily colonized by the donor and recipient bacteria. On the other hand, transfer was more difficult to detect in older animals or in certain animal species because of the difficulty in establishing the donors and recipients in the intestinal tract.

Sansonetti *et al.* (1980) investigated the transfer of wild type plasmid R64 and derepressed mutants (R64-11) in gnotoxenic (germ-free) chickens, using *E. coli* K-12 derivatives as donors and recipients. The derepressed plasmid transferred at higher frequencies than the wild type plasmid, presumably reflecting the higher transfer frequencies associated with derepressed plasmids *in vitro*. Similarly, Lafont *et al.* (1984) studied the transfer of chromosomal genes from an *E. coli Hfr* derivative in the intestine of gnotoxenic chicks. They observed multiple recombinants and relatively high transfer frequencies. The use of germ-free animals is even less representative of natural conditions than using young animals with an immature gut microbiota. Through the use of animal and chemostat models it has been demonstrated that the indigenous microbiota is a barrier to the colonization of invading *E. coli* and limits their ability to act as donors (Duval-Iflah *et al.* 1980; Freter *et al.* 1983; Freter 1986). Using a combination of continuous flow cultures and germ-free mice it was possible to mimic the natural caecal microflora using *E. coli* and 95 strictly anaerobic bacterial strains (Freter *et al.* 1983). The rate of plasmid transfer from an invading strain was highly dependent on colonization to prolong the residence-time of the plasmid carrying donor in the gut and to prevent the selection of 'fitter', plasmid-free segregants. This presumably explained the finding, particularly in human studies, that transfer frequencies were low unless there was some perturbation of the intestinal microbiota, such as fasting or antibiotic therapy (Smith 1970; Anderson *et al.* 1973; Jones *et al.* 1984; Freter 1986), although this may not always be the case (Williams 1977; Duval-Iflah *et al.* 1980).

The models of Freter (Freter *et al.* 1983) suggest that plasmid transfer does not occur between bacteria attached to the gut wall and those in the gut lumen. However, Duval-Iflah *et al.* (1980) found that a strain of *Serratia liquefaciens*, encoding resistance to 14 antibiotics on a 110 kilobase (kb) plasmid, could act as a donor to *E. coli* when transient in the gut of gnotobiotic mice containing a full human microflora. The presence of approx-imately 10^{10} *Bacteroides* spp. per gram of faeces did not entirely prevent conjugation while the administration of ampicillin promoted conjugation.

Gene transfer also occurs outside animals. Linton *et al.* (1981) suggested that the distribution of IncHII plasmids, which have a thermosensitive conjugation system in *E. coli* and *S. typhimurium*, during an outbreak of salmonellosis in calves was due to plasmid transfer, which occurred in the animals' environment and not in the intestine. In a related study (Timoney and Linton 1982), calves were fed cultures of donor strain, *E. coli* (pJT4), and genetically marked recipients. After 1–2 days the plasmid was detected in the recipients and in other *E. coli* recovered from the faeces. When muzzled calves were used transconjugants were not detected in the faeces, because the calves had had no contact with their environment. *In vitro* experiments showed that pJT4 transfer occurred in freshly voided faeces at 30°C but not at 37°C. This indicated that the transfer occurred in voided faeces in the calves' environment and that they subsequently ingested the transconjugants. Although antibiotics were used as a selective pressure it appeared that the enhanced colonization and persistence associated with the plasmid-containing bacteria provided a sufficient selective pressure to favour plasmid transfer (Timoney and Linton 1982).

The potential for gene transfer to occur on the skin of animals has been investigated by Naidoo and Lloyd (1984), who transferred resistance markers from multiresistant coagulase-negative *Staphylococci* spp., isolated from humans, to isolates from various animal species. They found that transfer occurred in broth and also on the skin of humans, mice and dogs, with transfer frequencies of between 10^{-6} and 10^{-8} per recipient. The transfer frequencies on skin were frequently higher than those in broth. Agarose gel electrophoresis showed the presence of a large number of plasmids, which together with the observation that transfer occurred in the presence of DNase, suggested that the transfer was by conjugation rather than transformation or transduction. The authors concluded that there was a potential for antibiotic resistance in *Staphylococcus* spp. to be transferred between animals and man, with pet dogs providing a link.

Indirect evidence of gene transfer

Epidemiological studies of plasmids from different bacterial isolates provide a large body of evidence to suggest that plasmids transfer between bacteria in the farm environment.

Chaslus-Dancla *et al.* (1987a) isolated eight trimethoprim-resistant *E. coli* from broiler chicks over a 6-week period. The resistance was mediated by a gene on a 65-kb plasmid, first isolated from a typical *E. coli* isolate. During the seventh week of the study the same plasmid was found in an unusual urease-negative cellobiose-positive *E. coli*. The authors concluded that the plasmid had transferred into a strain which itself was an efficient colonizer of the intestine.

In another study, a 100-kb plasmid encoding gentamicin and cipramycin resistance (AAC(3) IV) was isolated from healthy and sick lambs and calves (Chaslus-Dancla *et al.* 1987b). The distribution of the resistance marker was confirmed by DNA hybridization and this strongly suggested that plasmid transfer had occurred. Work by Campbell and colleagues (Campbell *et al.* 1986; Campbell and Mee 1987) on trimethoprim-resistance plasmids from *E. coli* isolated from humans and pigs also showed evidence of plasmid transfer. The plasmids consisted of one group with sizes of between 50 and 65 kb and one with sizes of >120 kb. The latter group all belonged to the IncFIV incompatibility group and could be further subdivided by restriction endonuclease analysis. Groups A, B and C were detected exclusively in human isolate, groups E and F were detected in porcine isolates and groups D and D′ were from both (Campbell and Mee 1987). The distribution of plasmid groups between *E. coli* biotypes suggested transfer of bacteria and plasmids both between and within human and porcine populations. In the case of the D and D′ plasmids it appeared that an ancestral plasmid initially spread into porcine *E. coli*, acquired mercury resistance and the type II dihydrofolate reductase (trimethoprim resistance) gene and was transferred back into *E. coli* colonizing the human population (Campbell *et al.* 1986). Similarly, different bacterial species or biotypes isolated from animal and human sources often contain closely related R-plasmids, which strongly suggested that plasmid transfer and subsequent bacterial colonization of intestinal tracts were common events (Towner *et al.* 1986).

An earlier study of antibiotic-resistance plasmids

in *Salmonella* spp. and in *E. coli* isolated from pigs in Japan (Ishiguro *et al.* 1980) showed the widespread distribution of IncHI, IncIα and IncFII plasmids with similar markers, in both bacterial species. Although restriction endonuclease analysis was not done, the authors concluded that plasmid transfer occurred in the pig or its surroundings. A study of IncFII plasmids encoding chloramphenicol resistance in *E. coli* isolated from pigs in Denmark (Jorgenson *et al.* 1980) indicated several different restriction patterns. That this was often coupled with additional antibiotic resistance determinants suggested that the plasmids had evolved from a common ancestor. The plasmids were also found in a variety of different serotypes of *E. coli*, indicating that the plasmids had evolved, spread by conjugation and subsequently moved between Danish farms.

As mentioned previously, the study of salmonella serovars isolated in the UK has shown the role of gene transfer in the evolution of bacteria. Prior to 1974 the predominant phage type of *S. typhimurium* isolated from calves was type 49 (Threlfall and Rowe 1984). In 1974 a new phage type (DT204) was isolated. This had evolved from the DT49 strain by the acquisition of a plasmid encoding tetracycline resistance and a 'typing phage restriction' phenotype. Additional drug resistance markers were subsequently acquired, carried on a conjugative IncHII plasmid. After 1–2 years a new phage type (DT204a) was recognized. This had evolved following the acquisition of a lysogenic bacteriophage that altered the phage type. Subtle shifts involving the loss of the 'typing phage restriction' phenotype from the small Tet^r plasmid, the acquisition of a transposon encoding trimethoprim resistance on the IncHII plasmid and a different lysogenic phage led to the appearance of DT204c, which became the predominant phage type infective calves for several years (Threlfall and Rowe 1984). More recently, *Salmonella enteritidis* phage type 4, which has been associated with an epidemic of food poisoning in humans in recent years, has been shown to be converted to a multiresistant strain of phage type 24 by the acquisition of a 37.5-kb IncN plasmid (Frost *et al.* 1989).

Gene transfer by transformation and transduction

The examples detailed above assume that the mechanism of gene transfer is plasmid-encoded conjugation. However, the potential also exists for gene transfer by transduction (involving bacteriophage) or transformation (uptake of exogenous DNA).

Transformation can be induced readily in the laboratory but probably occurs infrequently or not at all in nature because only a limited number of bacterial genera can become naturally competent. These include *Bacillus* spp., *Neissera* spp., *Acinetobacter* spp., *Streptococcus* spp. and certain *Pseudomonas* spp. (Stewart and Carlson 1986). Experiments by Griffith in the 1920s showed the transforming principle of DNA *in vivo* using rough and smooth variants of *Pneumococcus* (Griffith 1928). More recently it has been shown that natural *in vitro* transformation of the *Neisseria gonorrhoea* pilin gene DNA can be the basis of pilus antigenic variation (Seifert *et al.* 1988), with the suggestion that this may occur *in vivo*. The stability of DNA *in vivo* is unknown, although it is known to be rapidly degraded in freshwater ecosystems (Paul *et al.* 1989). However, in sediments it has been shown that DNA may absorb to sand and clay particles, where it is protected from enzymatic degradation while still accessible for transformation by species such as *Bacillus* (Lorenz *et al.* 1988). Because transformation is limited to certain genera, the stability of DNA is uncertain and the process is generally dependent on homologous DNA, transformation is unlikely to be a major mechanism of gene transfer.

Transduction involves the packaging of host DNA into the capsids of infecting bacteriophages, followed by release and infection of a new host. Transduction has been studied in fresh water (Saye *et al.* 1987) and in soil (Germida and Khachatourians 1988; Zeph *et al.* 1988) but not, apparently, in animals. Dhillon and Dhillon (1981) investigated the incidence of lysogenic enterobacteria from domesticated animals and showed that at least 21 per cent were lysogenic with their assay system. Similarly, Smith and Lovell (1985) showed that lysogenic phages were sufficiently common in *Salmonella* spp. to cause misleading results in triparental mating experiments designed to detect Tra$^+$ virulence plasmids. Lysogenic phages were frequently responsible for the transduction of a Tra$^-$ R-plasmid to both the final recipient and the original donor. In *Salmonella* spp., lysogenic phages are often associated with antigenic variation, virulence alteration and with a change in the phage type during epidemics (Threlfall and Rowe 1984).

Bacteriophages have been used to control diarrhoea in calves (Smith *et al.* 1987a,b). The dosing of calves with large numbers of bacteriophage may increase the potential for transduction in the intestine and the excretion of transducing phage into the environment. Phages are generally considered to package moderate amounts of either plasmid or chromosomal DNA and to survive for long periods, although electron microscopy has indicated that shear forces may damage tail and tail-fibre structures, perhaps resulting in reduced ability to infect bacteria (Reanney *et al.* 1983). However, the majority of phages have host ranges limited to species or subspecies, a principle used in phage-typing of bacteria, and that probably reduces the likelihood of widespread gene transfer occurring by this mechanism.

The use of genetically engineered micro-organisms

Probiotics

The term probiotic has been defined as a 'live microbial feed supplement which beneficially affects the host animal by improving its intestinal microbial balance' (Fuller 1989). Probiotics are intended for use in situations where the animal does not develop a normal gut microbiota, as possible may occur in calves, piglets and young broiler chicks reared intensively. Currently, probiotics consist of *Lactobacillus* spp., *Enterococcus* spp. and/or *Bifidobacterium* spp. (Fuller 1989). Once the principle behind probiotics, and the related practice of competitive exclusion (Impey and Mead 1989), has been fully accepted as a commercially viable husbandry practice, genetically manipulated probiotic organisms will probably be developed. Several possible traits could be included in a genetically engineered micro-organism (GEM) probiotic.

Certain forage feeds, such as barley, rye and oats contain pentosans and β-glucans, which interfere with digestion and nutrient absorption (Nasi 1988). Cloning genes encoding for pentosanase and β-glucanase production into suitable strains would allow the use of alternative cereals to corn and wheat. A wider understanding of the mechanism of adhesion could result in the use of GEM probiotics containing the 'virulence associated' adhesion genes from different bacterial species or for more than one tissue type in the intestinal tract. Recent work has shown that is is possible to manipulate *Lactobacillus* spp. in such a manner (McCarthy *et al.* 1988). It has also been suggested that probiotic organisms could be engineered so that they synthesize vitamins or produce antimicrobial and anti-helmintic substances (Tannock 1988).

Live vaccines

Genetic engineering techniques can be applied to the production of attenuated bacterial strains containing cloned antigens from organisms causing serious and costly diseases. An example of an attenuated strain has been described by Curtiss (1988). *Salmonella typhimurium* invades the gut-associated lymphoid tissue following colonization and this is followed by the invasion of other tissues and occasionally fatal bacteraemia. Curing the strain of a virulence plasmid did not affect the migration to lymphoid tissue, but prevented further invasion. Similarly, mutations in the *cya/crp* genes, which encode for adenylate cyclase and the cAMP receptor protein, also reduced virulence. A cloning vector that does not use antibiotic resistance genes has been constructed and can be used if the host is an aspartate semialdehyde dehydrogenase (*asd*) deletant. The vector encodes a functional *asd* gene, which complements the chromosomal mutation and allows virtually any antigenic determinant to be cloned and delivered to the lymphoid tissue in a highly immunogenic form. The use of such strains in agricultural animals or humans where plasmid transfer could occur would require additional safeguards to limit plasmid transfer. This could consist of the gene being cloned into a transposon and inserted into the chromosome, or by flanking the gene with chromosomal sequences and selecting for recombination into the chromosome.

Unintentional use of GEMs in animal husbandry

There are two other means by which GEMs may be consumed by animals. First, by the use of forages inoculated with engineered lactic acid bacteria to assist silage production (Seale 1986) and, secondly, by using as a feedstuff by-products of industrial fermentations using GEMs. The market for silage inoculants is increasing rapidly and although there are no engineered silage inoculants available at present, there are advantages to be accrued from manipulating strains to produce, or to overpro-

duce, enzymes and antimicrobial compounds. The silage will then be fed to ruminants and the survival and fate of the silage inoculants in the intestinal tract will have to be evaluated.

The likely incease in biotechnology industrial fermentor capacity will result in large amounts of microbial cell debris requiring disposal. It has been suggested that this would be suitable for use as an animal feedstuff. If engineered bacteria were used then it would be important to ensure that downstream processing, drying and any heating during feed compounding results in the killing of the engineered strains, although it may still be possible for any recombinant DNA molecules to transform rumen and gut bacteria and lead to the incorporation of their DNA in the intestinal microbiota.

Potential for the spread of GEMs by animals

By way of summary it may be useful to consider the potential for GEMs to be spread by agricultural and farm animals. The large-scale release of engineered micro-organisms in agriculture may lead to contact with domestic animals, either deliberately as part of the release programme or accidentally, after the animals have eaten treated crops or waste products. To gain permission to release a product containing GEMs, proof that any engineered genes will have been inserted into the chromosome or placed on non-mobilizable plasmid vectors will be needed (Molin *et al.* 1987; Bej *et al.* 1988). The bacterial hosts themselves may well be attenuated in the case of live vaccines. In any event, if the worst-case scenario is considered, the GEM may survive and colonize an animal host and transfer genes into indigenous bacteria. The consequences of this are unlikely to be serious. The genes in a live vaccine, for instance, will not encode for virulence, and if acquired by a native bacterium will merely make the transconjugants highly immunogenic and liable to be destroyed by the host animal. Similarly, the acquisition of catabolic genes for feed improvement will probably only make the process of feed conversion more efficient. Even the acquisition of antibiotic marker genes from the GEM will be negligible compared to the large gene pool of antibiotic resistance that is known to exist in the intestinal microbiota of farm animals, particularly those that are reared intensively (Linton 1977).

In conclusion, bacteria can survive in the

environment and are likely to continue to do so. Similarly, gene transfer is known to occur and the emergence of antibiotic resistance in animal pathogens illustrates this phenomenon particularly well. The challenge for those seeking to release GEMs will be to tailor their product such that the survival and spread of the bacteria and genotype is not of significance from the point of view of environmental pollution, public health and commerce.

References

Amyes, S.G.B. (1986). Epidemiology of trimethoprim resistance. *Journal of Antimicrobial Chemotherapy* (Suppl. C), **18**, 215–21.

Anderson, J.D., Gillespie, W.A., and Richmond, M.H. (1973). Chemotherapy and antibiotic resistance transfer between enterobacteria in the human gastro-intestinal tract. *Journal of Medical Microbiology*, **42**, 277–83.

Bale, M.J., Bennett, P.M., Hinton, M., and Beringer, J.E. (1990). The survival of GEMs and bacteria on inanimate surfaces and in animals. In *Bacterial genetics in natural environments*, (Ed. J.C. Fry and M.J. Day), pp. 231–9. Chapman and Hall, London.

Bej, A.K., Parhn, M.H., and Atlas, R.M. (1988). Model suicide vector for containment of genetically engineered micro-organisms. *Applied and Environmental Microbiology*, **54**, 2472–7.

Bennett, P.M., and Linton, A.H. (1986). Do plasmids influence the survival of bacteria. *Journal of Antimicrobial Chemotherapy* (Suppl. C), **18**, 123–6.

Beringer, J.E., and Bale, M.J. (1988). The survival and persistence of genetically-engineered micro-organisms. In *The release of genetically engineered micro-organisms*, (Ed. M. Sussman, C.H. Collins, F.A. Skinner, and D.E. Stewart-Tull), pp. 29–46. Academic Press, London.

Campbell, I.G., and Mee, B.J. (1987). Mapping of trimethoprim resistance genes from epidemiologically related plasmids. *Antimicrobial Agents and Chemotherapy*, **31**, 1440–1.

Campbell, I.G., Mee, B.J., and Nikoletti, S.M. (1986). Evolution and spread of IncFIV plasmids conferring resistance to trimethoprim. *Antimicrobial Agents and Chemotherapy*, **29**, 807–13.

Chaslus-Dancla, E., Gerbaud, G., Lagorce, M., Lafont, J.-P., and Courvalin, P. (1987a). Persistance of an antibiotic resistance plasmid in intestinal *Escherichia coli* of chickens in the absence of selective pressure. *Antimicrobial Agents and Chemotherapy*, **31**, 784–8.

Chaslus-Dancla, E., Lagorce, M., Lafont, J.-P., Gerbaud, G., Courvalin, P., and Martel, J.L. (1987b). Probable transmission between animals of a plasmid encoding aminoglycoside 3-*N*-acetyltransferase IV and dihydrofolate reductase I. *Veterinary Microbiology*, **15**, 97–104.

Colwell, R.R., Somerville, C., Knight, I., and Straub, W. (1988). Detection and monitoring of genetically engineered micro-organisms. In *The release of genetically engineered micro-organisms*, (Ed. M. Sussman, C.H. Collins, F.A. Skinner, and D.E. Stewart-Tull), pp. 47–60. Academic Press, London.

Cox, N.A., Bailey, J.S., and Thomson, J.E. (1983). Salmonella and other Enterobacteriacae found in commercial poultry feed. *Poultry Science*, **62**, 2169–75.

Curtiss, R. (1988). Engineering organisms for safety: what is necessary? In *The release of genetically engineered micro-organisms*, (Ed. M. Sussman, C.H. Collins, F.A. Skinner, and D.E. Stewart-Tull), pp. 7–20. Academic Press, London.

Dhillon, T.S., and Dhillon, E.K. (1981). Incidence of lysogeny, colicinogeny and drug resistance in enterobacteria isolated from sewage and rectums of humans and some domesticated species. *Applied and Environmental Microbiology*, **41**, 894–902.

Duval-Iflah, Y., Raibaud, P., Tancrede, C., and Rousseau, M. (1980). R-plasmid transfer from *Serratia liquefaciens* to *Escherichia coli* in vitro and in vivo in the digestive tract of gnotobiotic mice associated with human fecal flora. *Infection and Immunity*, **28**, 981–90.

Findlay, C.R. (1972). The persistence of *Salmonella dublin* in slurry tanks and on pasture. *Veterinary Record*, **91**, 233–5.

Freter, R. (1986). The need for mathematical models in understanding colonisation and plasmid transfers in the mammalian intestine. In *Banbury report 24: Antibiotic resistance genes: Ecology, transfer and expression*, (Ed. S.B. Levy and R.P. Novick), pp. 81–93. Cold Spring Harbor, New York.

Freter, R., Freter, R.R., and Brickner, H. (1983). Experimental and mathematical models of *Escherichia coli* plasmid transfer in vitro and in vivo. *Infection and Immunity*, **39**, 60–84.

Frost, J.A., Ward, L.R., and Rowe, B. (1989). Acquisition of a drug resistance plasmid converts *Salmonella enteritidis* phage type 4 to phage type 24. *Epidemiology and Infection*, **103**(2), 243–8.

Fuller, R. (1989). Probiotics in man and animals. *Journal of Applied Bacteriology*, **66**, 365–78.

Gauger, H.C., and Greaves, R.F. (1946). Isolation of *Salmonella typhimurium* from drinking water in an infected environment. *Poultry Science*, **25**, 476–8.

Germida, J.J., and Khachatourians, G.G. (1988). Transduction of *Escherichia coli* in soil. *Canadian Journal of Microbiology*, **34**, 190–3.

Griffith, F. (1928). The significance of pneumococcal types. *Journal of Hygiene (Cambridge)*, **27**, 113–58.

Harry, E.G. (1964). The survival of *Escherichia coli* in the dust of poultry houses. *Veterinary Record*, **76**, 466–70.

Hinton, M., Kaukas, A., and Linton, A.H. (1986). The

ecology of drug resistance in enteric bacteria. *Journal of Applied Bacteriology*, **61**, Symposium Supplement, 77S–92S.

Impey, C.S., and Mead, G.C. (1989). Fate of salmonellas in the alimentary tract of chicks pre-treated with mature caecal microflora to increase colonisation resistance. *Journal of Applied Bacteriology*, **66**, 469–75.

Ishiguro, N., Goto, J., and Sato, G. (1980). Genetical relationships between R-plasmids derived from *Salmonella* and *Escherichia coli* obtained from a pig farm, and its epidemiological significance. *Journal of Hygiene*, **84**, 365–79.

Jones, P.W. (1976). The effect of temperature, solids content and pH on the survival of salmonellas in cattle slurry. *British Veterinary Journal*, **132**, 284–93.

Jones, P.W. (1980). Health hazards associated with the handling of animal wastes. *Veterinary Record*, **106**, 4–7.

Jones, F.T., Langlois, B.E., Cromwell, G.L., and Hays, V.W. (1984). Effect of chlortetracycline on the spread of R-100 plasmid-containing *Escherichia coli* BEL15R from experimentally-infected pigs to uninfected pigs and chicks. *Journal of Animal Science*, **58**, 519–26.

Jorgenson, S.T., Grinstead, J., Bennett, P., and Richmond, M.H. (1980). Persistence and spread of a chloramphenical resistance-mediating plasmid in antigenic types of *Escherichia coli* pathogenic for piglets. *Plasmid*, **4**, 123–9.

Lafont, J.-P., Bree, A., and Platt, M. (1984). Bacterial conjugation in the digestive tracts of gnotoxenic chickens. *Applied and Environmental Microbiology*, **47**, 639–42.

Linton, A.H. (1977). Antibiotic resistance: the present situation reviewed. *Veterinary Record*, **100**, 354–60.

Linton, A.H., Timoney, J.F., and Hinton, M. (1981). The ecology of chloramphenical-resistance in *Salmonella typhimurium* and *Escherichia coli* in calves with endemic salmonella infection. *Journal of Applied Bacteriology*, **50**, 115–29.

Lorenz, M.G., Aardema, B.W., and Wackernagel, W. (1988). Highly efficient genetic transformation of *Bacillus subtilis* attached to sand grains. *Journal of General Microbiology*, **34**, 107–12.

McCarthy, D.M., Lin, J.H.C., Rinckel, J.A., and Savage, D.C. (1988). Genetic transformation in *Lactobacillus* spp. strain 100-33 of the capacity to colonize the non-secreting gastric epithelium in mice. *Applied and Environmental Microbiology*, **54**, 416–22.

McEldowney, S., and Fletcher, M. (1988). The effect of temperature and relative humidity on the survival of attached bacteria on dry solid surfaces. *Letters in Applied Microbiology*, **7**, 83–6.

Marshall, B., Flynn, P., Kamely, D., and Levy, S.B. (1988). Survival of *Escherichia coli* with and without ColE1::Tn5 after aerosol dispersal in a laboratory and a farm environment. *Applied and Environmental Microbiology*, **54**, 1776–83.

Molin, S., Klemm, P., Poulsen, R.K., Biehl, H.,

Gerdes, K., and Andersson, P. (1987). Conditional suicide system for containment of bacteria and plasmids. *BioTechnology*, **5**, 1315–18.

Naidoo, J., and Lloyd, D.H. (1984). Transmission of genes between staphylococci on skin. In *Antimicrobials and agriculture*, (Ed. M. Woodbine), pp. 282–95. Butterworths, London.

Nasi, M. (1988). Enzyme supplementation of laying hen diets based on barley and oats. In *Biotechnology in the feed industry*, (Ed. T.P. Lyons), pp. 199–204. Alltech, Nicholasville, Kentucky.

Paul, J.H., Jeffrey, W.H., David, A.W., Deflaun, M.F., and Cazares, L.H, (1989). Turnover of extracellular DNA in eutrophic and oligotrophic freshwater environments of southwest Florida. *Applied and Environmental Microbiology*, **55**, 1823–8.

Reanney, D.C., Gowland, P.C., and Slater, J.H. (1983). Genetic interactions among microbial communities. In *Microbes in their natural environment*, (Ed. J.H. Salter, R, Whittenbury, and J.W.T. Wimpenny), pp. 379–421. Cambridge University Press, Cambridge.

Sansonetti, P., Lafont, J.-P., Jaffee-Brachet, A., Guillot, J.-F., and Chaslus-Dancla, E. (1980). Parameters controlling interbacterial plasmid spreading in a gnotoxenic chicken gut system: influence of plasmid and bacterial mutations. *Antimicrobial Agents and Chemotherapy*, **17**, 327–33.

Saunders, J.R. (1984). Genetics and evolution of antibiotic resistance. *British Medical Bulletin*, **40**, 54–60.

Saye, D.J., Ogunseitan, O., Sayler, G.S., and Miller, R.V. (1987). Potential for transduction of plasmids in a natural freshwater environment: Effect of plasmid donor concentration and a natural microbial community on transduction in *Pseudomonas aeruginosa*. *Applied and Environmental Microbiology*, **53**, 987–95.

Seale, D. (1986). Bacterial inoculants as silage additives. *Journal of Applied Bacteriology*, **61**, Symposium Supplement, 9S–26S.

Seifert, H.S., Ajioka, R.S., Marchal, C., Sparling, P.F., and So, M. (1988). DNA transformation leads to pilin antigenic variation in *Neisseria gonorrhoeae*. *Nature*, **336**, 392–5.

Smith, H.W. (1970). The transfer of antibiotic-resistance between strains of enterobacteria in chicken, calves and pigs. *Journal of Medical Microbiology*, **3**, 165–80.

Smith, H.W., and Lovell, M.A. (1985). Transduction complicates the detection of conjugative ability in lysogenic Salmonella strains. *Journal of General Microbiology*, **131**, 2087–9.

Smith, H.W., Huggins, M.B., and Shaw, K.M. (1987a). The control of experimental *Escherichia coli* diarrhoea in calves by means of bacteriophages. *Journal of General Microbiology*, **133**, 1111–26.

Smith, H.W., Huggins, M.B., and Shaw, K.M. (1987b). Factors influencing the survival and multiplication of bacteriophages in calves and their environments. *Journal of General Microbiology*, **133**, 1127–35.

Stewart, G.J., and Carlson, C.A. (1986). The biology of natural transformation. *Annual Review of Microbiology*, **40**, 211–35.

Stott, J.A., Hodgson, J.E., and Chaney, J.C. (1975). Incidences of Salmonellae in animal feed and the effect of pelleting on content of Enterobacteriacae. *Journal of Applied Bacteriology*, **39**, 41–6.

Tannock, G.W. (1988). Molecular genetics: A new tool for investigating the microbial ecology of the gastrointestinal tract. *Microbial Ecology*, **15**, 239–56.

Threlfall, E.J., and Rowe B. (1984). Antimicrobial drug resistance in salmonellae in Britain – a real threat to public health? In *Antimicrobials and agriculture*, (Ed. M. Woodbine), pp. 513–24. Butterworths, London.

Timoney, J.F., and Linton, A.H. (1982). Experimental ecological studies on H2 plasmids in the intestines and faeces of the calf. *Journal of Applied Bacteriology*, **52**, 417–24.

Towner, K.J., Wise, P.J., and Lewis, M.J. (1986). Molecular relationship between trimethoprim R plasmids obtained from human and animal sources. *Journal of Applied Bacteriology*, **61**, 535–40.

Tucker, J.F. (1967). Survival of salmonellae in built-up litter for housing of rearing and laying fowls. *British Veterinary Journal*, **123**, 92–103.

Turnbull, P.C.B., and Snoeyenbos, G.H. (1973). The roles of ammonia, water activity and pH in the salmonellacidal effect of long-used poultry litter. *Avian Diseases*, **17**, 72–86.

Williams, P.H. (1977). Plasmid transfer in the human alimentary tract. *FEMS Microbiology Letters*, **2**, 91–5.

Williams, J.E. (1981a). Salmonellas in poultry feeds – a world wide review. Part I. *World's Poultry Science Journal*, **37**, 6–25.

Williams, J.E. (1981b). Salmonellas in poultry feeds – a world wide review. Part II. *World's Poultry Science Journal*, **37**, 91–5.

Williams, J.E., and Benson, S.T. (1978). Survival of *Salmonella typhimurium* in poultry feed and litter at three temperatures. *Avian Diseases*, **22**, 742–7.

Wray, C. (1975). Survival and spread of pathogenic bacteria of veterinary importance within the environment. *The Veterinary Bulletin*, **45**, 543–50.

Wray, C. (1986). Some aspects of the occurrence of resistant bacteria in the normal animal flora. *Journal of Antimicrobial Chemotherapy* (Suppl. C), **18**, 141–7.

Wray, C., and Sojka, W.J. (1977). Reviews of the progress of dairy science: *Salmonella* infections in cattle. *Journal of Dairy Research*, **44**, 383–425.

Zeph, L.R., Onaga, M.A., and Stotzky, G. (1988). Transduction of *Escherichia coli* by bacteriophage P1 in soil. *Applied and Environmental Microbiology*, **54**, 1731–7.

3

Prospects for the use of selected xenobiotic-degrading and genetically engineered micro-organisms in the treatment of chemical wastes

Nicholas C. McClure and Andrew J. Weightman

Introduction

IN 1980 THE UK GOVERNMENT published the report of a joint working party, under the chairmanship of Dr A. Spinks, that had been established to study the industrial applications of biotechnology. The Spinks report (Anonymous 1980), along with others published around that time, foreshadowed various developments in microbiology, including applications in waste treatment. More specifically, the potential for use of genetically engineered micro-organisms (GEMs) in the detoxification of recalcitrant chemical wastes was highlighted. Ten years later, and as the title of this chapter indicates, we are still discussing the potential practical applications of genetically engineered microbes in waste treatment. However, progress towards new, biotechnologically based 'clean technologies' has been far from slow and the problems of establishing specialized biodegradative functions in large-scale, open treatment plants, with or without the use of genetic engineering, are now more apparent.

The last decade has seen widespread advances in our understanding of biodegradation. Two important research developments have been the extension and application of recombinant techniques for analysis and manipulation of biodegradative genes, and an improved understanding of biodegradative processes, particularly those occurring in anoxic environments. In addition, new methodologies are being devised for the study of biodegradation using laboratory microcosms. Microbial inoculants are being used with variable degrees of success for *in situ* treatment of chemically contaminated ecosystems. Also an outline rationale for the assessment of risks and environmental impact associated with release of GEMs containing cloned catabolic genes is receiving international attention.

For the purpose of this review a distinction will be made between applications of microbiology in waste treatment-associated processes, such as nitrification, floc formation and general reduction of biological oxygen demand, which we shall not discuss, and applications in treatments of organic 'micropollutants', upon which we shall focus. In particular, we will consider advances in the research effort over the last decade, which are providing a clearer understanding of biodegradation, a better definition of pollution problems that are amenable to biological solutions and development of strategies for treatment.

Pollutants and pollution problems

An important factor in defining pollution problems has been the collation and analysis of data that have been used to identify and list 'priority pollutants'. The US Environmental Protection Agency's list, published in 1979 (Keith and Telliard 1979) has probably been the most influential. As a result of European Community directives, the UK Department of the Environment (DoE) is in the process of introducing legislation for controlling inputs of 'dangerous substances' into aquatic environments. This has resulted in the establishment of a so-called priority 'Red List' of pollutants and an additional 'First Priority Candidate List', the 'Black List' (Anonymous 1988), although exactly how comprehensive these, and other lists, are, is still a matter for debate. However, most importantly, they identify the major pollutants that we face and clearly demonstrate that, despite some local differences related to sites of chemical production, industrialized countries have to deal with the same pollutants.

The major agents of biodegradation in the environment are micro-organisms (Weightman and Slater 1988). Laboratory studies suggest that they are extraordinarily versatile and can adapt to degrade a wide range of xenobiotic compounds. This is shown by the variety of microbes, mainly bacteria, that have been isolated from laboratory enrichments with organic pollutants. Microbial biodegradation is frequently associated with incomplete breakdown of organic pollutants and biotransformations can produce limited, but often significant, changes in chemical structure. The extent to which biodegradation of a particular compound in the biosphere approaches mineralization (complete breakdown of the compound into oxidized products and elemental constituents) is primarily determined by the activities of microbial enzymes catalysing sequential, degradative reactions and which, together, constitute a complete metabolic (usually catabolic) pathway. As can be seen from Table 3.1, many of the priority pollutants identified on the UK DoE's Red and Black Lists are known to be catabolized by laboratory-selected bacteria. Paradoxically, recalcitrance to degradation in the environment is a major factor in the schemes used to identify priority pollutants in Table 3.1. Consequently, there is a need to distinguish between two types of recalcitrance. In 'environmental recalcitrance', factors other than the lack of appropriate microbial degradative enzymes may account for persistence of the pollutant in complex ecosystems, while with 'biological recalcitrance' the pollutant is resistant to bioconversion because the necessary biodegradative functions do not exist.

Table 3.1. *List of UK priority ('Red List') and first priority candidate ('Black List') pollutants*

Red List	Black List
Mercury and its compounds	2-amino-4-chlorophenol*
Cadium and its compounds	Anthracene*
Gamma-hexachlorocyclohexane*	Azinphos-ethyl
DDT*	Biphenyl*
Pentachlorophenol*	Chloroacetic acid*
Hexachlorobenzene*	2-chloroethanol*
Hexachlorobutadiene	4-chloro-2-nitrotoluene*
Aldrin	Cyanuric chloride*
Dieldrin	2,4-D*
Endrin	Demeton-O
PCBs*	1,4-dichlorobenzene*
Dichlorvos	1,1-dichloroethylene*
1,2-dichloroethane*	1,3-dichloropropan-2-ol*
Trichlorobenzene*	1,3-dichloropropene
Atrazine	Dimethoate
Simazine*	Ethylbenzene*
Tributyltin compounds	Fenthion
Triphenyltin compounds	Hexachloroethane
Trifuralin	Linuron*
Fenitrothion	Mevinphos
Azinphos-methyl	Parathion*
Malathion	Pyrazon (Chloridazon)*
Endosulfan	1,1,1-trichloroethane*
	Chloroprene
	3-chlorotoluene*
	Chloroform*
	Carbon tetrachloride

* Microbial degradation for these pollutants has been described in the literature.

Biological recalcitrance

Microbial infallability?

Given the range of xenobiotic compounds already shown to be biodegradable in the laboratory, and the rate at which reports of new biodegradative processes are being published, we might legitimately question whether any organic compound is biologically recalcitrant. The question of microbial infallability – whether micro-organisms have the potential to degrade all organic compounds – was first raised as a polemic by Rachel Carson (1962) in her book *Silent spring*. The response of biologists, provoked by her suggestion that micro-organisms possessed virtually an unlimited potential and could adapt to degrade any xenobiotic compound, was generally sceptical (Alexander 1965). However, Dagley (1984) has pointed out that as there are no energetic barriers to preclude such a capability, micro-organisms are restricted only by their ability to survive in the presence of the xenobiotic and to synthesize the necessary enzymes for transport and catabolism.

From the practical point of view, the number of xenobiotic compounds shown to be biodegradable under optimized laboratory conditions has increased at a steady rate during the last decade. Compounds such as pentachlorophenol (PCP), 2,4,5-trichlorophenoxyacetate (245-T) and highly substituted polychlorinated biphenyls (PCBs), were at one time thought to be biologically recalcitrant, but are now known to be biodegradable to the extent that the organic chlorines can be mineralized by microbial enzymes. Several bacteria that can utilize PCP, 245-T and dichlorobiphenyls as sole sources of carbon and energy have now been isolated.

It has been suggested that biodegradation of these xenobiotics is associated with newly evolved catabolic genes. An alternative explanation is that the intensity of research effort in this area has been such that methods for enrichment and isolation of micro-organisms with unusual biodegradative functions are better. A striking example of how research effort coupled with effective methodologies can completely alter our impression of biodegradative potential is provided in the literature on anaerobic biodegradation of xenobiotic compounds. Ten years ago, an uninformed reader of this literature might easily have supposed that the breakdown of xenobiotics in the biosphere was primarily a function of aerobic bacteria, of which the most important were 'pseudomonads'. It now appears that such a supposition is wrong and that the dearth of information reflected our ignorance of the catabolic versatility of other microbial groups, notably the anaerobes (see p. 27). It is unlikely that the degradative capabilities seen in the representatives of all the major groups of anaerobic eubacteria are newly evolved or recently acquired traits. However, the origins of genes encoding catabolic enzymes involved in aerobic and anaerobic degradation are only poorly understood.

Microbial adaptation and biodegradative potential

As pointed out by Chapman (1979), attempts to correlate chemical structure with biodegradative potential are prone to error. For example, a recent study (Parsons and Govers 1990) reported that determinations of 'quantitative structure–activity relationships' between biodegradation rates of organic compounds and chemical structure parameters were generally precluded because of lack of data about mechanisms and 'rate-limiting' steps of biodegradation. However, such data would not help to predict biodegradation rates in the biosphere, where recalcitrance can be a function of many factors that are independent of mechanisms and rates of catabolism (see p. 27). Acclimation (or acclimatization) of natural microbial communities is a commonly observed response to periodic and prolonged application of pollutants. Initially, biodegradation of the pollutant is absent or occurs only at a slow rate but, with time, the rate may increase dramatically (Barkey and Pritchard 1988). Thus, acclimation reflects adaptation of the indigenous microflora to degrade xenobiotic compounds. Barkey and Pritchard (1988) listed three mechanisms to account for adaptation.

1. Induction of (or relief from catabolite repression for) required biodegradative enzymes.
2. Genetic changes producing new biodegradative and associated physiological capabilities.
3. Selection of specific organisms.

Mechanisms 1 and 2 are both amenable to genetic manipulation.

Adaptation is also seen in laboratory enrichment cultures, where samples from the enrichment are inoculated into a culture medium containing a recalcitrant pollutant as a potential source of carbon and energy. Invariably, biodegradation of the pollutant in such enrichment cultures does not occur immediately but requires weeks or even months of incubation. Several authors have speculated as to what is happening during this period of adaptation. For example, there is a widely held belief that adaptation is the result of 'genetic condensation' during enrichment, whereby genetic rearrangements, recombinations and gene transfer result in the assembly of genes encoding the requisite biodegradative pathway. An extreme exposition of this idea was given by Kellogg et al. (1982) who suggested that isolation of a 245-T utilizer from a chemostat culture originally inoculated with soil and microbial cultures containing several different catabolic plasmids was the result of genetic condensation or, as they called it, 'plasmid-assisted molecular breeding'. It should be noted that no genetic evidence has yet been published to support a recombinational mechanism for this particular adaptation of xenobiotic degrading functions. However, mutational changes have been shown to be involved in microbial adaptation and degradation of xenobiotic compounds. For example, activation of silent or cryptic genes encoding dehalogenases has been suggested to explain the adaptation of *Pseudomonas putida* strain PP1 to utilize chlorinated alkanoate herbicides (Senior *et al.* 1975; Slater *et al.* 1985). Furthermore, Wyndham (1986) showed that adaptation of *Acinetobacter calcoaceticus* to the utilization of aniline depended on the dynamics of two related genotypes, designated Ami^0 and Ami^+. The former had a selective advantage at low aniline concentrations (in the micromolar range) but was outcompeted by the latter at high aniline concentrations (in the millimolar range).

Alexander and his colleagues have offered alternative explanations for the acclimation period before mineralization of nitrophenols in acquatic environments. Wiggins *et al.* (1987) suggested that it reflected the time taken for biodegrading microorganisms to multiply to a level commensurate with detectable utilization of the xenobiotic (approximately 10^3–10^4 MPN (most probable number) per ml). Later, Wiggins and Alexander (1988) proposed additional mechanisms involving effects of xenobiotic concentration on the degradative microorganisms and the presence or absence of alternative substrates.

Some investigators have proposed that the development of DNA amplification methods and gene probe technology currently being used experimentally to monitor the fate of catabolic GEMs in complex ecosystems (Steffan and Atlas 1988; Steffan *et al.* 1989; King *et al.* 1990) may be useful for evaluating biodegradative potential. Thus, specific catabolic gene probes could be used to determine whether or not environmental samples contained homologous sequences. Such applications of the technology are, however, likely to be limited by the high specificity of the gene probes which would have to be used.

Anaerobic biodegradation

Anaerobiosis is a common phenomenon in many polluted ecosystems, e.g. groundwaters, sediments, landfills, lagoons and flooded soils. However, progress in our understanding of anaerobic biodegradation has been slow mainly because of the practical problems often encountered in culturing and isolating anaerobic bacteria. Most investigations focus on the biotransformation of xenobiotics by mixtures of anaerobes either in samples taken directly from the environment or consortia isolated by enrichment (Berry *et al.* 1987). These studies clearly show that anaerobic biodegradation, although qualitatively different from aerobic biodegradation, results from the activities of a wide range of catabolically versatile bacteria. Thus, halogenated compounds such as 24-D, 245-T (Suflita *et al.* 1984; Mikesell and Boyd 1985; Gibson and Suflita 1990), di- and trichlorobenzenes (Bosma *et al.* 1988), chlorophenols (Boyd and Shelton 1984; Genthner *et al.* 1989; Häggblom and Young 1990), chloroalkanes (Bouwer and McCarty 1983; Vogel and McCarty 1985; Bagley and Gossett 1990) and polychlorinated biphenyls (Quensen *et al.* 1990) have been shown to be dechlorinated by anaerobic mixed cultures. Efforts to examine catabolic activities associated with xenobiotic degradation in pure cultures have only recently been successful. A striking example is the characterization of a broad specificity aromatic dehalogenating activity in cell extracts of *Desulfomonile tiedjei* (originally designated strain DCB-1) (Shelton and Tiedje 1984; DeWeerd and Suflita 1990). Other studies of pure cultures of xenobiotic-degrading anaerobes include those by Criddle *et al.* (1990), who described the enrichment and isolation of *Pseudomonas* sp. strain KC, capable of mineralizing carbon tetrachloride under denitrifying conditions. Zeyer *et al.* (1990), Lovley *et al.* (1989) and Lovely and Lonergan (1990) have also described anaerobic biodegradation of aromatic hydrocarbons by pure cultures of anaerobes under denitrifying and Fe(III)-reducing conditions, respectively.

Environmental recalcitrance

There is a wide range of biodegradative functions in the biosphere and great potential for adaptation. Thus, it seems reasonable to assume that microorganisms that can degrade almost all organic pollutants, even without genetic manipulation, can be isolated. It follows, therefore, that environmental recalcitrance of pollutants results from the inability of polluted ecosystems to select such micro-organisms and support their growth and catabolic activity. The major factors affecting environmental recalcitrance are considered below.

Concentration effects

Many organic pollutants present problems in contaminated ecosystems at low and high concentrations. At high concentrations, toxic effects causing inhibition are frequently reported, whereas low concentrations, below the K_s for uptake and degradation of the xenobiotic, may persist in the presence of degrading organisms (Boethling and Alexander 1979; Alexander 1984; Leahy and Colwell 1990; Lamar et al. 1990). In oligotrophic and nutritionally fluctuating environments, resistance to starvation stress would be a prerequisite for survival (Sinclair and Alexander 1984; Lappin-Scott and Costerton 1990). Strategies such as the selection and manipulation of oligotrophic and starvation-resistant bacteria have been suggested for bioremediation in such ecosystems, but many difficulties associated with this approach remain to be overcome. The selection of xenobiotic degraders resistant to the toxic effects of the xenobiotic may provide a way forward for concentrated waste streams (Ewers et al. 1990). However, this approach can also present problems, in that degradative genes may be unstable at high concentrations of xenobiotic (Weightman et al. 1985; Strotmann et al. 1990).

Temperature

Temperature affects both the selection of xenobiotic-degrading bacteria and rates of degradation. However, the literature on this important aspect of biodegradation is almost non-existent and considerations of temperature are mostly confined to studies of hydrocarbon degradation and clean-up of oil spills at sea (Leahy and Colwell 1990). The few papers so far published on the subject indicate that the biochemistry of xenobiotic degradation by psychrophilic and thermophilic micro-organisms is not qualitatively different to that observed in mesophiles. What does differ is the composition of the microbial communities involved. For example, at temperatures in the thermophilic range Bacillus spp. predominate in enrichments for aerobic xenobiotic-degrading organisms.

Alternative nutrients and energy sources

As indicated earlier, xenobiotic biodegradation in the environment can be a fortuitous activity when broad specificity enzymes catalyze partial breakdown of the pollutant compound. If such activities do not provide nutrients or energy for the organisms involved, alternative sources are required to sustain them. Thus, biodegradation of pollutants, as secondary substrates (Lapat-Polasko et al. 1984; Schmidt et al. 1987; Topp et al. 1988; Hess et al. 1990) in some environments has been stimulated by the addition of alternative nutrient and energy sources (Atlas 1988; Jones and Alexander 1988; Song et al. 1990). Supply of: oxygen to aerobic environments; reducing equivalents (to anaerobic environments); and carbon, nitrogen and phosphorus to contaminated environments (in forms that are readily available to the indigenous micro-organisms), are currently being used for in situ bioremediation projects, most notably for the clean-up of the crude oil spilt in Prince William Sound by the tanker Exxon Valdez (Leahy and Colwell 1990; Fox 1990).

The presence of alternative growth substrates can also cause deleterious effects by interfering with xenobiotic biodegradation.

Compartmentation and availability

The availability of xenobiotics for microbial degradation in polluted ecosystems is affected by chemical reactions, e.g. polymerization, adsorption (to humic matter and clays) and solubility (Alexander 1984; Stucki and Alexander 1987). Thus, physical and chemical factors can restrict pollutant biodegradation in some environments despite the presence of micro-organisms with the appropriate catabolic potential. In some cases, physical compartmentation may not present an insurmountable barrier and mechanisms such as chemotaxis and water percolation (Harwood et al. 1990; Trevors et al. 1990) could transport xenobiotic-degrading populations to the contaminated site.

Interactions between micro-organisms

Interactions such as predation, competition and mutualism are major influences on the survival and activity of bacteria involved in xenobiotic biodegradation. In most ecosystems, biodegradation seems to be a function of microbial associations, involving poorly understood physiological and biochemical interactions (Weightman and Slater 1988). Secondary utilizers, which are unable to initiate breakdown of xenobiotics, can be essential in other ways for maintaining the integrity and activity of a xenobiotic-degrading community. For example, they may be involved in channelling degradation products from or providing additional nutrients to the primary utilizers (Lewis *et al.* 1984; Slater and Lovatt 1984).

Competition for available nutrient sources and predation are factors that can affect survival and growth of xenobiotic-degrading micro-organisms in ecosystems to which they are introduced exogeneously (Scheuerman *et al.* 1988). Once again, these factors are poorly understood and, while recognizing that they should be taken into account, investigations developing strategies for the use of microbial inoculants in bioremediation usually ignore them.

Genetic analysis of xenobiotic-degrading micro-organisms and the construction of catabolic GEMs

Involvement of transmissible catabolic plasmids in xenobiotic biodegradation and their use for *in vivo* genetic manipulation

Since the identification of the SAL and OCT plasmids (Chakrabarty 1972) a wide range of aerobic catabolic functions have been found to be encoded on naturally transmissible plasmids (Weightman and Slater 1988).

The first report of the genetic construction of a xenobiotic-degrading micro-organism by Friello *et al.* (1976) did not involve the use of *in vitro* recombinant methods but exploited the natural transfer ability of several unrelated catabolic plasmids to establish a mixture of catabolic pathways associated with hydrocarbon catabolism, e.g. TOL, NAH, CAM and OCT, in one strain of *Pseudomonas*. The next decisive step towards genetic manipulation for improvement of xenobiotic-degrading bacteria was reported by Reineke and Knackmuss (1979). In an elegant series of experiments they combined parts of two different plasmid-encoded 3CB pathways, one, pWR1, for 3-chlorobenzoate (3CB) catabolism and the other, pWWO (TOL), for methylbenzoate catabolism, to create a novel pathway for degradation of 3,5-dichlorobenzoate (35DCB) and 4-chlorobenzoate (4CB). These authors recognized that the inability of the plasmid pWR1-encoded 3CB pathway in *Pseudomonas* sp. B13 to catabolized 4CB was due to the narrow specificity of just one enzyme, benzoate 1,2-dioxygenase. Plasmid pWWO carried the

*xyl*XYZ gene cluster (formerly designated *xyl*D as part of the *meta*-cleavage pathway operon), which encoded a broad-specificity benzoate, 1,2-dioxygenase, that could also catalyse conversion of methybenzoates and chlorobenzoates to corresponding dihydrodiol products. Conjugal transfer of pWWO from *P. putida* to strain B13 produced transconjugants capable of mineralizing 4CB and 35DCB, as plasmids pWR1 and pWWO were compatible. Reineke and Knackmuss (1979) showed that the new catabolic capabilities of the transconjugants, produced by *in vivo* genetic manipulation, was primarily due to two factors. First, the recruitment of TOL plasmid genes encoding the broad-specificity benzoate 1,2-dioxygenase by the plasmid pWR1-encoded pathway. Secondly, inactivation of the pWWO plasmid gene encoding the *meta*-ring cleavage enzyme catechol 2,3-dioxygenase.

Reineke, Knackmuss and their colleagues used the principle of recruiting broad-specificity enzymes to augment the plasmid pWR1-encoded pathway for the *in vivo* construction of pseudomonads capable of mineralizing chlorosalicylate, chlorophenol, chloroaniline, chloronitrophenol, chlorobiphenyl and further chlorobenzoate isomers (Reineke *et al.* 1982; Rubio *et al.* 1986a,b; Latorne *et al.* 1984; Brahn *et al.* 1988; Hartmann *et al.* 1989; Mokoss *et al.* 1990). Constructions usually involved the transfer of plasmid pWR1 to host strains producing broad-specificity enzymes that were able to catalyse conversion of chloroaromatic substrates to corresponding chlorocatechols. Expression of the 3CB catabolic genes on plasmid pWR1 in these strains provided an *ortho*-ring cleavage pathway for mineralization of the chlorocatechols. The inactivation of attenuation of the host catechol 2,3-dioxygenase (a *meta*-cleavage enzyme), seen in all of these *in vivo* constructs, removed an interfering activity and effectively avoided misrouting of chlorocatechols down the unproductive *meta*-pathway. A similar rationale was used by Chapman (1988) for the expansion of a 3CB pathway in *P. putida* strain 633 to obtain biodegradation of 3- and 4-chlorotoluenes. Transfer of plasmid pWWO to strain 633 facilitated selection of transconjugants on chlorotoluene. Chapman (1988) suggested this was due to the induction of the 'upper pathway' broad-specificity enzymes encoded by the pWWO plasmid by the presence of chlorotoluenes, and that these enzymes converted them to chlorocatechols. The enzymes were normally associated with conversion of toluene and xylenes to their corresponding catechols. The *ortho*-cleavage pathway in strain 633 (pWWO) was able to mineralize these chlorocatechol products.

In vitro genetic manipulation of xenobiotic catabolic pathways

The system described above was used by Timmis and colleagues to develop *in vitro* recombinant methods for genetic manipulation of xenobiotic catabolic pathways (Ramos *et al.* 1986, 1987; Rojo *et al.* 1987). Their studies provided a more detailed genetic analysis of adaptation by catabolic pathway augmentation and, in the process, they constructed GEMs with novel degradative capabilities. Their experiments confirmed that 4-chlorobenzoate (4CB)-utilizing derivatives of *Pseudomonas* sp. B13 (pWR1) containing the *xyl*XYZ (*xyl*D) genes, and without other pWWO plasmid genes, were able to mineralize 4CB and utilize it as a sole source of carbon and energy. The *in vitro* construction of 35DCB utilizers was more complex and mutations altering the regulation of expression of these genes seemed to be required in addition to recruitment of the *xyl*XYZ-encoded benzoate 1,2-dioxygenase. Ramos *et al.* (1986) proposed that mutation in the *xyl*S gene, which encodes the activator of the *xyl*XYZ (and other *meta*-pathway) genes, was required to allow it to recognize and respond to 35DCB. Thus, whereas the wild type XylS activator would not activate the *xyl*XYZ genes in the presence of 35DCB, a mutant $XylS_{352}$ was able to do this, and 4CB-utilizing strains containing plasmids pWR1 and pWWO were also able to grow on 35DCB after the introduction of the $xylS_{352}$ mutant gene.

The role of plasmid stability in function of GEMs *in situ*

Bacteria with beneficial plasmid-born catabolic functions may suffer from loss of associated activities *in situ* due to plasmid instability. Plasmid instability can be caused by two major mechanisms:

1. Segregational instability, which results from defective partitioning of plasmids between daughter cells during division.
2. Structural instability resulting in physical changes such as deletions, insertions and re-arrangements.

Complex mechanisms ensure the stable segregation of natural plasmids during cell growth (Cohen *et al.* 1988; Nordström and Austin 1989; Thomas and Helinski 1989). This means that even for low-copy-number plasmids, plasmid-free segregants arise at low frequency under normal conditions. However, recombinant plasmids including high-copy-number plasmids used for the production of heterologous proteins in industrial fermentations may exhibit serious stability problems (Ensley 1986). Environmental growth conditions can affect plasmid stability, copy-number and expression of cloned genes (Sayadi *et al.* 1989) and these effects may differ with free or immobilized host cells. Structural instability may result from illegitimate recombination (between sequences with little homology) or as a consequence of recombination between repeated genetic sequences flanking genes of interest (Meulien *et al.* 1981; Eaton and Timmis 1986; Wyndham *et al.* 1988; Saint *et al.* 1990). Many natural catabolic plasmids exhibit structural instability, resulting in loss of specific catabolic genes in the absence of selection. The genes for complex catabolic pathways may be carried on transposons, which can also give rise to a complex series of rearrangements and cointegration products with other replicons, such as those observed with the TOL plasmid pWWO (Lehrbach *et al.* 1982; Tsuda and Jino 1988).

Several studies have investigated the stability of plasmids in bacteria in experimental microcosms and natural systems. Bacterial strains released to natural ecosystems may experience elevated rate of mutation at chromosomal loci, and in addition transferred plasmid DNA may be more prone to deletion and rearrangement in recipient strains (Miller *et al.* 1990). Plasmid stability can be affected by a wide range of environmental parameters, such as nutrient availability and oxygen concentration (Trevors *et al.* 1989). The nature of the host strain can influence both plasmid stability and expression of plasmid-borne genes (Caldwell *et al.* 1989).

When considering the application of GEMs for the degradation of environmental pollutants it is important to assess the stability of the catabolic phenotype both in the presence and absence of the primary substrate. Jain *et al.* (1987) reported that (*P. putida* harbouring the TOL plasmid pWWO was stably maintained in groundwater aquifer microcosms throughout an 8-week incubation period, in the presence and absence of toluene.

Fulthorpe and Wyndham (1989) studied the survival and activity of a 3CB catabolic genotype (plasmid, pBR60) in flowthrough lake microcosms and found that the catabolic genotype declined to very low levels after 8 weeks in the absence of 3CB. Discrepancies in estimates of 3CB-degrading bacteria by viable plate count and MPN–DNA hybridization were attributed to instability of the catabolic phenotype and the possibility of gene transfer to indigenous bacteria. Concentrations of 3CB of ≥ 1 μM provided a strong selective pressure for the catabolic phenotype. Golovleva *et al.* (1988) introduced a *Pseudomonas aeruginosa* strain BS827 carrying a naphthalene catabolic plasmid, pBS3, into a soil ecosystem. Again, the presence of the catabolic substrate, in this case kelthane, greatly enhanced the retention of its biodegradative ability *in situ*. In the absence of kelthane, 92 per cent of the introduced strain had lost their biodegradative ability after 41 days in the soil ecosystem (compared to 0–30 per cent in the presence of kelthane). The stability of recombinant catabolic plasmids in GEMs introduced into both soil and activated sludge microcosms has also been investigated (McClure *et al.* 1989, 1991; Ramos *et al.* 1991). A recombinant TOL plasmid (pWWO-EB62), encoding *p*-ethylbenzoate catabolism, was stably maintained in *P. putida* strain EEZ15, in the presence and absence of the substrate, in soil microcosms (Ramos *et al.* 1991). In our laboratory we have investigated the survival and function of a 3CB-degrading *P. putida* strain harbouring a recombinant catabolic plasmid, pD10, in a model activated sludge unit. In laboratory culture, plasmid pD10 was stably maintained in *P. putida* strain UWC1 and in a long-term experiment *P. putida* UWC1(pD10) maintained this plasmid in the unit for 8 weeks, this included a period of over consecutive 30 days in the absence of 3CB (McClure *et al.* 1989). This was confirmed by separate monitoring of the host strain and plasmid-borne phenotypes and by restriction endonuclease digestion of plasmid DNA from strains reisolated from the activated sludge.

Both the recombinant catabolic plasmids described above, the low-copy-number pWWO-EB62, and the high-copy-number plasmid pD10, were stably maintained in their host bacteria following introduction into experimental microcosms. However, studies with natural catabolic plasmids show that stability can be influenced by a wide variety of factors. Laboratory studies may lead to a greater understanding of the basis for plasmid instability such that stability of recombinant plasmids can be enhanced under environmental conditions.

Broad specificity enzymes associated with xenobiotic degradation

The studies referred to in the preceding section have clearly shown that genetic manipulation offers a means of accelerating the evolution of catabolic pathways for the biodegradation of xenobiotic and pollutant compounds. An effective development strategy is to identify and clone genes encoding broad substrate specificity enzymes, which may be recruited in a modular fashion to augment existing pathways. Manipulation of these genes facilitates pathway expansion by increasing the number of pollutant substrates that a biodegradative pathway can catabolize. Additional manipulation of cloned regulatory genes provides a method of releasing pathways from tight controls so that biodegradative functions are expressed in response to a wider variety of pollutant compounds and environmental stimuli.

The problem remains that only relatively few biodegradative pathways have been characterized in sufficient detail to permit this type of approach and the genetic study of many other potentially interesting biodegradative functions has hardly begun. Furthermore, most of the well characterized biodegradative pathways are for model compounds rather than priority pollutants. However, recent work has resulted in the identification of a number of exceptional broad specificity enzymes, which are priority candidates for genetical analysis and manipulation. Some examples, most relevant to bioremediation, are detailed below.

Broad specificity catabolic oxygenases and degradation of trichloroethylene

Studies over the past 5 years on the biodegradation of trichloroethylene (TCE) have uncovered several types of broad specificity oxygenases (Henry and Grbic-Galic 1991), which catalyse dechlorination of the compound. For example, toluene mono- and dioxygenases associated with methylaromatic catabolism, phenol and chlorophenol catabolic hydroxylases, and oxygenases such as the soluble methane mono-oxygenase (sMMO). Nelson *et al.*

(1986) first described a pure culture *Pseudomonas cepacia* strain G4 able to degrade TCE. Later work by the same group (Nelson *et al.* 1987; Shields *et al.* 1989; Folsom *et al.* 1990) implicated enzymes involved in toluene and phenol catabolism, specifically strain G4's toluene dioxygenase (Nelson *et al.* 1988). Wackett and Gibson (1988) and Wackett and Householder (1989) reported TCE degradation by a toluene dioxygenase produced by *Pseudomonas aeruginosa* strain F1. Winter *et al.* (1989) cloned the gene(s) encoding an unusual toluene mono-oxygenase from *Pseudomonas mendocina*, which was also expressed in *Escherichia coli* and resulted in TCE biodegradation by the recombinant strain. In this case, an advantage of cloning and expressing the catabolic gene(s) heterologously was that toluene was no longer required as a cosubstrate for TCE breakdown. Thus, the recombinant *E. coli* strain containing an IPTG-inducible toluene mono-oxygenase showed better TCE degradation characteristics than the *P. mendocina* from which the gene was originally isolated. Harker and Kim (1990) reported that *Alcaligenes eutrophus* strain JMP134 produced at least two oxygenases capable of degrading TCE. One, tentatively identified as a chlorophenol hydroxylase involved in 2,4-dichlorophenoxyacetate catabolism, was encoded by the plasmid pJP4. The other was a chromosomally encoded enzyme, probably a phenol hydroxylase. TCE degradation by methanotrophs and alkane-utilizing bacteria has been reported by several authors. For example, Wackett *et al.* (1989) surveyed a range of micro-organisms producing oxygenases and found TCE degradation only in propane utilizers (all *Mycobacterium* spp.). Mixed and pure cultures of methanotrophs have also been found to catalyse fortuitous oxygenation of TCE (Fogel *et al.* 1986; Oldenhuis *et al.* 1989; Tsien *et al.* 1989; Henry and Grbic-Galic 1991). These studies amply illustrate that even before exposure to a problem xenobiotic such as TCE, the biosphere contains a number of broad-specificity enzymes which, through adaptation and/or genetic manipulation may be recruited to provide a biodegradative route.

Oxidation of priority pollutants by *Phanerochaete chrysosporium*

The lignin-degrading fungus *Phanerochaete chrysosporium* produces a non-specific oxidizing system, which normally catalyses the breakdown of the persistent biopolymer. In addition, this system is thought to be the catalyst that allows the fungus to degrade a remarkable range of recalcitrant xenobiotics (Hammel 1989). The first indications of the biodegradative potential of *P. chrysosporium* were published by Bumpus *et al.* (1985) and Eaton (1985). Since then, the list of priority pollutants that can be oxidized by the fungus has increased steadily and now includes: 245-T (Bumpus *et al.* 1985), polychlorinated biphenyls (Eaton 1985), DDT (Bumpus and Aust 1987), pentachlorophenol (Mileski *et al.* 1988), thianthrene (Schreiner *et al.* 1988), 2,4,6-trinitrotoluene (Fernando *et al.* 1990), lindane and chlordane (Kennedy *et al.* 1990), dyes (Cripps *et al.* 1990) and 2,4-dichlorophenol (Valli and Gold 1991). At the moment, relatively little is known about the pathway(s) involved in biodegradation, although some details are now emerging regarding catabolism and dechlorination of chlorophenols (Mileski *et al.* 1988; Valli and Gold 1991). The potential application of *P. chrysosporium* in environmental detoxification is already under consideration and may be rationalized when more information on the biochemistry and genetics of its xenobiotic catabolic capacity is available.

Biotechnology and bioremediation

Bioremediation strategies

Two basic approaches are currently under investigation for bioremediation. The first is *in situ* treatment of contaminated sites, and may involve inoculation with degradative micro-organisms or introduction of nutrients to stimulate indigenous activities (see p. 28). The second involves construction of reactors that allow contained treatment of chemical wastes (modified activated sludge treatment systems, digestors, sequencing batch reactors, fluidized bed and other biofilm systems, and immobilized bacteria). Contained, reactor-type treatments are generally preferred to *in situ* bioremediation because of the complexities and heterogeneity that are often associated with the latter. For example, St John and Sikes (1988) and McCarty (1988) describe the considerable challenges facing biotechnologists in the bioremediation of contaminated soils and groundwaters, and point out that well-organized, interdisciplinary efforts are required to devise effective biological treatments for complex industrial waste sites. Conventional activated sludge systems and digestors do not provide effective treatment for chemical pollutants (Hill *et al.* 1986) but may be adapted to handle specific effluents, as in the case of the model system for treatment of morpholine described by Brown and Knapp (1990). The bioreactor design is often empirical, involving the use of selected xenobiotic-degrading bacteria inoculated into biofilms, on a variety of support media (Frick *et al.* 1988; Phelps *et al.* 1990), or immobilized at high cell densities (Heitkamp *et al.* 1990).

Testing the application of GEMs for *in situ* waste treatment

Despite the extensive achievements made in the construction of bacterial strains capable of the breakdown of toxic and previously recalcitrant compounds, there are few reports describing the activity of such strains under environmentally relevant conditions. Concern has been expressed over potentially deleterious environmental effects following the release of GEMs into natural ecosystems (Anonymous 1988). As a result, attention has been focused on the use of accurate microcosms, which mimic target ecosystems and can be used to test both the efficacy and the effects of genetically-engineered microbial inoculants (Sojka and Ying 1987; Trevors 1988; Fry *et al.* 1992). There are numerous reports on the successful use of non-engineered bacterial inoculants for enhanced degradation of target pollutants under natural or simulated natural conditions. The few examples where GEMs have been used to investigate pollutant degradation in complex microcosms have resulted in varying degrees of success. Hansen *et al.* (1984) showed that an *E. coli* strain with a recombinant plasmid pRR130 carrying mercury resistance genes was effective at removing mercury from raw sewage in continuous culture. Dwyer *et al.* (1988) studied the survival and activity of *Pseudomonas* sp. strain B13 and genetically engineered derivatives with extended catabolic ranges. One derivative, strain FR1 (pFRC20P), survived well under most conditions and was able to degrade mixtures of 3CB and 4-methylbenzoate in a simple activated sludge microcosm. However, when addition of these substrates was staggered, the substrate added second was only partially degraded (about 50 per cent). The addition of both substrates (5 mM) to an unacclimatized system had a deleterious effect on the total bacterial and GEM population and 3CB degradation did not occur.

In our studies on the survival and function of genetically engineered 3CB degraders in a complex activated sludge microcosm and batch cultures, we have demonstrated that different *Pseudomonas* host strains carrying plasmid pD10 varied in their ability to degrade 3CB in waste water. Although *P. putida* UWC1(pD10) survived well in the activated sludge unit, this strain did not enhance 3CB breakdown when the unit was run in a continuous mode. Results from batch culture experiments showed that both alternative substrates and temperature affected the breakdown of 3CB by strain UWC1 and activated sludge-derived bacteria to which plasmid pD10 was transferred. A laboratory-selected transconjugant strain, *P. putida* ASR2.8(pD10) showed a high rate of 3CB breakdown at 15°C in batch culture. This strain also enhanced 3CB breakdown in a fully functional activated sludge unit and was maintained in the unit at a high level (about 10^6 CFU/ml; McClure *et al.* 1991). However, only approximately 20 per cent of the 3CB was degraded in the unit following inoculation with strain ASR2.8. Using lower 3CB concentrations and inoculation with a 3CB-degrader isolated by enrichment from the target ecosystem, we have observed almost total 3CB breakdown in the activated sludge unit.

As with laboratory-selected, non-engineered bacteria, the efficacy of GEMs for *in situ* biodegradation will vary and many factors may affect the ability of a particular strain to function in target ecosystems. Microcosms that accurately mimic such environments are essential to assess potential inoculants and may also provide a means of selecting strains that are more likely to prove effective *in situ*.

Conclusion

In response to the higher priority currently being given to environmental issues and clean technologies, the last decade has seen an upsurge of research into microbial biodegradation. While many investigations seem concerned primarily with the isolation and characterization of bacteria showing novel degradative capabilities, the wider issue of bioremediation is being considered and real applications pursued. Biodegradation by groups of micro-organisms, such as the fungi and anaerobes, which received scant attention at the beginning of the 1980s, is now gaining wider interest. However, more basic research into the biochemistry and genetics of xenobiotic catabolism by these groups is needed. Some investigators are now using well characterized xenobiotic-degrading organisms to address the problem of environmental recalcitrance. In many cases, the use of aerobic bacteria in bioremediation applications is dependent not so much on the isolation of strains producing the appropriate catabolic functions, but more on developing technologies for treatment of chemical wastes, both *in situ* and with bioreactors. Priorities for the next decade will be to push forward technology transfer from the laboratory to develop effective waste treatment systems, and is likely to involve specific solutions related to each particular pollution problem.

Applications of genetic engineering in this area are more difficult to predict. The regulatory position remains unclear, although the release of catabolic GEMs to the environment would not seem to pose any great risk. However, the major problem will be to demonstrate a need for GEMs to solve a problem that cannot be achieved using selected, non-engineered micro-organisms alone. It will be interesting to see if the first application of GEMs for bioremediation of a real pollution problem will have to await the new century.

Acknowledgements

We gratefully acknowledge the Water Research Centre for funding some of the research described here carried out in our laboratory.

References

Alexander, M. (1965). Biodegradation: problems of molecular recalcitrance and microbial fallibility. *Advances in Applied Microbiology*, 7, 47–52.

Alexander, M. (1984). Ecological constraints on genetic engineering. In *Genetic control of environmental pollutants*, (Ed. G.S. Omenn and A. Hollaender), p. 151. Plenum Press, New York.

Anonymous (1980). *Biotechnology. Report of a joint working party*. HMSO, London.

Anonymous (1988). *Inputs of dangerous substances to water: proposals for a unified system of control*. Department of the Environment Circular 7/89, HMSO, London.

Atlas, R.M. (1988). Biodegradation of hydrocarbons in the environment. In *Environmental biotechnology: reducing risks from environmental chemicals through biotechnology*, (Ed. G.S. Omenn). Plenum Press, New York.

Bagley, O.M., and Gossett, J.M. (1990). Tetrachlorethene transformation to trichloroethene and *cis*-1, 2-dichloroethylene by sulphate-reducing enrichment cultures. *Applied and Environmental Microbiology*, 56, 2511–16.

Barkay, J., and Pritchard, H. (1988). Adaptation of aquatic microbial communities to pollutant stress. *Microbiological Sciences*, 5, 165–9.

Berry, D.F., Francis, A.J., and Bollag, J.-M. (1987). Microbial metabolism of homocyclic and heterocyclic aromatic compounds under anaerobic conditions. *Microbiological Reviews*, 51, 43–59.

Boethling, R.S., and Alexander, M. (1979). Effect of concentration of organic chemicals on their biodegradation by natural microbial communities. *Applied and Environmental Microbiology*, 37, 1211–16.

Bosma, T.N.P., van der Meer, J.R., Schraa, G., Tros, M.E., and Zehnder, A.J.B. (1988). Reductive dechlorination of all trichloro- and dichlorobenzene isomers. *FEMS Microbiology Ecology*, 523, 223–9.

Bouwer, E.J., and McCarty, P.L. (1983). Transformations of 1- and 2-carbon halogenated aliphatic organic compounds under methanogenic conditions. *Applied and Environmental Microbiology*, 45, 1286–94.

Boyd, S.A., and Shelton, D.R. (1984). Anaerobic biodegradation of chlorophenols in fresh and aclimated sludge. *Applied and Environmental Microbiology*, 47, 272–7.

Brahn, C., Bayly, N.C., and Knackmuss, H.-J. (1988). The *in vivo* construction of 4-chloro-2-nitrophenol assimilatory bacteria. *Archives of Microbiology*, 150, 171–7.

Brown, V.R., and Knapp, G.I. (1990). The effect of withdrawal of morpholine from the influence and its reinstatement on the performance and microbial ecology of a model activated sludge plant treating a morpholine containing influent. *Journal of Applied Bacteriology*, 69, 43–53.

Bumpus, J.A., and Aust, S.D. (1987). Biodegradation of DDT [1,1,1-trichloro-2-2-bis (4-chlorophenyl) ethane]. *Applied and Environmental Microbiology*, 53, 2001–8.

Bumpus, J.A., Tien, M., Wright, D., and Aust, S.D. (1985). Oxidation of persistent environmental pollutants by a white rot fungus. *Science*, 228, 1434–6.

Caldwell, B.A., Ye, C., Griffiths, R.P., Mayer, C.L., and Morita, R.Y. (1989). Plasmid expression and maintenance during long-term starvation-survival of bacteria in well water. *Applied and Environmental Microbiology*, 55, 1860–4.

Carson, R. (1962). *Silent spring*. Houghton Mifflin, USA.

Chakrabarty, A.M. (1972). Genetic basis of the biodegradation of salicylate in *Pseudomonas*. *Journal of*

Bacteriology, **112**, 815–23.

Chapman, P.J. (1979). Degradation mechanisms. In *Workshops: microbial degradation of pollutants in marine environments*, (Ed. A.W. Bouquin and P.H. Pritchard), p. 28. US Environmental Protection Agency, Florida.

Chapman, P.J. (1988). Constructing microbial strains for degradation of halogenated aromatic hydrocarbons. In *Environmental biotechnology: reducing risks from environmental chemicals through biotechnology*, (Ed. G.S. Omenn), pp. 81–96. Plenum Press, New York.

Cohen, S.N., Miller, C.A., Beancage, S., and Biek, D.P. (1988). Stable inheritance of bacterial plasmids: Practical considerations in the release of organisms into the environment. In *Environmental biotechnology: reducing risks from environmental chemicals through biotechnology*, (Ed. G.S. Omenn), pp. 97–104. Plenum Press, New York.

Criddle, C.S., DeWitt, J.T., Grbic-Galic, D., and McCarty, P.L. (1990). Transformation of carbon tetrachloride by *Pseudomonas* sp. strain KC under denitrification conditions. *Applied and Environmental Microbiology*, **56**, 3240–6.

Cripps, C., Bumpus, J.A., and Aust, S.D. (1990). Biodegradation of A20 and heterocyclic dyes by *Pharnerochaete chrysporium. Applied and Environmental Microbiology*, **56**, 1114–18.

Dagley, S. (1984). Introduction. In *Microbial degradation of organic compounds*, (Ed. D.T. Gibson), p. 1. Marcel Dekker, New York.

DeWeerd, K.A., and Suflita, J.M. (1990). Anaerobic aryl reductive dehalogenation of halobenzoates by cell extracts of *Desulfomonile tiedjei. Applied and Environmental Microbiology*, **56**, 2995–3005.

Dwyer, D.F., Rojo, F., and Timmis, K.N. (1988). Fate and behaviour in an activated sludge microcosm of a genetically engineered micro-organism designed to degrade substituted aromatic compounds. In *The release of genetically engineered micro-organisms*, (Ed. M. Sussman, C.H. Collins, F.A. Skinner, and D.E. Stewart-Tull), pp. 77–89). Academic Press, London.

Eaton, D.C. (1985). Mineralization of polychlorinated biphenyls by *Phanerochaete chrysosporium. Enzyme and Microbial Technology*, **7**, 194–6.

Eaton, R.W., and Timmis, K.N. (1986). Spontaneous deletion of a 20-kilobase DNA segment carrying genes specifying isopropylbenzene metabolism in *Pseudomonas putida* ne204. *Journal of Bacteriology*, **168**, 428–30.

Ensley, B.D. (1986). Stability of recombinant plasmids in industrial microorganisms – methods of increasing plasmid stability. *Critical Reviews in Biotechnology*, **4**, 763–77.

Ewers, J., Freier-Schröder, D., and Knackmuss, H.-J. (1990). Selection of trichloroethene (TCE) degrading bacteria that resist inactivation by TCE. *Archives of Microbiology*, **156**, 410–13.

Fernando, T., Bumpus, J.A., and Aust, S.D. (1990). Biodegradation of TNT (2,4,6-trinitrotoluene) by *Phanerochaete chrysosporium. Applied and Environmental Microbiology*, **56**, 1666–71.

Fogel, M.M., Taddeo, A.R., and Fogel, S. (1986). Biodegradation of chlorinated ethemers by a methane-utilizing mixed culture. *Applied and Environmental Microbiology*, **51**, 720–4.

Folsom, B.R., Chapman, P.J., and Pritchard, P.H. (1990). Phenol and trichloroethylene degradation by *Pseudomonas cepacia* G4: Kinetics and interactions between substrates. *Applied and Environmental Microbiology*, **56**, 1279–85.

Fox, J.L. (1990). More confidence about degrading work. *Biotechnology*, **8**, 604.

Frick, T.D., Crawford, R.L., Martinson, M., Chresand, T., and Bateson, G. (1988). Microbiological cleaning of groundwater contaminated by pentachlorophenol. In *Environmental biotechnology: reducing risks for environmental chemicals through biotechnology*, (Ed. G.S. Ommen), pp. 173–93. Plenum Press, New York.

Friello, D.A., Mylroie, J.R., and Chakrabarty, A.M. (1976). Use of genetically-engineered multi-plasmid microorganisms for rapid degradation of fuel hydrocarbons. In *Proceedings of 3rd International Biodegradation Symposium*, (Ed. J.M. Sharpely), pp. 205–14. Applied Science Publications, Essex, UK.

Fry, J.C., McClure, N.C., and Weightman, A.J. (1992). Survival of genetically engineered microorganisms (GEMs) in activated sludge. In *Genetic interactions between microorganisms in the natural environment*, (Ed. E.M. Willington and J.D. Van Elsas), pp. 198–215. Pergamon Press, London.

Fulthorpe, R.R., and Wyndham, R.C. (1989). Survival and activity of a 3-chlorobenzoate-catabolic genotype in a natural system. *Applied and Environmental Microbiology*, **55**, 1584–90.

Genthner, B.R.S., Allen Price II, W., and Pritchard, P.H. (1989). Anaerobic degradation of chloroaromatic compounds in aquatic sediments under a variety of enrichment conditions. *Applied and Environmental Microbiology*, **55**, 1466–71.

Gibson, S.A., and Suflita, J.M. (1990). Anaerobic biodegradation of 2,4,5-trichlorophenonyacetic acid in samples from a methanogenic aquifier: stimulation by short-chain organic acids and alcohols. *Applied and Environmental Microbiology*, **56**, 1825–32.

Golovleva, L.A., Pertsova, R.N., Boromin, A.M., Travkim, V.M., and Kozlovsky, S.A. (1988). Kelthane degradation by genetically engineered *Pseudomonas aeruginosa* BS 817 in a soil ecosystem. *Applied and Environmental Microbiology*, **54**, 1587–90.

Häggblom, M.M., and Young, L.V. (1990). Chlorophenol degradation coupled to sulfate reduction. *Applied and Environmental Microbiology*, **56**, 3255–60.

Hammel, K.E. (1989). Organopollutant degradation by

ligninolytic fungi. *Enzyme and Microbial Technology*, **11**, 776–7.

Hansen, C.L., Zwolinski, G., Martin, D., and Williams, J.M. (1984). Bacterial removal of mercury from sewage. *Biotechnology and Bioengineering*, **26**, 1330–3.

Harker, A.R., and Kim, Y. (1990). Trichloroethylene degradation by two independent aromatic degrading pathways in *Alcaligenes eutrophus* JMP134. *Applied and Environmental Microbiology*, **56**, 1179–81.

Hartman, J., Engelberts, K., Nordham, B., Schmidt, E., and Reineke, W. (1989). Degradation of 2-chlorobenzoate by *in vivo* constructed hybrid pseudomonads. *FEMS Microbiology Letters*, **61**, 17–22.

Harwood, C.S., Parker, R.E., and Dispensa, M. (1990). Chemotaxis of *Pseudomonas putida* towards chlorinated benzoates. *Applied and Environmental Microbiology*, **56**, 1501–3.

Heitkemp. M.A., Camel, V., Reuter, T.J., and Adams, W.J. (1990). Biodegradation of *p*-nitrophenol in an aqueous waste stream by immobilized bacteria. *Applied and Environmental Microbiology*, **56**, 2967–73.

Henry, S.M., and Grbic-Galic, D. (1991). Influence of endogenous and exogenous electron donors and trichloroethylene oxidation toxicity on trichloroethylene oxidation by methanotrophic cultures from a groundwater aquifier. *Applied and Environmental Microbiology*, **57**, 236–44.

Hess, T.F., Schmidt, S.K., Silverstein, J., and Howe, B. (1990). Supplemental substrate enhancement of 2,4-dinitrophenol mineralization by a bacterial consortium. *Applied and Environmental Microbiology*, **56**, 1551–8.

Hill, N.P., McIntyre, A.E., Perry, R., and Lester, J.N. (1986). Behaviour of chlorophenoxy herbicides during the activated sludge treatment of municipal wastewater. *Water Research*, **20**, 45–52.

Jain, R.K., Sayler, G.S., Wilson, J.T., Houston, L., and Pacia, D. (1987). Maintenance and stability of introduced genotypes in groundwater aquifier material. *Applied and Environmental Microbiology*, **53**, 998–1002.

Jones, S.H., and Alexander, M. (1988). Effect of inorganic nutrients on the acclimation period preceding mineralization of organic chemicals in lake water. *Applied and Environmental Microbiology*, **54**, 3177–9.

Keith, I.H., and Telliard, W.A. (1979). Priority pollutants – a perspective. *Environmental Science and Technology*, **13**, 416–23.

Kellogg, J.J., Chatterjee, D.K., and Chakrabarty, A.M. (1982). Plasmid-assisted molecular breeding: new techniques for enhanced biodegradation of persistant toxic chemicals. *Science*, **214**, 1133–5.

Kennedy, D.W., Aust, S.D., and Bumpus, J.A. (1990). Comparative biodegradation of alkyl halide insecticides by the white rot fungus, *Phanerochaete chrysosporium* BKM-F-1767). *Applied and Environmental Microbiology*, **56**, 2347–458.

King, J.M.H., Di Grazia, P.M., Applegate, B., Burlage, R., Sansevarino, J., Dunbar, P., Lariner, F., and Sayler, G.S. (1990). Rapid sensitive bioluminescent reporter technology for naphthalene exposure and biodegradation. *Science*, **249**, 778–81.

Lamar, R.T., Larsen, M.J., and Kirk, T.K. (1990). Sensitivity to and degradation of pentachlorophenol by *Phanerochaete* spp. *Applied and Environmental Microbiology*, **56**, 3519–26.

LaPat-Polasko, L.T., McCarty, P.L., and Zehnder, A.J.B. (1984). Secondary substrate utilization of methylene chloride by an isolated strain of *Pseudomonas* sp. *Applied and Environmental Microbiology*, **47**, 825–30.

Lappin-Scott, H.M., and Costerton, J.W. (1990). Starvation and penetration of bacteria in soils and rocks. *Experientia*, **46**, 807–12.

Latorne, J., Reineke, W., and Knackmuss, H.-J. (1984). Microbial metabolism of chloroanilines: enhanced evolution by natural genetic exchange. *Archives of Microbiology*, **140**, 159–65.

Leahy, J.G., and Colwell, R.R. (1990). Microbial degradation of hydrocarbons in the environment. *Microbiological Reviews*, **54**, 305–15.

Lehrbach, P.R., Ward, J., Meulien, P., and Broda, P. (1982). Physical mapping of Tol plasmids pWWO and pND2 and various R-plasmid-TOL derivatives from *Pseudomonas* spp. *Journal of Bacteriology*, **152**, 1280–3.

Lewis, D.L., Hudson, R.E., and Freeman III, L.F. (1984). Effects of microbial community interactions on transformation rates of xenobiotic chemicals. *Applied and Environmental Microbiology*, **48**, 561–5.

Lovley, D.R., and Lonergan, D.J. (1990). Anaerobic oxidation of toluene, phenol and *p*-cresol by the dissimilatory non-reducing organism, GS-15. *Applied and Environmental Microbiology*, **56**, 1858–64.

Lovley, D.R., Baedecker, M.J., Lonergan, D.J., Cozzarelli, I.M., Phillips, E.J.P., and Seigel, D.I. (1989). Oxidation of aromatic contaminants coupled to microbial iron reduction. *Nature*, **339**, 297–9.

McCarty, P.L. (1988). Bioengineering issues related to *in situ* remediation of contaminated soils and groundwater. In *Environmental biotechnology: reducing risks from environmental chemicals through biotechnology*, (Ed. G.S. Omenn), pp. 143–63. Plenum Press, New York.

McClure, N.C., Weightman, A.J., and Fry, J.C. (1989). Survival of *Pseudomonas putida* UWC1 containing cloned catabolic genes in a model activated-sludge unit. *Applied and Environmental Microbiology*, **55**, 2627–34.

McClure, N.C., Fry, J.C., and Weightman, A.J. (1991). Survival and catabolic activity of natural and genetically engineered bacteria in a laboratory-scale activated-sludge unit. *Applied and Environmental*

Microbiology, **57**, 366–73.

Meulien, P., Downing, R.G., and Broda, P. (1981). Excision of the 40n segment of the TOL plasmid from *Pseudomonas putida* mt-2 involves direct repeats. *Molecular and General Genetics*, **186**, 97–101.

Mikesell, M.D., and Boyd, S.A. (1985). Reductive dechlorination of the pesticides 2,4-D and 2,4,5-T, and pentachlorophenol anaerobic sludges. *Journal of Environmental Quality*, **14**, 337–40.

Mileski, G.J., Bumpus, J.A., Jurek, M.A., and Aust, S.D. (1988). Biodegradation of pentachlorophenol by the white rot fungus *Phanerochaete chrysosporium*. *Applied and Environmental Microbiology*, **54**, 2888–9.

Miller, R.V., Kokjohn, T.A., and Sayler, G.S. (1990). Environmental and molecular characterization of systems which affect genome alteration in *Pseudomonas aeruginosa*. In *Pseudomonas biotransformations, pathogenesis and evolving biotechnology*, (Ed. S. Silver, A.M. Chakrobarty, B. Iglewski, and S. Kaplan). American Society for Microbiology, Washington D.C.

Mokoss, H., Schmidt, E., and Reineke, W. (1990). Degradation of 3-chlorobiphenyl by *in vivo* constructed hybrid pseudomonads. FEMS Microbiology Letters, **71**, 179–85.

Nelson, M.J.K., Montgomery, S.O., O'Neill, E.J., and Pritchard, P.H. (1986). Aerobic metabolism of trichloroethylene by a bacterial isolate. *Applied and Environmental Microbiology*, **52**, 383–4.

Nelson, M.J.K., Montgomery, S.O., Mahaffey, W.R., and Pritchard, P.H. (1987). Biodegradation of trichloroethylene and involvement of an aromatic biodegradative pathway. *Applied and Environmental Microbiology*, **53**, 949–54.

Nelson, M.J.K., Montgomery, S.O., and Pritchard, P.H. (1988). Trichloroethylene metabolism by microorganisms that degrade aromatic compounds. *Applied and Environmental Microbiology*, **54**, 604–6.

Nordström, K., and Austin, S.S. (1989). Mechanisms that contribute to the stable segregation of plasmids. *Annual Review of Genetics*, **23**, 37–69.

Oldenhuis, R., Vink, R.L.J.M., Janssen, D.B., and Withalt, B. (1989). Degradation of chlorinated aliphatic hydrocarbons by *Methylosinus trichosporium* OB36 expressing soluble methane monooxygenase. *Applied and Environmental Microbiology*, **55**, 2819–26.

Parsons, J.R., and Govers, H.A.J. (1990). Quantitative structure-activity relationships for biodegradation. *Ecotoxicology and Environmental Safety*, **19**, 212–27.

Phelps, J.J., Niedzielski, J.J., Schram, R.M., Herbes, S.E., and White, D.C. (1990). Biodegradation of trichloroethylene in continuous-recycle expanded bed bioreactors. *Applied and Environmental Microbiology*, **56**, 1702–9.

Quensen III, J.F., Boyd, S.A., and Tiedje, J.M. (1990). Dechlorination of four commercial polychlorinated biphenyl mixtures (aroclors) by anaerobic microorganisms from sediments. *Applied and*

Environmental Microbiology, **56**, 2360–9.

Ramos, J.L., Duque, E., and Ramos-Gonzalez, M.-I. (1991). Survival in soils of an herbicide-resistant *Pseudomonas putida* strain bearing a recombinant TOL plasmid. *Applied and Environmental Microbiology*, **57**, 260–6.

Ramos, J.L., Stolz, A., Reineke, W., and Timmis, K.N. (1986). Altered effector specificities in regulators of gene expression: TOL plasmid *XylS* mutants and their use to engineer expansion of the range of aromatics degraded by bacteria. *Proceedings of the National Academy of Sciences (USA)*, **83**, 8467–71.

Ramos, J.L., Wasserfallen, A., Rose, K., and Timmis, K.N. (1987). Redesigning metabolic routes: manipulation of TOL plasmid pathway for catabolism of alkylbenzoates. *Science*, **235**, 593–6.

Reineke, W., and Knackmuss, H.-J. (1979). Construction of haloaromatic utilising bacteria. *Nature*, **277**, 385–6.

Reineke, W., Jeenes, D.J., Williams, P.A., and Knackmuss, H.-J. (1982). TOL plasmid pWWO in constructed halobenzoate-degrading *Pseudomonas* strains: prevention of *meta*-pathway. *Journal of Bacteriology*, **150**, 195–201.

Rojo, F., Pieper, D.H., Engesser, K.H., Knackmuss, H.-J., and Timmis, K.N. (1987). Assemblage of *ortho*-cleavage route for simultaneous degradation of chloro- and methylaromatics. *Science*, **238**, 1395–8.

Rubio, M.A., Engesser, K.-H., and Knackmuss, H.-J. (1986a). Microbial metabolism of chlorosalicylates: accelerated evolution by natural genetic exchange. *Archives of Microbiology*, **145**, 116–22.

Rubio, M.A., Engesser, K.-H., and Knackmuss, H.-J. (1986b). Microbial metabolism of chlorosalicylates: effect of prolonged subcultivation on constructed strains. *Archives of Microbiology*, **145**, 123–5.

Saint, C.P., McClure, N.C., and Venables, W.A. (1990). Physical map of the aromatic amine and *m*-tolute catabolic plasmid pTDN1 in *Pseudomonas putida*: location of a unique *meta*-cleavage pathway. *Journal of General Microbiology*, **136**, 615–25.

Sayadi, S., Nasri, M., Barbotin, J.N., and Thomas, D. (1989). Effects of environmental growth conditions on plasmid stability, plasmid copy number, and catechol 2,3,-dioxygenase activity in free and immobilised *Escherichia coli* cells. *Biotechnology and Bioengineering*, **33**, 801–8.

Scheuerman, P.R., Schmidt, J.P., and Alexander, M. (1988). Factors affecting the survival and growth of bacteria introduced into lake water. *Archives of Microbiology*, **150**, 320–5.

Schmidt, S.K., Scow, K.M., and Alexander, M. (1987). Kinetics of *p*-nitrophenol mineralisation by a *Pseudomonas* sp. effects of second substrates. *Applied and Environmental Microbiology*, **53**, 2617–23.

Schreiner, R.P., Stevens, S.E., and Tien, M. (1988). Oxidation of thianthrene by the ligninase of

Phaenerochaete chrysosporium. Applied and Environmental Microbiology, **54**, 1858–60.

Senior, E., Bull, A.T., and Slater, J.H. (1975). Enzyme evolution in a microbial community growing on the herbicide dalapon. *Nature*, **263**, 476–9.

Shelton, D.R., and Tiedje, J.M. (1984). Isolation and partial characterization of bacteria in an aerobic consortium that mineralizes 3-chlorobenzoic acid. *Applied and Environmental Microbiology*, **48**, 840–8.

Shields, M.S., Montgomery, S.O., Chapman, P.J., Cuskey, S.M., and Pritchard, P.H. (1989). Novel pathway of toluene catabolism in the trichloroethylene-degrading bacterium G4. *Applied and Environmental Microbiology*, **55**, 1624–9.

Sinclair, J.L., and Alexander, M. (1984). Role of resistance to starvation in bacterial survival in sewage and lake water. *Applied and Environmental Microbiology*, **48**, 410–15.

Slater, J.H., and Lovatt, D. (1984). Biodegradation and the significance of microbial communities. In *Microbial degradation of organic compounds*, (Ed. D.T. Gibson), p. 439. Marcell Dekker, New York.

Slater, J.H., Weightman, A.J., and Hall, B.G. (1985). Dehalogenase genes of *Pseudomonas putida* PP3 on chromosomally located transposable elements. *Molecular Biology and Evolution* **2**, 557–67.

Sojka, S.A., and Ying, W.C. (1987). Genetic engineering and process technology for hazardous waste control. *Developments in Industrial Microbiology*, **27**, 129–33.

Song, H.-G., Wang, X., and Bartha, R. (1990). Bioremediation potential of terrestrial fuel spills. *Applied and Environmental Microbiology*, **56**, 652–6.

Steffan, R., and Atlas, R.M. (1988). DNA amplification to enhance detection of genetically engineered bacteria in environmental samples. *Applied and Environmental Microbiology*, **54**, 2185–91.

Steffan, R., Breen, A., Atlas, R.M., and Sayler, G.S. (1989). Application of gene probe methods for monitoring specific microbial populations in freshwater ecosystems. *Canadian Journal of Microbiology*, **35**, 681–5.

St John, W.D., and Sikes, D.J. (1988). Complex industrial waste sites. In *Environmental biotechnology: reducing risks from environmental chemicals through biotechnology*, (Ed. G.S. Ommenn), pp. 237–53. Plenum Press, New York.

Strotmann, U.J., Pentenga, M., and Janssen, D.B. (1990). Degradation of 2-chloroethanol by wild type and mutants of *Pseudomonas putida* US2. *Archives of Microbiology*, **154**, 294–300.

Stucki, G., and Alexander, M. (1987). Role of dissolution rate and solubility on biodegradation of aromatic compounds. *Applied and Environmental Microbiology*, **53**, 292–7.

Suflita, J.M., Stobt, J., and Tiedje, J.M. (1984). Dechlorination of 2,4,5-trichlorophenoxyacetic acid (2,4,5-T) by anaerobic microorganisms. *Journal of Agricultural and Food Chemistry*, **32**, 218–21.

Thomas, C.M., and Helinski, D.R. (1989). Vegetative replication and stable inheritance of IncP plasmids. In *Promiscuous plasmids of gram-negative bacteria*, (Ed. C.M. Thomas), pp. 1–25. Academic Press, London.

Topp, E., Crawford, R.L., and Hanson, R.S. (1988). Influence of readily metabolizable carbon on pentachlorophenol metabolism by a pentachlorophenol-degrading *Flavobacterium* sp. *Applied and Environmental Microbiology*, **54**, 2452–9.

Trevors, J.T., (1988). Use of microcosms to study genetic interactions between microorganisms. *Microbiological Science*, **5**, 132-6.

Trevors, J.T., van Elsas, J.D., Starodut, M.E., and Van Overbeek, L.S. (1989). Survival of and plasmid stability in *Pseudomonas* and *Klebsiella* spp. introduced into agricultural drainage water. *Canadian Journal of Microbiology*, **35**, 675–80.

Trevors, J.T., van Elsas, J.D., van Overbeek, L.S., and Starodub, M. (1990). Transport of a genetically engineered *Pseudomonas fluorescens* strain through a soil microcosm. *Applied and Environmental Microbiology*, **56**, 401–8.

Tsien, H.-C., Brusseau, G.A., Hansen, R.S., and Wackett, L.P. (1989). Biodegradation of trichloroethylene by *Methylosinus trichosporium* OB36. *Applied and Environmental Microbiology*, **55**, 3155–61.

Tsuda, M., and Jino, T. (1988). Identification and characterisation of Tn*4653*, a transposon covering the toluene transposon Tn*4651* on TOL plasmid pWWO. *Molecular and General Genetics*, **213**, 72–7.

Valli, K., and Gold, M.H. (1991). Degradation of 2,4-Dichlorophenol by the lignin-degrading fungus *Phanerochaete chrysosporium. Journal of Bacteriology*, **173**, 345–52.

Vogel, T.M. and McCarty, P.L. (1985). Biotransformation of tetrachloroethylene to trichloroethylene, dichloroethylene, vinyl chloride and carbon dioxide under methanogenic conditions. *Applied and Environmental Microbiology*, **49**, 1080–3.

Wackett, L.P., and Gibson, D.T. (1988). Degradation of trichloroethylene by toluene dioxygenase in whole-cell studies with *Pseudomonas putida* F1. *Applied and Environmental Microbiology*, **54**, 1703–8.

Wackett, L.P., and Householder, S.R. (1989). Toxicity of trichloroethylene to *Pseudomonas putida* F1 is mediated by toluene dioxygenase. *Applied and Environmental Microbiology*, **55**, 2723–5.

Wackett, L.P, Brussean, G.A., Householder, S.R., and Hansen, R.S. (1989). Survey of microbial oxygenases trichloroethylene degradation by propane-oxidising bacteria. *Applied and Environmental Microbiology*, **55**, 2960–6.

Weightman, A.J., and Slater, J.H. (1988). The problem of xenobiotics and recalcitrance. In *Micro-organisms in action: concepts and applications in microbial ecology*, (Ed. J.M. Lynch and J.E. Hobbie), pp. 322–47.

Blackwell Scientific Publications, Oxford.

Weightman, A.J., Weightman, A.L., and Slater, J.H. (1985). Toxic effects of chlorinated and brominated alkanoic acids on *Pseudomonas putida* PP3: selection at high frequencies of mutations in genes encoding dehalogenases. *Applied and Environmental Microbiology*, **49**, 1496–501.

Wiggins, B.A., and Alexander, M. (1988). Role of chemical concentration and second carbon sources in acclimination of microbial communities for biodegration. *Applied and Environmental Microbiology*, **54**, 2803–7.

Wiggins, B.A., Jones, S.H., and Alexander, M. (1987). Explanations for the acclimation period preceding the mineralization of organic chemicals in aquatic environments. *Applied and Environmental Microbiology*, **53**, 791–6.

Winter, R.B., Yen, K.-M., and Ensley, B.D. (1989). Efficient degradation of trichloroethylene by a recombinant *Escherichia coli*. *Biotechnology*, **7**, 282–5.

Wyndham, R.C. (1986). Evolved aniline catabolism in *Acinetobacter calcoaceticus* during continuous culture of river water. *Applied and Environmental Microbiology*, **51**, 781–9.

Wyndham, R.C., Singh, R.K., and Straus, N.A. (1988). Catabolic instability, plasmid gene deletion and recombination in *Alcaligenes* sp. BR60. *Archives of Microbiology*, **150**, 237–43.

Zeyer, J., Eicher, P., Dolfing, J., and Schwarzenbach, R.P. (1990). Anaerobic degradation of aromatic hydrocarbons. In *Biotechnology and biodegradation*, (Ed. D. Kakely, A. Chakrabarty and G.S. Omenn), pp. 33–40. Gulf Publishing Company, Houston.

4

Gene transfer in the environment: conjugation

M. Day and J.C. Fry

Introduction

OUR UNDERSTANDING of the mechanisms of conjugation is still at an elementary stage. Although the 'end-product' of the process is largely predictable only two aspects need to be universal:

1. All donors must express both internal regulatory factors (required for the orientation of DNA transfer) and surface components, which allow for the recognition.
2. Penetration of the recipient.

Conjugation is one of three systems that permit genes to be exchanged between bacteria. It is a parasexual process requiring cell-to-cell contact between donor and recipient cells (Willetts and Skurray 1980). The genetic information required for self-transfer is encoded on transfer-proficient plasmids (Tra$^+$) and cells may be host to more than one plasmid (Bender and Cooksey 1986). These plasmids can also mobilize other non-self-transmissible plasmids (Tra$^-$) and chromosomal genes. Plasmids are circular, double-stranded, autonomously replicating DNA molecules, which are widely distributed through the bacterial kingdom (Day 1987). They are generally regarded as dispensable to their host cell because they do not often encode genetic characters (phenotypes) essential for the more routine aspects of cell growth and survival. Phenotypes encoded by plasmids include those for catabolism, (naphthalene and cellulolysis), resistance (UV and mercury), biosynthesis (nitrogen fixation) and a range of other miscellaneous types, e.g. tumourigenicity), modulation (Day 1987) and mutability (Frigo 1985).

Conjugation mechanisms

Conjugative pili have been identified for plasmids from various incompatibility groups tested in *Escherichia coli* and *Pseudomonas* spp. (Paranchych and Frost 1987). These pili are essential for conjugation (Achtman *et al*. 1978) and have been broadly divided into two morphological groups, 'long flexible' (1 μm) and 'short rigid' (0.1 μm) (Bradley 1980). Long flexible pili are well characterized and those expressed by the F plasmid allow exchange to take place in liquids and on surfaces at about the same frequency. Plasmids capable of this are said to exhibit a universal mating phenotype. The short and rigid pili expressed by plasmids of the Inc N, P and W groups are 10^3–10^5 times more efficient on solid surfaces (Bradley 1980). It is not known if the coincidence of rigid pili with the plasmids in these Inc groups, which are all in the broad host range, has any significance. In addition, some plasmids in the Inc groups I_1, I_2, I_5, B, K and Z determine both rigid and flexible pili (Bradley 1984).

The number of conjugative pili per cell is typically small and their lengths vary considerably with host growth phase and cultural conditions (Curtiss 1969; Frost *et al*. 1985). The tips of pili initiate contact with the recipient cell and allow stable cell-to-cell aggregates to form. Current models for conjugation suggest that, on contact, pilus dissembly 'pulls' the surface of the cells together, a contact that develops to form a DNA transport pore between the cells (Ippen-Ihler 1989).

Conjugation in Gram-positive bacteria is very different because pili are not found. Instead, donor cells form a DNA transport pore directly between cells (Ippen-Ihler 1989). *Streptococcus faecalis* can form aggregates both with recipient cells and in the presence of the culture supernatants of recipient cells. Investigations into this phenomenon have shown these plasmid conjugation systems to be induced by pheromones (Dunney *et al.* 1985). These pheromones (also known as clumping inducing agents – CIAs) are small peptides excreted by the recipient cells. When plasmid-carrying donor cells are exposed to an appropriate CIA, plasmid-encoded genes are induced to produce new surface proteins, which promote cell clumping and allow conjugal transfer to proceed. Individual *S. faecalis* cells excrete multiple pheromones, each of which can specifically induce a mating response from a donor cell carrying a particular type of plasmid (Clewell *et al.* 1987).

Two further variations on the conjugational process have been established. First, transposon-mediated conjugation has been observed. The most extensively studied element is Tn*916*, a 16.4-kb transposon encoding tetracyclin resistance, isolated in *S. faecalis* (Franke and Clewell 1982; Clewell and Gawron-Burke 1986), which can transpose between plasmids and the chromosome. It is capable of conjugal transfer between cells with or without plasmids at frequencies of 10^{-6}–10^{-8} in filter matings.

Retrotransfer describes a second variation, and one that is a process of reciprocal genetic exchange (Powell *et al.* 1989). In matings with an IncP1 plasmid, donor cells can inherit markers from the recipient at frequencies similar to those transferred from the donor (Schoonejans and Tousant 1983; Mergeay and Gerits 1978). IncQ plasmids could also be transferred from a *Desulfovibrio desulphuricans* recipient to the donor if the donor possessed an IncP plasmid (Thiry *et al.* 1984). Plasmids with this property are termed hermaphoriditic, as they endow the interacting host cells with the ability to exchange genes reciprocally. So far, this ability has been observed in IncP1, IncM and some IncN plasmids, but not in IncW, IncFI, IncFII or IncV (Thiry *et al.* 1984).

Cellular factors affecting conjugal transfer

A variety of genetic and physiological factors influences the free flow of genes in and between bacterial communities.

Surface (entry) exclusion

The phenomenon of surface (entry) exclusion is exhibited by many conjugative plasmids, which encode products that interfere with the capacity of the cell to pair and/or for DNA to move between cells carrying the same type of conjugative plasmid (Ou 1980). A possible advantage to sibling cells growing in close association is that this mechanism will reduce energy expenditure on unproductive self-matings.

Mobilization

Mobilization is a process by which smaller plasmids, under 30 kb (Thiry *et al.* 1984). which are not large enough to encode a self-conjugative system, may be transferred from one cell to another by the transfer genes of another plasmid. Recent work suggests that the minimum size of conjugal plasmids may be well below 20 kb (Hopwood *et al.* 1986), and in the well-characterized streptomycete plasmid pIJ101 (8.9 kb) the transfer region is only 2.1 kb (Keiser *et al.* 1982). In the broad host range plasmid pEV1 (11 kb) the Tra functions are encoded in less than 3.5 kb and in other plasmids it is commonly less than 10 kb (McNeil and Gibbon 1986).

Transfer may be achieved in two different ways. Donation describes the process by which a plasmid incapable of self-transmission uses the transfer system of a coresident conjugal plasmid to move from its host cell into the recipient. This can only occur if the non-conjugal plasmid possesses an *ori*T or *bom* site (basis of mobility) and Mob genes. The Mob genes of RSF1010 permit transfer not only among Gram-negative bacteria but also, surprisingly, from *Agrobacterium* sp. into plant cells (Buchanan-Wollaston *et al.* 1987).

Plasmids may also be mobilized by conduction (Clarke and Warren 1979). The plasmid forms a cointegrate with a conjugal plasmid and is then transferred. After transfer into the new host, the cointegrate resolves into the two plasmids.

Genther *et al.* (1988) found that 26 out of 68 natural freshwater isolates could act as recipients for the broad host range (BHR) cloning vector R1162 when mobilized by plasmid R68. Mobilization, from a strain of *E. coli* carrying the small non-transmissible plasmid pBR325 in a sewage treatment plant microcosm has been used to demonstrate the presence of natural mobilizing

Table 4.1. *Effect of recipient species on frequency of transfer of three natural plasmids pQM1, pQM3 and pQM4*

Recipient bacteria	Strain	Plasmid transfer frequency (per recipient)		
		pQM1[1]	pQM3[2]	pQM4[2]
Pseudomonas aeruginosa	PAO348	8×10^{-1}	3×10^{-8}	2×10^{-5}
	PU21	3×10^{-1}	2×10^{-6}	3×10^{-10}
Pseudomonas putida	KT2440	4×10^{-4}	2×10^{-2}	3×10^{-4}
	PP3	7×10^{-8}	4×10^{-7}	ND
Pseudomonas fluorescens	AR41	5×10^{-6}	5×10^{-6}	ND
Pseudomonas maltophilia	AR27	4×10^{-8}	1×10^{-9}	5×10^{-9}
Alteromonas putrifaciens	10695	ND	ND	3×10^{-7}
Escherichia coli	LE392	ND	1×10^{-6}	5×10^{-3}
Acinetobacter calcoaceticus	BD413	ND	2×10^{-9}	2×10^{-9}
Proteus vulgaris	6213	ND	6×10^{-9}	2×10^{-8}
Klebsiella pneumoniae	8172	ND	9×10^{-9}	7×10^{-8}*

[1] Bale *et al.* 1988b; [2] Rochelle *et al.* 1988b; * Lowest detectable frequency; ND, no detectable transfer.

plasmids (Gealt *et al.* 1985; McPherson and Gealt 1986; Mancini *et al.* 1987). McClure *et al.* (1989), working in an activated sludge microcosm, have shown mobilization of a recombinant plasmid pD10 from an added *Pseudomonas putida* host by indigenous plasmids to sludge bacteria. Native soil streptomycetes harbour BHR conjugative plasmids. Ross-McQueen *et al.* (1985) and Raffii and Crawford (1988) have shown mobilization of non-conjugative plasmids in triparental matings with *Streptomyces lividans* in sterile soil. Although this was at a lower frequency in microcosms than in the laboratory, their data clearly show the potential for mobilization in soil exists.

As mobilization is achieved by conjugal plasmids, those factors that effect conjugation will logically also effect mobilization. At present we do not know the extent of mobilization in natural environments or many of the factors that may effect its success. These examples, however, show that there is widespread potential for the mobilization of plasmids in nature.

Incompatibility

Incompatibility reflects a relationship between the replication and/or segregation systems of two plasmids such that they cannot be reliably inherited by the daughter cells at division (Bukhari *et al.* 1977; Datta 1979). Incompatibility can also be caused by the host cell being lysogenic for a phage. For example, phage B3 lysogens of *Pseudomonas aeruginosa* are non-permissive for the maintenance of IncP plasmids (Stanisich and Ortiz 1976).

Restriction and modification

Restriction and modification (RM) classically describes a double enzyme system with the potential to limit the establishment of exogenous DNA acquired through any transfer process (Linn and Arber 1968). The first enzyme is an endonuclease, which cleaves target DNA at a specific sequence, rendering it biologically inactive. The second enzyme, a methylase, modifies the target site and protects it from cleavage (Roberts 1983). There is a wide range of such enzymes and some cells may have more than one RM system, encoded on plasmids and phage, as well as on the chromosomes. No work of this type has yet been reported in microcosms or *in situ*.

BHR plasmids

BHR plasmids are those that can be transferred and stably maintained in bacteria belonging to different genera (Datta and Hedges 1972).

How broad is broad host range? Plasmids considered to have really broad host range, as measured by their ability to enter most Gram-negative bacteria, belong to the IncP, W and Q groups. We have examined the transfer range of three naturally occurring plasmids, which have narrow or BHR properties. The host range of pQM1 (Bale *et al.* 1988b) was narrow, confined to fluorescent pseudomonads, while that of pQM3 and pQM4 was broader (Table 4.1).

BHR plasmids vary greatly in the transfer frequencies observed between different hosts. For

Table 4.2. *Transfer characteristics of pQM3*

Recipient bacteria	Transfer frequency (per recipient) from the following donor bacteria		
	P. cepacia	*P. fluorescens*	*P. putida*
Pseudomonas aeruginosa PAO348	8×10^{-4} (28, 34, 44, 78, 156)	6×10^{-3} (91)	2×10^{-3} (28, 34, 44, 78)
Pseudomonas fluorescens AR41	5×10^{-6} (91)	8×10^{-2} (91)	5×10^{-3} (91)
Pseudomonas putida KT2440	2×10^{-2} (78)	1×10^{-3} (91)	6×10^{-1} (78)
Escherichia coli LE392	1×10^{-6} (78)	ND	ND

Transfer frequencies were the means of four determinations (Rochelle *et al.* 1988a).
Figures in brackets are the sized (kb) of plasmid bands detected in the transconjugants following each mating. ND, not done.

example, the host and recipient strains can be seen to exert a considerable influence (over 1000-fold) on the transfer frequency of pQM3 (Table 4.2). An even greater effect (over 10^8-fold) on transfer success between a smaller group of bacteria, was seen with the narrow host range plasmid pQM1 (Bale *et al.* 1988a).

Self-maintenance and the ability to replicate are major features required by BHR plasmids for persistence in diverse microbial populations. Soil streptomycetes harbour BHR conjugative plasmids (Ross-McQueen *et al.* 1985), which can mobilize smaller plasmids (Raffi and Crawford 1988). BHR plasmids like RP1 (IncP1) and R300B (IncQ) both have such maintenance and gene expression qualities. In fact, the only species known to be non-permissive for IncP plasmids are *Bacteriodes* spp. (Guiney *et al.* 1984). Although RP1 can transfer into *Myxococcus xanthus*, it is unable to replicate unless it becomes integrated into the host cell chromosome (Breton *et al.* 1985). A similar observation has been made for other narrow host range plasmids of the IncF1 and IncIa groups (Mergeay and Gerits 1978; Boulnois *et al.* 1985). Thus, transfer range does not equate with maintenance range. In addition, conjugal plasmid transfer can occur from *E. coli* to Gram-positive bacteria (Trieu-Cout *et al.* 1987) and even to yeast (Heinemann and Sprague 1989). Smaller BHR plasmids (e.g. RSF1010) need to be mobilized to realize their potential and such plasmids have been shown to enter plant tissue after mobilization by the *Agrobacterium* plasmid pTi (Buchanan-Wollaston *et al.* 1987).

Expression

After a plasmid is transferred to a new host, a minimal level of expression is required for it to be maintained in the host. If it is to alter the host cell phenotype then further genes will have to be expressed. For transcription to occur the host RNA polymerases must be able to recognize the plasmid-encoded promoters and initiate RNA synthesis. Promoter sequences are different in different species and their efficiency may vary considerably in novel genetic backgrounds. For example, many *P. aeruginosa* genes are not expressed in *E. coli* because the promoters are not recognized (Chakrabarty *et al.* 1978; Kokjohn and Miller 1985). Other genes are expressed at different levels, for example tetracyclin resistance mediated by Tn*10* is about 10-fold lower in *E. coli* than in *Haemophilus* spp. (Levy *et al.* 1984). Tn5 expresses only kanamycin resistance in *E. coli* but in *Klebsiella* spp. and a range of other Gram-negative bacteria it also expresses streptomycin and bleomycin resistances (Marshall *et al.* 1988). It is obvious, then, that the failure of a gene to express prevents phenotypic detection. However, this does not mean that the gene is absent from the gene pool and nor does it eliminate the possibility of it becoming active on transfer to another strain or species.

Mutation rate

The average mutation rate per nucleotide in *E. coli* is about 10^{-9}–10^{-10} per generation (Lewin 1987). Simple calculations show that a large proportion of these will be neutral or silent, and will therefore have little effect on the stability of the plasmid or expression of its genes. As most plasmids exist in cells at a predetermined copy-number, any mutations that adversely affect the reproductive fitness of the plasmid will result in its loss.

Other types of mutation affect the structural integrity of plasmids. For example, the frequency of RP1 transfer in soil isolates of *Pseudomonas fluorescens* is improved with large deletions in the plasmid (Voisard *et al.* 1988). A similar observation was made by Tu *et al.* (1989), who have identified a range of point mutations and deletions in the plasmid RP4 that are necessary for its stable maintenance in *Xanthomonas campestris* var. *citri*. They suggest that the increase in stability was a result of various mutations selected to inactivate the *kil* and *kor* functions, which interfere with its stability in the host cell.

Rochelle *et al.* (1988a) report changes in the size of a mercury resistance plasmid, pQM3, which occurred on its transfer to a range of recipients (Table 4.2). Similar modifications in plasmid size after transfer have been reported by others (Beckman *et al.* 1982; Szabo and Mills 1984; Lessie and Gaffney 1986) for pseudomonad plasmids.

Genera involved and potential for spread

Conjugal gene transfer has been widely demonstrated in both Gram-positive and Gram-negative bacteria. The diversity of bacteria in which plasmids have been identified indicates that genetic isolation *in situ* will only occur when an individual population is physically isolated. The degree to which any organism participates in gene transfer (or is genetically isolated) will depend on the host and recipient (their densities, genotypes, phenotypes, physiology, etc.), the plasmid (its transmissibility, genotype, etc.) and environmental factors (pH, temperature, nutrient status, etc.).

Survival

Liang *et al.* (1982) reported that the survival and growth of bacteria were reduced if inoculated into a non-sterile environment like sewage, soil or water. Conversely, if they were placed in a sterile environment survival and growth was higher (Stotsky and Babich 1986).

The survival of two laboratory strains and three natural soil isolates with BHR plasmids (pRD1 and RP4) all decreased in non-sterile soil (Schilf and Klingmuller 1983). However, Krasovsky and Stotzky (1987) showed that the survival of *E. coli* transconjugants was better than either of the parental strains. These apparently conflicting observations illustrate the paucity of our knowledge about the biological characteristics needed to persist *in situ* and we should not, therefore, expect all plasmids (recombinant or not) to have similar potential for ecological effect or to confer the same survival characteristics on all their hosts.

More work has been done on the survival in soil of plasmid-free strains of *P. fluorescens* and *B. subtilis* (Van Elsas *et al.* 1987) and *E. coli* (Trevors *et al.* 1987) than in water. Golovleva *et al.* (1988) examined the survival of a plasmid-bearing *P. aeruginosa* strain growing on kelthane. In contaminated soil the plasmid was maintained in over 70 per cent of the population.

The transfer of chromosomal genes and survival of both donor and recipient cells was raised at pH values approaching neutrality in soil (Krasovsky and Stotzky 1987; Weinberg and Stotzky 1972). None of these observations consider the potential contribution of transfer from non-viable or non-recoverable cells or by the uptake of free DNA into the resident indigenous microbial population.

In situ transfer

In situ work is rather loosely defined in the literature because very few experiments have been performed unenclosed or under natural conditions. Most work has been done in laboratory microcosms and outside in some form of enclosure, which isolates the experiment from indigenous microflora. Sometimes these have been added to the test system, but more often they have been omitted.

In the experimental assessment of transfer, by whatever genetic mechanism, judgements of frequency are made on the recovered transcipients. At present a successful transfer but non-survival, non-expression or a failure to recover transcipients are all practically observed as a reduction in frequency and are classed as failure to transfer. Stotzky (1989) has also noted that the type of bacteria and plasmids involved, together with the physiochemical nature of the environment, all contribute to the survival and hence observed success of gene transfer. The total number of cells and their presentation is another important aspect to be borne in mind. The total number of cells and ratio of potential donor cells to recipients in the *milieu* will influence the frequency of transfer and consequently must be taken into account in any assessment strategy. Studies in the laboratory (Rochelle *et al.* 1989b), in enclosures in lake water (O'Morchoe *et al.* 1988) and *in situ* (Bale *et al.* 1988a) all demonstrate that this is indeed a primary factor in successful transfer.

Environmental factors affecting gene transfer

Habitat – soil

Stotzky (1989) stressed the importance of the types of bacteria and their associated plasmids, together with the physiochemical nature of the soil environment, to the survival of, and gene transfer into, the recipient. Logically this should apply to all other environments. The first studies on transfer in soil were done under sterile conditions (Weinberg and Stotzky 1972) and were recognized as having severe limitations in their applicability to the natural situation. The apparent inability to detect transfer at high rates *in situ* in bulk soil may be due to low population densities of donor and recipient populations, low fitness and competitiveness values and to various barriers affecting transfer. Chromosomal gene transfer only occurred between *E. coli* strains in sterile soil (Krasovsky and Stotzky 1987), which indicates that the indigenous microbial population inhibits transfer in some way(s).

Devanas and Stotzky (1987) give as an example a case where RP4 transferred in non-sterile soil but two other plasmids did not. They comment on the observation that as the nutrient status of the non-sterile soil rose so did RP4 transfer frequency and, conversely, as the relative humidity (rH) value fell, plasmid transfer frequency also dropped. Neither of these factors was so important in sterile soil. The BHR plasmid pRD1 transferred from *E. coli*, when added to non-sterile soil, into indigenous bacteria at the lower frequency of 10^{-9} compared to a transfer frequency of about 10^{-5} in laboratory experiments (Schilf and Klingmuller 1983). Van Elsas *et al.* (1988b) were unable to detect transfer of RP4 into the indigenous soil population from the rhizosphere or into individual soil isolates, even when the soil was amended. A similar observation

was made by Trevors and Oddie (1986), who were unable to show plasmid transfer in unamended soil. This was unexpected for two reasons. Pertsova *et al.* (1984) had shown transfer of 3-chlorobenzoate degradative capacity between *P. aeruginosa* and *P. putida* strains and to indigenous pseudomonads, while Schilf and Klingmuller (1983) had found, in laboratory matings, that 1.3 per cent of their soil isolates were recipients of RP4. Van Elsas *et al.* (1987) report what appears to be conjugal exchange of a tetracyclin-resistance plasmid between *Bacillus* spp. in sterile soil microcosms. Nutrients and bentonite additions increased the transfer frequency. Similar experiments in non-sterile soil were unsuccessful unless bentonite was added. This amendment also increased the survival of the recipients. No *in situ* experiments have yet been done in soil (J.D. Van Elsas, personal communication).

Plasmid transfer between streptomycetes appears to occur more freely in soil than between the non-mycelial bacteria discussed above. Large numbers of transconjugants were isolated from intraspecific and interspecific crosses in soil microcosms using a multiple antibiotic-resistance recombinant plasmid (pIJ673), *Streptomyces lividans* and *Streptomyces violaceolatus* (Wellington *et al.* 1990). Transfer occurred in sterile and non-sterile soil, whether or not it was amended with starch and chitin.

Habitat – aqueous

There are two other habitats in which gene transfer has been studied, both of which are aqueous. First, there are the waste treatment units and secondly, there are various river and lake systems.

A common feature of each of these investigations is that a large proportion of the bacteria exist enclosed in a polysaccharide matrix in flocs, particles or as a biofilm on surfaces, like those of submerged stones (epilithon). Although universal mating systems do occur, as indicated by a high-plate-to-broth ratio of 1 : 1.9 for pQM1 (Bale *et al.* 1987) many other plasmids have surface-preferred mechanism(s), e.g. RP1 (1 : 700), pQM3 (1 : 662) and pQM4 (1 : 1016) (Rochelle *et al.* 1989a).

To our knowledge the only unenclosed *in situ* transfer experiments to have been done are those reported by Bale *et al.* (1987, 1988a,b). Although their protocol is based on a biofilm it offers a tool by which *in situ* transfers can be more widely studied.

The currently available evidence suggests that conjugative plasmid transfer occurs more readily and at higher frequencies in aquatic habitats than in soil. Transfer also occurs whether the environment is amended with nutrients or not (Bale *et al.* 1987). From *in situ* experiments, done in epilithon, frequencies between about 10^{-10} and 10^{-1} have been obtained; the actual value dependent mainly on the temperature and donor-to-recipient ratio (Fry and Day 1990). However, conjugation frequencies of about 10^{-5} have been obtained in wastewater systems (Altherr and Kasewick 1982; Mach and Grimes 1982).

pH

In laboratory matings between strains of *E. coli* (Curtiss 1976) transfer was confined between pH 6 and 8.5.

Two soil studies (Weinberg and Stotzky 1972; Krasovsky and Stotzky 1987) have shown that raising the pH to neutrality also increased the frequency of gene exchange and the survival of donors, recipients and exconjugants. Chromosomal gene transfer and survival of donors and recipients in soil was also raised at pH values approaching neutrality (Weinberg and Stotzky 1972; Krasovsky and Stotzky 1987).

Laboratory matings with natural plasmid pQM1 (Fig. 4.1) show pH to have had little effect at 25°C, but it did have an effect at 37°C. Other natural plasmids, and RP1, show that pH affects transfer efficiency in the range 4.5–8.5, but in no case was transfer totally inhibited (Rochelle *et al.* 1988a). Also included are data from laboratory experiments showing that transfer of natural plasmids from fresh epilithon is little effected by pH.

Temperature

Mobilization of a small mercury resistance plasmid pQM17 by RP1 had a broad temperature optimum, between 15 and 37°C (Fig. 4.2), with the frequency falling rapidly below 15°C. Transfer still occurred at 5°C. What is surprising is that the self-transfer of RP1 is 4000-fold lower than the frequency with which it mobilizes pQM17; this effect is seen at all temperatures (Rochelle *et al.* 1988b). If this differential between self-transfer and mobilization is typical and is found for other small plasmids, then *in situ* mobilization frequencies will be significantly higher than transfers of larger self-transmissible plasmids.

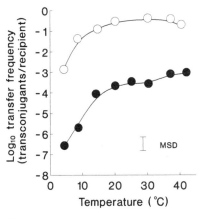

Fig. 4.2. The effect of temperature on the mobilization of pQM17 by RP1 (○), compared with RP1 self-transfer (●) (Rochelle *et al.* 1988b, 1989a). MSD, minimum significant difference.

Figure 4.3 shows the effect of temperature on the transfer of RP1 and four natural exogenously isolated plasmids (Fry and Day 1990). Both pQM1 and pQM3 had optimal transfer frequencies at 20–25°C. This is similar to optima reported by others for plasmids in soil bacteria (Kelly and Reanney 1984), a marine *E. coli* (Gauthier *et al.* 1985), the transfer of antibiotic resistance from a sewage isolate (Altherr and Kasewick 1982) and R1*drd*19 (Singleton 1983). The transfer frequency of RP1 reached a plateau between 20 and 42°C.

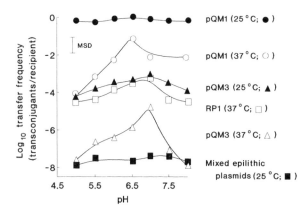

Fig. 4.1. The effect of pH on plasmid transfer in laboratory experiments (Rochelle *et al.* 1989a).

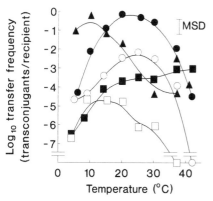

Fig. 4.3. The effect of temperature on the frequency of conjugation between various pseudomonads. The plasmids pQM1 (●), PR1 (■), pQM3 (○), pQM4 (□) and pQM85 (▲) (Rochelle *et al.* 1989a; Fry and Day 1990). MSD, minimum significant difference.

Plasmids pQM1 and pQM3 show similar transfer profiles to RP1 at lower temperatures, but are about 100 to 1000 times higher at their optimum (Fig. 4.3). Plasmid pQM85 was isolated during the winter at incubation temperatures of 10°C. The results show that pQM85 has a very high and sharp peak of transfer at 10°C. These results clearly suggest that plasmids exist in the epilithon that transfers best during winter conditions. Finally, the transfer frequency of pQM1 at 6°C *in situ* is about 500 times lower than is seen in these laboratory experiments (Bale *et al.* 1987).

We have also examined the transfer of pQM1 between a natural psychotroph (*P. fluorescens* SAM1) and mesotroph (*P. putida* (MES1). Both bacteria grew well at 20°C but had different optimum transfer temperatures of 20–25°C (SAM1) and 25–30°C (MES1). Transfer of pQM1 from SAM1 to MES1 was high (1.5×10^{-2} per recipient) at 20°C, similar to the SAM1 self-cross. The transfer frequency was about 180 times lower when transferred from MES1 to SAM1 (J. Daiper, J.C. Fry and M.V. Day unpublished results). These data suggest that the transfer of plasmids like pQM1 could show directionality in the environment.

An increase in the temperature of nutrient-amended soil has a general promotional influence on transfer frequencies. For example, with RP4 the transfer frequency was detectable (Van Elsas *et al.* 1988b), and increased with temperature above 15°C.

In a wastewater treatment plant microcosm

study, Altherr and Kasewick (1982) examined the intrastrain transfer of a tetracyclin and streptomycin resistance plasmid from a sewage *E. coli* host to a recipient isolated from a local creek. Transfer was optimal at 25°C. In a microcosm containing raw sewage the frequency was optimal at 22.5°C and lower at 29.5°C. Khalil and Gealt (1987) examined transfer of a recombinant plasmid between *E. coli* in synthetic wastewater. Exconjugants were detected after incubation at 10 and 50°C, with a maximum between 30 and 40°C.

Using teflon-coated test chambers, held underwater in a lake, O'Morchoe *et al.* (1988) found little reduction in the conjugal transfer frequency of R68.45 between temperatures of 20 and 28°C. Figure 4.4 compares the transfer frequencies of pQM1 obtained from *in situ* and microcosm experiments (Fry and Day 1990). It is clear that the laboratory experiments give higher transfer frequencies than the microcosms or *in situ* experiments at equivalent temperatures. These results show there to be a remarkable agreement between the microcosm and *in situ* experiments, suggesting that the microcosms are a good simulation of nature.

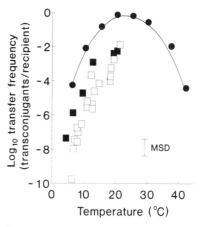

Fig. 4.4. Comparison of the transfer frequency of pQM1 obtained *in situ* (□) (Bale *et al.* 1987; Bale *et al.* 1988a,b) and in microcosms (■) (Bale *et al.* 1987; Rochelle *et al.* 1989b) with laboratory experiments (●) (Rochelle *et al.* 1989a). MSD, minimum significant difference.

Singleton and Anson (1981) have reported a synergistic effect between non-optimum temperature and pH. Their results with Rl*drd*19 and are supported by those of Rochelle *et al.* 1989a) with RP1 and both show an optimum transfer frequency at about pH 7.0.

Anaerobosis

In one laboratory study it was shown that the expression of resistance phenotypes encoded by 45 different *E. coli* plasmids was identical, whether the cells were incubated anaerobically or not (Stallions and Curtiss 1972). However, Burman (1977) showed that the expression of pili was much reduced under reduced oxygen tension. This may explain the observation by Anderson (1975) that the transfer of R1 was inhibited by anaerobosis. Burman (1975) showed this mobilization to be solely a donor function. Moodie and Woods (1973) found that the transfer frequency of two drug resistance plasmids between *E. coli* strains was reduced 100–1000 times by anaerobic incubation. Rochelle *et al.* (1988a) have shown that anaerobosis affected both expression and transfer. They also found that RP1 would transfer as efficiently aerobically as anaerobically, a report that conflicts with that of Graves and Riggs (1980), who stated that RP1 frequencies were depressed 400 times under anaerobosis. It is possible that this difference may not be plasmid encoded, because Graves and Riggs (1980) used a *Pseudomonas* intraspecies cross, while Rochelle *et al.* (1989a) used a *Pseudomonas* strain as donor and an *E. coli* as recipient. Trevors and Starodub (1987) have shown that for one plasmid, transfer was higher in aerobic soils than in those with reduced oxygen tension.

Nutrient status

The transfer of RP4 between pseudomonads introduced into non-sterile loamy soils, or into indigenous soil populations from the rhizosphere or into indigenous isolates, was mostly below the level of detection (Van Elsas *et al.* 1987, 1988a). When amendments to the soil, such as additions of nutrients (and bentonite) were made, the transfer frequency became detectable (Van Elsas *et al.* 1987), but only at temperatures above 15°C.

Transfer frequencies in non-sterile microcosms with soil of rhizosphere origin or, if not, amended by the addition of nutrients, were raised above those observed in bulk soil (Trevors and Oddie, 1986; Van Elsas *et al.* 1988c). Nutrient amendment had little or no effect in sterile soils. Figure 4.5 illustrates the effect of nutrient concentration, as a function of the distance from the root, on the transfer frequency in soil microcosms.

Rochelle *et al.* (1989a) have demonstrated that

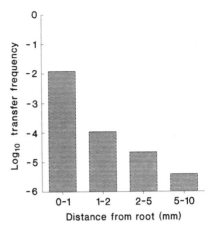

Fig. 4.5. The effect of distance from the root mat on RP1 transfer frequency in laboratory experiments performed in root chambers (Van Elsas *et al.* 1988b).

pQM1, pQM3 and RP1 transfer frequencies are each reduced 100-fold between 0 and 2 gC/litre compared with media with 10 gC/litre. The transfer of a natural plasmid from epilithon into laboratory recipients is similarly effected.

Time

Time, or the duration needed for the optimal exchange of a plasmid(s), maybe an important criterion. Laboratory studies shows that most plasmid transfers reach a maximum within 4 hours (Fig. 4.6) however, the transfer frequency of both pQM4 and natural unknown plasmids, transferring from a suspension of epilithic bacteria, increased with incubation time.

Value of microcosms

It is probable that most releases of recombinant organisms will occur deliberately for agricultural use, but that some will come via accidental spillage, which results in the contamination of soil, and then water (through run-off). Some accidental releases into waterways will also inevitably come from industrial plants using GEMs to produce high value pharmaceutical products. For this reason, it is critical that knowledge of GEM behaviour is obtained both in soils and in water microcosms. A microcosm mimics, to some extent, one environment and enables the experimental evaluation of the potential effects that will occur in nature.

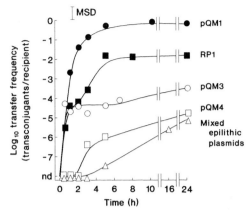

Fig. 4.6. The effect of mating time on the transfer of RP2, pQM3, pQM4 and natural plasmids from a mixed natural suspension to *Pseudomonas aeruginosa* recipient (Rochelle *et al.* 1989a). MSD, minimum significant difference.

Results from microcosms must provide reliable predictions of *in situ* results to be of maximum value to a regulatory authority. To ensure this it is essential that data from *in situ* and microcosm experiments are compared. In particular, GEM survival and plasmid transfer should be similarly influenced by environmental variables in both types of experiment. As GEMs cannot yet be easily released into the environment, *in situ*/microcosm comparisons must be done with non-engineered plasmids and organisms. Unfortunately, such comparisons have rarely been undertaken. Detection of chromosomal gene transfer and recombination *in situ* between indigenous populations of bacteria is probably impossible at present. Circumstantial evidence from a study of *Rhizobium* populations (Harrison *et al.* 1989) suggests that this is indeed a rare event, even between closely related bacteria. This conclusion is very different from that obtained from work done on plasmid exchange (Fry and Day 1990). Perhaps the difference is one peculiar to the bacteria in these ecological systems.

Risk assessment

Risk assessment must take many factors into account. Several of these are ecological, for example the movement of bacteria, whether they be plasmid-bearing or not, through soil and water. This may be a significant parameter and Smith *et al.* (1985) have shown that over 90 per cent of bacteria were translocated through soil columns.

However, Drahos *et al.* (see Chapter 11) have shown little movement through the soil surface. It is also necessary to quantify the contribution to transfer that viable but non-culturable (Roszak *et al.* 1984) plasmid-bearing bacteria, e.g. GEMs, might be able to make to the assessment equation.

The biological survival, persistence, replication and success of both natural and recombinant DNA in bacterial communities will depend on many factors. These include the host and its genotype, the novel genes and their effect on the cell's phenotype, its frequency of transfer, adaptability/mutability and the selective pressure imposed by the cellular environment.

Is gene transfer by conjugation a primary mechanism? Laboratory studies are generally done in a highly defined manner with a non-indigenous plasmid/host system. When the same plasmid/host transfer system is placed either in a microcosm or *in situ*, transfer frequencies are often found to be far lower, often by several orders of magnitude. This observation has reinforced the idea that *in situ* transfer frequencies will also be low. There is now evidence to show that this assumption is invalid (Bale *et al.* 1988b) and that our ideas of genetic exchange in and between microbial communities need to be revised.

In laboratory experiments done in soil Walter *et al.* (1987) found transfer to be more apparent than real, in that it occurred on the selective agar plate, used for the isolation and enumeration of recombinants, and not in the soil. This is certainly not the case in the *in situ* riverwater studies of Bale *et al.* (1987, 1988a,b); these experiments used controls that clearly showed that no transfer occurred on the selective medium. *In situ* transfer in soil has not yet been demonstrated. In a different type of investigation, one based on population studies, Schofield *et al.* (1987) present circumstantial evidence for exchange of, and recombination between, symbiotic plasmids in *Rhizobium*.

The predictability of transfer needs to take into account the fluctuating environment. Studies and extrapolations from laboratory studies, under optimal conditions, are not necessarily any indication or prediction of the degree of successful gene transfer *in situ*.

Our work has analysed the effect of temperature on the transfer of the natural plasmid pQM1 between strains of *P. aeruginosa* under controlled laboratory conditions, in simple microcosms and *in situ*. Although it is clear that in a defined situation transfer frequency of pQM1 is highest, those from microcosms were only slightly higher than those from *in situ* experiments. Results at 20°C were indistinguishable (see Fig. 4.4).

Plasmids capable of retrotransfer and/or mobilization of other plasmids add a further dimension to the assessment of risks from the release of GEMs. If such plasmids are widely distributed in nature then there is evidence to show that transfer is not likely to be the directional process nor is it the low likelihood event it has been thought to be.

Conclusion

Bacteria do not exist in physical isolation in any environment and so it is unlikely that they are genetically isolated. Gene transfer processes do not exist in isolation either, their influence of the evolution of the microbial community is cumulative. The potential for change is described by the physiochemical nature of the environment and a combination of the efficiency of the transfer mechanisms operating and densities of the members of that community. The fact that identifiable and individual species do exist shows that there is a strong selective pressure to maintain species diversity and individual specialism. Conjugation is a universal process and we still have much to learn about its biology.

The original question of whether gene transfer occurs *in situ* has been positively answered by several workers. As a result there are further questions. First, to what extent does gene transfer contribute to and influence the evolution of natural bacterial populations and their genetic response to natural and man-made challenges? Secondly, what are the regulatory factors that promote and inhibit this exchange? We will begin to appreciate the genetics, development and biological design of gene exchange systems by answering these questions. From the application of this knowledge a more realistic assessment of the potential ecological consequences on the release of GEMs will come.

Acknowledgements

Part of the work reported in this chapter was supported by Natural Environment Research Council and by the European Economic Community Biotechnology Action Program (BAP 0379). Thanks are also due to our research staff whose work is discussed here.

References

Achtman, M., Morelli, G., and Schwuchow, S. (1978). Cell–cell interactions in conjugating *Escherichia coli*: role of pili and the fate of mating aggregates. *Journal of Bacteriology*, **135**, 1055–61.

Altherr, M.R., and Kasewick, K.L. (1982). *In situ* studies with membrane diffusion chambers of antibiotic resistance in *Escherichia coli*. *Applied and Environmental Microbiology*, **44**, 838–43.

Anderson, J.D. (1975). Factors that may prevent transfer of antibiotic-resistance between Gram-negative bacteria in the gut. *Journal of Medical Microbiology*, **8**, 83–8.

Bale, M.J., Fry, J.C., and Day, M.J. (1987). Plasmid transfer between strains of *Pseudomonas aeruginosa* on membrane filters attached to river stones. *Journal of General Microbiology*, **133**, 3099–107.

Bale, M.J., Day, M.J., and Fry, J.C. (1988a). Novel method for studying plasmid transfer in undisturbed river epilithon. *Applied and Environmental Microbiology*, **54**, 2756–8.

Bale, M.J., Fry, J.C., and Day, M.J. (1988b). Transfer and occurrences of large mercury resistance plasmids in river epilithon. *Applied and Environmental Microbiology*, **54**, 972–8.

Beckman, W., Gaffney, T., and Lessie, T.G. (1982). Correlation between auxotrophy and plasmid alteration in mutant strains of *Pseudomonas cepacia*. *Journal of Bacteriology*, **149**, 1154–8.

Bender, C.L., and Cooksey, D.A. (1986). Indigenous plasmids in *Pseudomonas syringe* p.v. *tomato*: conjugative transfer and role in copper resistance. *Journal of Bacteriology*, **165**, 534–41.

Boulnois, G.J., Varley, J.M., Sharpe, G.S., and Franklin, F.C.H. (1985). Transposon donor plasmid, based on ColEb-P9, for use in *Pseudomonas putida* and a variety of other Gram-negative bacteria. *Molecular and General Genetics*, **200**, 65–7.

Bradley, D.E. (1980). Morphological and seriological relations of conjugal pili. *Plasmid*, **4**, 155–69.

Bradley, D.E. (1984). Characterisation and functions of thick and thin conjugative pili determined by transfer derepressed plasmids of Inc groups, Il 12, 15, B, K and Z. *Journal of General Microbiology*, **130**, 1489–502.

Breton, A.M., Taoua, S., and Guespin, M.J. (1985). Transfer of RP4 to *Myxococcus xanthus* and evidence of its integration into the chromosome. *Journal of Bacteriology*, **161**, 523–8.

Buchanan-Wollaston, V., Passiatore, J.E., and Cannon, F. (1987). The *mob* and *ori*T mobilisation functions of a bacterial plasmid promote its transfer to plants. *Nature*, **328**, 172–5.

Bukhari, A.I., Shapiro, J.A., and Adhya, S.L. (1977). *DNA insertion elements, plasmids and episomes*, Appendix B, pp. 601–71. Cold Spring Harbor Laboratories, Cold Spring Harbor, New York.

Burman, L.G. (1975). Amplification of sex repressor function of one fi^+ R factor during anaerobic growth of *Escherichia coli*. *Journal of Bacteriology*, 123, 265–71.

Burman, L.G. (1977). Expression of R-plasmid functions during anaerobic growth of *Escherichia coli* K12 host. *Journal of Bacteriology*, 131, 69–75.

Charkabarty, A.M., Freillo, D.A., and Bopp, L.H. (1978). Transposition of plasmid DNA sequences specifying hydrocarbon degradation and their expression in various microorganisms. *Proceedings of the National Academy of Sciences, USA*, 75, 3109–12.

Clarke, A.J., and Warren, G.J. (1979). Conjugal transmission of plasmids. *Annual Review of Genetics*, 13, 69–125.

Clewell, D.B., and Gawron-Burke, C. (1986). Conjugative transposons and the dissemination of antibiotic resistance in *Streptococci*. *Annual Review of Microbiology*, 40, 635–59.

Clewell, D.B., An, F.Y., Mori, M., Ike, Y., and Susuki, A. (1987). *Streptococcus facaelis* sex pheromone (cADI) response: Evidence that the peptide inhibitor excreted by pADI containing cells maybe plasmid determined. *Plasmid*, 17, 65–8.

Curtiss, R. III (1969). Bacterial conjugation. *Annual Review Microbiology*, 23, 69–136.

Curtiss, R. III (1976). Genetic manipulation of microorganisms: Potential benefits and biohazards. *Annual Review of Microbiology*, 30, 507–33.

Datta, N. (1979). Plasmid Classification: Incompatibility grouping. In *Plasmids of medical, environmental and commercial importance*, (Ed. K.N. Timmis and A. Puhler), pp. 3–12. Elsevier, Amsterdam.

Datta, N., and Hedges, R.W. (1972). Host ranges of R factors. *Journal of General Microbiology*, 70, 453–60.

Day, M.J. (1987). The biology of plasmids. *Science Progress*, 71, 203–20.

Devanas, M.A., and Stotzky, G. (1987). Survival of Genetically engineered microbes in the environment: Effect of host/vector relationship. In *Developments in industrial microbiology*, Vol. 29, (Ed. G. Pierce), pp. 287–96. Elsevier, Amsterdam.

Dunney, G.M., Zimmerman, D.L., and Tortorello, M.L. (1985). Induction of surface exclusion by *Streptococcus faecalis* sex pheromones: Use of monoclonal antibodies to identify an inducible surface antigen involved in the exclusion process. *Proceedings of the National Academy Sciences, USA*, 82, 8582–6.

Franke, A.E., and Clewell, D.B. (1982). Evidence for a chromosomal borne resistance transposon (Tn*916*) in *Streptococcus faecalis* that is capable of 'conjugal' transfer in the absence of a conjugal plasmid. *Journal of Bacteriology*, 145, 495–502.

Frigo, S.M. (1985). Influencia de plasmidio R na frequentcia de mutacao em *Escherichia coli*. *Revista de Microbiologia*, (São Paulo) 16, 255–9.

Frost, L.S., Finlay, B.B., Opgenorth, A., Paranchych, W., and Lee, J. S. (1985). Characterisation and sequence analysis of pilin from F-like plasmids. *Journal of Bacteriology*, 164, 1238–47.

Fry, J.C., and Day, M.J. (1990). Plasmid transfer in the epilithon. In *Bacterial genetics in natural environments*. (Ed. J.C. Fry and M.J. Day), pp. 172–81. Chapman and Hall, London.

Gauthier, M.J., Cauvin, F., and Briettmayer, J.-P. (1985). Influence of salts and temperature on the transfer of mercury resistance from a marine pseudomonad to *Escherichia coli*. *Applied and Environmental Microbiology*, 50, 38–40.

Gealt, M.A., Chai, M.D., Alpert, K.B., and Boyer, J.C. (1985). Transfer of plasmids pBR322 and pBR325 in wastewater from laboratory strains of *Escherichia coli* to bacteria indigenous to the waste disposal system. *Applied and Environmental Microbiology*, 49, 836–41.

Genther, F.J., Chatteerjee, P., Barkay, T., and Bourquin, A.W. (1988). Capacity of aquatic bacteria to act as recipients of plasmid DNA. *Applied and Environmental Microbiology*, 54, 115–17.

Golovleva, L.A., Pertsova, R.N., Boronin, A.M., Travkin, V.M., and Kozlovlky, S.A. (1988). Kelthane degradation by genetically engineered *Pseudomonas aeruginosa* BS827 in a soil ecosystem. *Applied and Environmental Microbiology*, 54, 1587–90.

Graves, J.F., and Riggs, H.G. (1980). Anaerobic transfer of antibiotic resistance in *Pseudomonas aeruginosa*. *Applied and Environmental Microbiology*, 40, 1–6.

Guiney, D.G., Hasegawa, P., and Davis, C.E. (1984). Plasmid transfer from *E. coli* to *Bacteroides fragilis*: differential expression of antibiotic resistance plasmid. *Proceedings of the National Academy Sciences, USA*, 81, 7203–6.

Harrison, S.P., Jones, D.G., and Young, J.W. (1989). Rhizobium population genetics: genetic variation within and between populations from diverse locations. *Journal of General Microbiology*, 135, 1061–9.

Heinemann, J.A., and Sprague, G.F. (1989). Bacterial conjugative plasmids mobilise DNA transfer between bacteria and yeast. *Nature*, 319, 205–9.

Hopwood, D.A., Keiser, T., Lydiate, D.J., and Bibb, M.J. (1986). *Streptomyces plasmids*: Their biology and use as cloning vectors. In *The bacteria*, Vol. IX. (Ed. S.W. Queener and L.E. Day), pp. 159–229. Academic Press, Orlando, Florida.

Ippen-Ihler, K. (1989). Bacterial conjugation. In *Gene transfer in the environment*, (Ed. S.B. Levy and R.V. Miller), pp. 33–72. McGraw Hill, New York.

Khalil, T.A., and Gealt, M.A. (1987). Temperature, pH and cations affect the ability of *Escherichia coli* to mobilise plasmids in L broth and synthetic wastewater. *Canadian Journal of Microbiology*, 33, 733–7.

Keiser, T., Hopwood, D.A., Wright, H.M., and Thompson, C.J. (1982). pLJ101, a multicopy broad host range Streptomyces plasmid: Functional analysis and development of DNA cloning vectors. *Molecular and General Genetics*, **185**, 223–38.

Kelly, W.J., and Reanney, D.C. (1984). Mercury resistance among soil bacteria: Ecology and transferability of genes encoding resistance. *Soil Biology and Biochemistry*, **61**, 1–8.

Kokjohn, T.A., and Miller, R.V. (1985). Cloning and characterisation of the *rec*A gene of *Pseudomonas aeruginosa*. *Journal of Bacteriology*, **163**, 568–72.

Krasovsky, V.N., and Stotzky, G. (1987). Conjugation and genetic recombination in sterile and non sterile soil. *Soil Biology and Biochemistry*, **19**, 631–8.

Lessie, T.G., and Gaffney, T. (1986). Catabolic potential of *Pseudomonas cepacia*. In *The bacteria*, Vol. X, (Ed. J.R. Sokatch), pp. 439–81. Academic Press, London.

Levy, S.B., Buu, A., and Marshall, B. (1984). Transposon Tn*10*-like tetracyclin resistance determinants in *Haemophilus parainfluenzae*. *Journal of Bacteriology*, **160**, 87–94.

Lewin, B. (1987). *Genes 3*. John Wiley and Sons, London.

Liang, L.N., Sinclair, J.L., Mallory, L.M., and Alexander, M. (1982). Fate in model ecosystems of microbial species of potential use in genetic engineering. *Applied and Environmental Microbiology*, **44**, 708–14.

Linn, S., and Arber, W. (1968). Host specificity of DNA produced by *Escherichia coli*, X. *In vitro* restriction of phage fd replicative form. *Proceedings of the National Academy Sciences, USA*, **59**, 1300–6.

Mach, P.A., and Grimes, D.J. (1982). R plasmid transfer in a wastewater treatment plant. *Applied and Environmental Microbiology*, **44**, 1395–403.

Mancini, P., Fertels, S., Nave, D., and Gealt, M.A. (1987). Mobilisation of pH3V106 from *Escherichia coli* HB101 in a laboratory scale waste treatment facility. *Applied and Environmental Microbiology*, **53**, 665–71.

Marshall, B.W., Flynn, A., Kamely, D., and Levy, S.B. (1988). Survival of *Escherichia coli* with and without col E1: Tn 5 after aerosol dispersal in a laboratory and farm environment. *Applied and Environmental Microbiology*, **54**, 1776–83.

McClure, N.C., Weightman, A.J., and Fry, J.C. (1989). Survival of *Pseudomonas putida* UWC1 containing cloned catabolic genes in a model activated sludge unit. *Applied and Environmental Microbiology*, **55**, 2627–34.

McNeil, T., and Gibbon, P.H. (1986). Characterisation of the *Streptomyces* plasmid pVE1. *Plasmid*, **16**, 182–94.

McPherson, P., and Gealt, M.A. (1986). Isolation of indigenous wastewater bacterial strains capable of mobilising plasmid pBR325. *Molecular and General Genetics*, **180**, 501–10.

Mergeay, M., and Gerits, J. (1978). F′ plasmid transfer from *Escherichia coli* to *Pseudomonas fluorescens*. *Journal of Bacteriology*, **135**, 18–28.

Moodie, H.L., and Woods, D.R. (1973). Anaerobic R-factor transfer in *Escherichia coli*. *Journal of General Microbiology*, **76**, 437–40.

O'Morchoe, S., Ogunseitan, O., Sayler, G.S., and Miller, R.V. (1988). Conjugal transfer of R68.45 and FP5 between *Pseudomonas aeruginosa* in a natural freshwater environment. *Applied and Environmental Microbiology*, **54**, 1923–9.

Ou, J.T. (1980). Mating due to loss of surface exclusion as a cause for thermosensitive growth of bacteria containing the Rtsl plasmid. *Molecular and General Genetics*, **180**, 501–10.

Paranchych, W., and Frost, L.S. (1987). The physiology and biochemistry of pili. *Advances in Bacterial Physiology*, **29**, 53–114.

Pertsova, R.N., Kunc, R., and Golovleva, L.A. (1984). Degradation of 3-chlorobenzoate in soil by pseudomonads carrying biodegradative plasmids. *Folia Microbiologica*, **29**, 242–7.

Powell, B., Mergeay, M., and Christofi, N. (1989). Transfer of broad host-range plasmids to sulphate-reducing bacteria. *FEMS Microbial Letters*, **59**, 269–74.

Raffi, F., and Crawford, D.L. (1988). Transfer of conjugative plasmids and mobilisation of a nonconjugative plasmid between *Streptomyces* strains on agar and in soil. *Applied and Environmental Microbiology*, **54**, 1334–40.

Roberts, R.J. (1983). Restriction and modification enzymes and their recognition sequences. *Nucleic Acid Research*, **11**, 135–67.

Rochelle, P.A., Day, M.J., and Fry, J.C. (1988a). Structural rearrangements of a broad host range plasmid encoding mercury resistance from an epilithic isolate of *Pseudomonas cepacia*. *FEMS Microbial Letters*, **52**, 245–50.

Rochelle, P.A., Day, M.J., and Fry, J.C. (1988b). Occurrence, transfer and mobilisation in epilithic strains of *Acinetobacter* of mercury-resistance plasmids capable of transformation. *Journal of General Microbiology*, **134**, 2933–41.

Rochelle, P.A., Fry, J.C., and Day, M.J. (1989a). Factors affecting conjugal transfer of plasmids encoding mercury resistance from pure cultures and mixed natural suspensions of epilithic bacteria. *Journal of General Microbiology*, **135**, 409–24.

Rochelle, P.A., Fry, J.C., and Day, M.J. (1989b). Plasmid transfer between *Pseudomonas* spp. within epilithic films in a rotating disc microcosm. *FEMS Microbiology Ecology*, **62**, 127–36.

Ross-McQueen, D.A., Anderson, N.A., and Schottel, J.L. (1985). Inhibitory reactions between natural isolates of streptomyces. *Journal of General Microbiology*, **131**, 1149–55.

Roszak, D.B., Grimes, D.J., and Colwell, R.R. (1984). Viable but non recoverable stages of *Salmonella enteriditis* in aquatic systems. *Canadian Journal of Microbiology*, **30**, 334–8.

Schilf, W., and Klingmuller, W. (1983). Experiments with *Escherichia coli* on dispersal of plasmids in environmental samples. *Recombinant DNA technology Bulletin*, **6**, 101–2.

Schofield, P.R., Gibson, A.H., Dudman, W.F., and Watson, J.M. (1987). Evidence for gene exchange and recombination of Rhizobium symbiotic plasmids in a soil population. *Applied and Environmental Microbiology*, **53**, 2942–7.

Schoonejans, E., and Tousant, A. (1983). Utilisation of plasmid pULB113 (RP4::miniMu) to construct a linkage map of *Erwinia caratovora* subsp. *chrysanthemi*. *Journal of Bacteriology*, **154**, 1489–92.

Singleton, P. (1983). Colloidal clay inhibits conjugal transfer of R-plasmid R1*drd*-19 in *Escherichia coli* below 22°C. *Applied and Environmental Microbiology*, **46**, 756–7.

Singleton, P., and Anson, A.E. (1981). Conjugal transfer of R-plasmid R1*drd*19 in *E. coli* below 22°C. *Applied and Environmental Microbiology*, **42**, 789–91.

Smith, M.S., Thomas, G.W., White, R.E., and Rotonga, D. (1985). Transport of bacteria though intact and disturbed soil columns. *Journal of Environmental Quality*, **14**, 87–91.

Stallions, R.D., and Curtiss, R. III. (1972). Energy requirements for specific pair formation during conjugation in *Escherichia coli* K12. *Journal of Bacteriology*, **94**, 490–2.

Stanisich, V.A., and Ortiz, J.M. (1976). Similarities between plasmids of the IncP incompatibility group derived from different genera. *Journal of General Microbiology*, **94**, 281–9.

Stotzky, G. (1989). Gene transfer among bacteria in soil. In *Gene transfer in the environment*, (Ed. S.B. Levy and R.V. Miller), pp. 165–222. McGraw Hill, New York.

Stotzky, G., and Babich, H. (1986). Survival of, and genetic transfer by, genetically engineered bacteria in natural environments. *Advances in Applied Microbiology*, **31**, 93–138.

Szabo, L.J., and Mills, D. (1984). Integration and excision of pMC7105 in *Pseudomonas syringe* pv. *phaesolicola*: involvement of repetitive sequences. *Journal of Bacteriology*, **157**, 821–7.

Thiry, G., Mergeay, M., and Faelen, M. (1984). Back mobilisation of Tra⁻ Mob⁺ plasmids mediated by various IncM. IncN and IncP1 plasmids. *Archives of the International Journal of Physiology and Biochemistry*, **92**, 64–5.

Trevors, J.T., and Oddie, K.M. (1986). R-plasmid transfer in soil and water. *Canadian Journal of Microbiology*, **32**, 610–13.

Trevors, J.T., and Starodub, M.E. (1987). R plasmid transfer in non-sterile agricultural soil. *Systematics in Applied Microbiology*, **9**, 312–15.

Trevors, J.T., Barkay, T., and Bourquin, A.W. (1987). Gene transfer among bacteria in soil and aquatic environments: a review. *Canadian Journal of Microbiology*, **33**, 191–8.

Trieu-Cout, P., Carlier, C., Martin, P., and Couvalin, P. (1987). Plasmid transfer by conjugation from *Escherichia coli* to Gram-positive bacteria. *FEMS Microbiology Letters*, **48**, 289–94.

Tu, J., Wang, H.-R., Chou, H.-C., and Tu, C.Y.-L. (1989). Mutation(s) necessary for the residence of RP4 in *Xanthomonas campestris* p.v. *citri*. *Current Microbiology*, **19**, 217–22.

Van Elsas, J.D., Govaart, J.M., and Van Veen, J.A. (1987). Transfer of plasmid pFT30 between bacilli in soil influenced by bacterial population dynamics and soil conditions. *Soil Biology and Biochemistry*, **19**, 639–47.

Van Elsas, J.D., Nikkel, M., and Van Overbeek, L.S. (1988a). Detection of RP4 transfer in soil and rhizosphere, and the occurrence of homology to RP4 in soil bacteria. *Current Microbiology*, **19**, 1–7.

Van Elsas, J.D., Trevors, J.T., and Starodub, M.E. (1988b). Bacterial conjugation between Pseudomonas in the rhizospheres of wheat. *FEMS Microbiology Ecology*, **53**, 299–306.

Van Elsas, J.D., Trevors, J.T., and Starodub, M.E. (1988c). Plasmid transfer in soil and rhizosphere. In *Risk assessment for deliberate releases*, (Ed. W. Klingmuller), pp. 89–99. Springer-Verlag, Berlin.

Voisard, C., Rella, M., and Haas, D. (1988). Conjugative transfer of plasmid RP1 to soil isolates of *Pseudomonas fluorescens* is facilitated by certain large RP1 deletions. *FEMS Microbial Letters*, **55**, 9–14.

Walter, M.V., Porteous, A. and Seidler, R.J. (1987). Formation of false positive transconjugants following *in vitro* conjugation experiments. *Abstracts of the Annual Meeting of the American Society for Microbiology*, p. 302.

Weinberg, S.R., and Stotzky, G. (1972). Conjugation and gene recombination of *Escherichia coli*. *Soil Biology and Biochemistry*, **41**, 171–80.

Wellington, E.M.H., Cresswell, N., and Saunders, V.A. (1990). Growth and survival of streptomycete inoculants and extent of plasmid transfer in sterile and non-sterile soil. *Applied and Environmental Microbiology*, **56**, 1413–19.

Willetts, N., and Skurray, R. (1980). The conjugation system of F-like plasmids. *Annual Review of Genetics*, **14**, 41–76.

5

Gene transfer in the environment: transduction

Tyler A. Kokjohn and Robert V. Miller

Historical perspective

AFTER THEIR DISCOVERY of conjugation in *Escherichia coli* in the 1940s (Lederberg and Tatum 1946a,b), Lederberg's group turned their attention to the study of the closely related bacterial species *Salmonella typhimurium* in the hope of finding that this parasexual process was operating in that microorganism as well. They discovered that genetic transfer could be demonstrated between certain genetically marked strains of this species. However, it soon become obvious that, unlike conjugation, gene transfer between these bacteria could take place not only when the cells of the two mutants were in contact but also when one strain was exposed to a cell-free extract of the other (Lederberg *et al*. 1951). Characterization of the system revealed that the transfer of genetic elements from one cell to the other by this process, which they termed transduction, was mediated by the packaging of bacterial DNA into viral particles (Zinder and Lederberg 1952).

The mediators of this gene transfer system – bacteriophages (or, more simply, phages) – are obligate intracellular parasites of bacteria and manifest no independent metabolic activity. Bacteriophages can propagate only within their bacterial hosts and use the host's ribosomes and macromolecular synthetic machinery to do so. When free in the environment, phage particles (virions) are metabolically inert and cannot multiply. The systematic study of bacteriophages in their own right has resulted in enormous strides in the understanding of basic molecular biology and genetics.

Bacteriophages were first described in 1919 by two independent investigators, F.W. Twort and T. D'Herelle (Duckworth 1976). Priority in discovery of bacteriphages has remained a matter of controversy. Twort referred to a 'glassy transformation' of bacterial cells, which seemed to be infective, while D'Herelle reported on an agent antagonistic to bacterial growth. Initially, phages were thought to have potential for disease control, although efforts towards this end have met with little success (Stent 1963; Duckworth 1987).

Bacteriophages are extraordinarily diverse both in physical structure (Fig. 5.1) and in their reproductive strategies. Phages may have either RNA or DNA as their genetic material. The RNA phages have not been shown to be transducing and will not be considered in this review.

Consequences of phage–host interaction

Bacteriophages are recognized by their effects on the growth of bacterial cells. As first observed by D'Herelle (Stent 1963), the most easily identified symptom of phage infection of a susceptible bacterium is lysis of the host. Lysis is caused by the dissolution of the bacterial cell and is accompanied by the release of newly produced progeny virions. If appropriate amounts of phage are inoculated on a solid surface, such as an agar plate with enough bacteria to produce a continuous layer of growth, a plaque results, which is due to localized lysis of the bacterial lawn (Fig. 5.2). A plaque is simply a colony of phages that arises from a single virion.

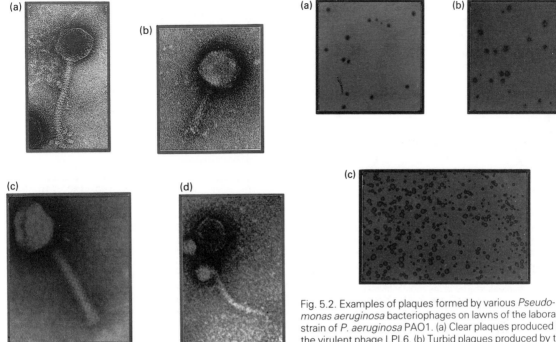

Fig. 5.1. Virions of several *Pseudomonas aeruginosa* bacteriophages. (a) The specialized transducing phage D3 (Miller *et al.* 1974). (b) The generalized transducing phage F116 (Miller *et al.* 1974). (c) The generalized transducing phage B86 (Kilbane and Miller 1988). (d) ø1407, a temperate phage isolated from the lung of a cystic fibrosis patient.

Fig. 5.2. Examples of plaques formed by various *Pseudomonas aeruginosa* bacteriophages on lawns of the laboratory strain of *P. aeruginosa* PAO1. (a) Clear plaques produced by the virulent phage LPL6. (b) Turbid plaques produced by the temperate phage FII6L. The bull's eye appearance is due to the growth of immune lysogens in the centre of the plaque. (c) Plaques of bacteriophages present in an environmental sample from a freshwater pond. Note the heterogeneity of plaque morphology and the predominance of turbid plaques.

Bacterial viruses are normally enumerated by determining the concentration (titre) of plaque-forming units (PFU) in a given volume.

Lysis of the host with production of new virions is not the only possible consequence of phage infection. Various species of bacteriophages (termed temperate) are capable of an alternative response in which they establish a long-term symbiotic relationship with their host (lysogeny). Bacteriophages that are not capable of lysogenizing their hosts, and that can only elicit a lytic response to infection, are referred to as virulent.

While not limited to temperate bacteriophages, successful transduction is most often mediated by this type of virus. Therefore, a short discussion of both the temperate and lytic life-cycles of bacteriophages is appropriate before we examine transduction and its potential as a mechanism for gene transfer in the environment.

The lytic or vegetative life-cycle

When a bacterial virus infects a susceptible host, the most frequent outcome is the expression of the viral genome leading to the replication of the viral nucleic acid and synthesis of viral proteins. The phage genome is subsequently packaged into shells (capsids) composed of virally-encoded proteins to produce infectious virions. After a latent period, during which the production of virions takes place, the host cell bursts, releasing a number of progeny phages. This process is referred to as the vegetative or lytic response as, for most phages, it results in the lysis and death of the host cell.

To produce these virions, the infecting virus confiscates the host's metabolic machinery. During lytic growth, the phage will utilize macromolecular pools of the cell to carry out virally-encoded functions while suppressing host-directed synthetic activity. The host cell must be metabolically active to support phage replication (Freifelder 1983). The

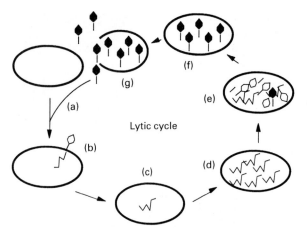

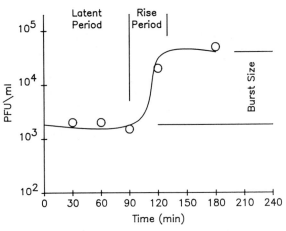

Fig. 5.3. The lytic or vegetative life-cyle of a bacteriophage. Adsorption (a): initial contact is probably random and is followed by an irreversible binding stage. Injection (b): the phage genome is injected into the host cell's cytoplasm (see Fig. 5.5). Latent period (c–f): (c) early functions on the phage genome are expressed. These are mostly associated with replication of the phage nucleic acid; (d) as the number of phage genomes increases late functions begin to be expressed. These mostly code for capsid proteins; (e) as the various structural proteins of the phage particle are produced, they are assembled into capsids. Phage DNA is then packaged into the empty heads to produce a virion; (f) as the number of virions in the infected cell increases, expression of lysozyme is initiated. Rise period (g): the host cell lyses releasing a burst of progeny virions.

Fig. 5.4. A one-step growth curve for the *Pseudomonas aeruginosa* bacteriophage LPL6. A suspension of growing bacterial cells is infected with phage and then diluted. After various periods of incubation, samples are taken and the number of infectious particles determined by mixing with an indicator bacterium and counting plaques (see Fig. 5.2). During the latent period (here 90 min), each infected cell acts as a single infectious centre (plaque). During the rise period the infected cells lyse and each progeny virion is now capable of producing an infectious centre. The burst size is the average number of virions produced by an infected cell and is the ratio of the number of infectious centres at the end of the rise period to the number at the beginning of the rise period. In this example the burst size is approximately 25. PFU, plaque-forming unit.

replication of progeny phages in an infected cell follows a somewhat typical and reproducible pattern, the general outline of which is described below and illustrated in Figures 5.3 and 5.4.

Adsorption
Phage adsorption to bacterial cells appears to follow first order kinetics, at least under idealized conditions in the laboratory (Stent 1963). Initial contact between phage and bacterium is probably the result of random collision. However, phage particles become attached to bacterial cells at defined receptors and the ability of a phage to infect a bacterium is limited to one or a few species due to the specificity of the receptor. The ability of individual cells within a given bacterial species to adsorb phages is also not uniform. It is always possible to find clones derived from cells that are resistant to specific phages (Stent 1963; Hayes 1968).

The adsorption process is complex and clearly separable into distinct phases. At first, binding to

the receptor is reversible. Phage particles that desorb at this stage are capable of infecting new hosts and producing progeny phages. This stage is followed by an irreversible binding state, when structural changes occur in the receptor and phage particle.

The host cell does not usually participate actively in the adsorption process. Phage particles are bound by metabolically inactive cells, or even by cell fragments, if the receptor is present (Stent 1963). Adsorption is frequently quite sensitive to environmental conditions. The ionic composition and pH of the medium and the environmental temperature are significant factors affecting phage adsorption (Stent 1963). Cations are often important to effective adsorption and have been postulated to neutralize repulsive electrostatic forces between phage and bacterium (Stent 1963). A detailed discussion of these factors and their potential for affecting transduction in the environmental setting will follow (see p. 69).

Injection

Following adsorption, the phage genome is introduced into the host cell's cytoplasm to begin the process of progeny production. To facilitate this process, the tail of the phage capsid may contract, thus acting as a microsyringe to inject the phage nucleic acid directly into the host (Fig. 5.5). In addition, phage tails often contain enzymes that aid in the penetration process by weakening the cell envelope.

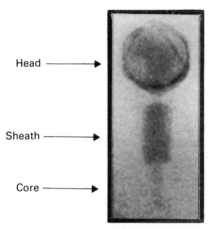

Head

Sheath

Core

Fig. 5.5. An empty capsid of the generalized transducing *Pseudomonas aeruginosa* phage E79. Note that the tail sheath has contracted. This action injects the tail core through the host cell's envelope allowing the introduction of the phage DNA into the cytoplasm.

The latent period

Upon injection, the viral genes begin to be expressed in a distinct, temporally-regulated pattern. As there is no release of virions from the infected cell, this period is referred to as the latent period. During this time, phage proteins and chromosomes are produced and assembled into nascent phage particles. The highly coordinated nature of this process is emphasized by the fact that premature lysis of phage-infected cells reveals that there are no virions present within the cell early in the latent period (this is termed the eclipse period) (Doermann 1952). Progeny phages are produced in an essentially assembly-line fashion within the infected cells, and the rate of increase of virions during the latent period is not exponential but linear.

The length of the latent period is characteristic of the phage species, the metabolic state of the host and the environmental conditions. It is not directly affected by the number of phage particles infecting a single cell (Ellis and Delbruck 1939).

The rise period

The latent period is followed by a period when infected cells begin to lyse, releasing mature virions. This period of rapid increase is referred to as the rise period and corresponds to the end of the phage maturation cycle. A single infected cell will release a large number of virions when it bursts. The average number of progency phages released by an infected cell is referred to as the burst size, and is characteristic of the phage (Ellis and Delbruck 1939; Stent 1963; Hayes 1968).

The temperate life-cycle

In addition to the lytic mode of development, infection by temperate phages may lead to lysogenization of the host. In the temperate or lysogenic response, the lytic functions of the viral genome are repressed by a phage-encoded DNA binding protein (Ptashne 1986) and the viral genome is carried symbiotically as a prophage in the host cell. Because the repressor protein is present in the host cell, the lysogenized cell is immune to superinfection by the same (and sometimes other closely related) phages (Stent and Calender 1978). In many cases, immunity does not inhibit absorption of the phage DNA but simply inhibits its expression. This form of phage–host interaction is referred to as lysogeny, and the infected host bacterium as a lysogen, because it has the potential for producing phage particles through induction of the prophage to lytic growth (Lwoff 1953). Such induction occurs spontaneously at a low frequency in each cell generation. Many prophages can also be induced to the lytic response at high frequency after exposure of the host cell to various stresses.

The physical condition of the prophage varies depending on the species of virus. For example, phage lambda of *E. coli*, P22 of *S. typhimurium*, and D3 of *Pseudomonas aeruginosa* insert directly into the host chromosome at defined positions (Susskind and Botstein 1978; Cavenagh and Miller 1986; Weisberg 1987). On the other hand, prophages P1 of *E. coli* and F116 of *P. aeruginosa* are maintained as low-copy-number plasmids (Miller *et al.* 1977; Sternberg and Hoess 1983).

Lysogeny is common in the bacterial world, in

fact it has been suggested that it represents the norm and not the exception for bacteria in nature (Hayes 1968; Holloway and Krishnapillai 1979). In certain species of bacteria, most, if not all isolates, are polylysogenic for several distinct types of phages.

Establishment and maintenance of lysogeny

The establishment and maintenance of lysogeny is controlled both by bacterial and bacteriophage genes (Herskowitz and Hagen 1980). Temperate phages allow the expression of only a small subset of their genes when in the lysogenic state (Hayes 1968). The best-studied system of lysogenization is that of bacteriophage lambda, which infects the enteric organism *E. coli*. During lysogeny, the expression of prophage lambda genes is inhibited by a repressor protein (the product of the *c*I gene) encoded by the phage genome (Ptashne 1967), which binds to the phage chromosome and inhibits transcription of most of the phage's genes. Similar repressors have been identified for a number of phages, including P22 of *S. typhimurium* (Levine 1957). D3 of *P. aeruginosa* (Miller and Kokjohn 1987), ø16-3 of *Rhizobium meliloti* (Dallman *et al.* 1987), and D108 (Levin and DuBow 1987) and 434 (Levine *et al.* 1979) of *E. coli*.

The establishment of lysogeny in bacteriophage lambda is a highly regulated process whose main feature is the regulation of expression of the *c*I gene (Echols and Green 1971). A detailed description of this decision-making process can be found in Ptashne (1986). A significant aspect of this system is its ability to assess the physiological condition of the host cell (Herskowitz and Hagen 1980). It appears that upon infecting an *E. coli* cell, phage lambda determines the nutritional state of the host and responds appropriately. If the cell is starving, an elevated level of cAMP will be found and will set in motion a chain of events that favours the establishment of lysogeny by increasing expression of the *c*I gene. If, on the other hand, the energy state of the host cell is high, a low concentration of cAMP will be found and the system will favour lytic growth by reducing the synthesis of *c*I protein.

The system is also set-up to sense when the available population of potential host cells is being exhausted (Kourilsky 1973). If a cell is infected simultaneously by several phages, the establishment of lysogeny is favoured. A high ratio of phage to bacteria (multiplicity of infection – MOI) will favour multiple infections and suggest that the

numbers of suitable hosts is becoming depleted. Similar abilities to monitor host availability have been reported for the lysogeny-establishment systems of phage D3 of *P. aeruginosa* (Kokjohn and Miller 1988) and P22 of *S. typhimurium* (Levine 1957).

Once established, lysogeny is maintained by the constant production of the prophage-encoded repressor of lytic functions. The expression of this protein makes lysogens immune to superinfection by phages of the same species as the resident prophage. Transcription from any superinfecting phage genome is immediately inhibited by the presence of the repressor protein.

Lysogenic induction

The repressor protein must be present at all times for lysogeny to be maintained. If its production ceases or it is inactivated, the prophage will initiate vegetative growth resulting in cell lysis. This condition is continually occurring in a subpopulation of lysogenized cells, leading to the continuous production of virions at low levels (Stent 1963; Ptashne 1986). The induction of lytic growth of a prophage may be caused by a number of environmental factors. The stress associated with the exposure of the lysogen to various agents can lead to the induction of the lytic response at high frequency. Inducing agents share the common property of directly damaging cellular DNA or inhibiting DNA replication (Lwoff 1953; Roberts and Devoret 1983; Walker 1984). In addition, starvation conditions may induce certain prophages of *E. coli* (Melechen and Go 1980).

While the molecular mechanisms of prophage induction have been studied in detail in only a few phages, it appears that these stresses produce a signal that ultimately leads to the destruction of the repressor protein. Studies of phage lambda have revealed that prophage induction is initiated by interaction of the *c*I repressor protein with the host cell *recA* gene product (Clark 1973). In response to the inducing agent, the RecA protein becomes activated to a state in which it promotes cleavage of the *c*I protein at a precise position in the peptide chain, yielding two distinct polypeptide fragments (Roberts and Roberts 1975; Roberts and Devoret 1983). This cleavage eliminates the DNA binding affinity of the repressor protein for the phage promoters. When enough repressor protein is cleaved, vegetative growth is initiated (Bailone *et al.* 1979).

It is clear that a functional RecA protein is required for induction of a number of prophages in *E. coli* (Clark 1973). Moreover, stress stimulation of induction requires that the RecA protein be specifically activated (Quillardet *et al.* 1982). This activation occurs when normal DNA replication is inhibited (Roberts and Devoret 1983). For example, RecA protein will be activated in a cell that has suffered damage to its DNA through irradiation with ultra-violet (UV) light. Only when in this activated state will the RecA protein promote the cleavage of the phage repressor. The induction of the prophage will only occur when the DNA damage is extensive: more than 90 per cent of the *cI* repressor must be inactivated before lytic growth will ensue (Bailone *et al.* 1979). This response has been hypothesized to be a survival adaptation of the phage. It allows the prophage to escape from a severely damaged host and maximizes its chances of survival.

Stress-induced lytic growth is a common property of temperate phages (Miller 1992, see also p. 70). Approximately half the species of temperate bacteriophages of *E. coli* are UV-inducible (Roberts and Devoret 1983). A number of other bacterial species have been shown to contain analogues of the *E. coli recA* gene and to host *recA*-dependent, UV-inducible prophages (Kokjohn 1989; Miller and Kokjohn 1990). These species include, among others, *Bacillus subtilis* (Lovett *et al.* 1988). *P. aeruginosa* (Kokjohn and Miller 1987), *S. typhimurium* (Sauer *et al.* 1982; Downs and Roth 1987), *Rhizobium meliloti* (Better and Helinski 1983; Dallman *et al.* 1987), and *Vibrio cholerae* (Goldberg and Mekalanos 1986).

Lysogenic conversion

In addition to expressing superinfection immunity, cells lysogenic for certain prophages will exhibit modification in their biochemical properties. Lysogenic conversion is the acquisition by a lysogen of new, phage-encoded phenotypic properties (Low and Porter 1978). Such new traits may be significant in the ecology of the phage (Reanney 1976) and have an effect on the potential for environmental gene transmission by transduction. For example, lysogenization of *P. aeruginosa* with the specialized-transducing (Cavenagh and Miller 1986) temperate phage D3 results not only in

superinfection immunity but also in resistance to adsorption of this page to the cell envelope (Kuzio and Kropinski 1983). At the same time, adsorption of the unrelated virulent phage E79 is also inhibited. Upon lysogenization with D3, the lipopolysaccharide of the host is modified in such a way that the affinity of these phages for their receptors is reduced significantly.

Archived prophages

Recently, Downs and Roth (1987) reported that variants of prophage P22 of *S. typhimurium* can exist in a unique quiescent state, termed archived, which is distinctly different from classic lysogeny. Cells containing an archived prophage do not express superinfection immunity to P22, nor do they spontaneously release phage particles. However, when these investigators perturbed purine metabolism of strains containing archived viruses, these cryptic prophages were activated and phage particles were released from the cells. This induction was independent of the *recA* gene and the archived prophages were not, unlike classic P22 prophages, UV-inducible. The exact physical condition of the archived prophage DNA is not known but it appears to be modified in some unknown way. To date, archiving has only been demonstrated in *S. typhimurium*, and the signifance of this phenomenon to the microbial world is still to be determined.

Transduction

Transduction is defined as the transfer of genetic information, mediated by bacteriophage particles, between bacterial cells (Masters 1985). Two types of transduction are recognized. In the first – generalized (or unrestricted) transduction – any genetic element within the host cell has an equal probability of being transduced by the phage vector (Zinder and Lederberg 1952). Both chromosomal and plasmid genes can be transduced (Ozeki and Ikeda 1968). The second type – specialized (or restricted) transduction (Morse 1954) – allows only the transfer of specific genetic elements. The molecular mechanisms for the production of specialized and generalized transducing particles are fundamentally different.

Generalized transduction

Generalized transducing particles are produced by the inappropriate packaging of host DNA into a phage capsid in place of the phage genome (Fig. 5.6). Either chromosomal or plasmid DNA can be packaged and transferred. Transducing particles contain only bacterial or plasmid DNA (Ikeda and Tomizawa 1965; Yamagishi and Takahashi 1968; Chakrabarty and Gunsalus 1969; Ebel-Tsipis *et al.* 1972; de Lancastre and Archer 1981). They do not contain any phage DNA and the transductants formed are not immune to superinfection by the mediating phage (Saye *et al.* 1987b). Transducing particles can be produced either during primary lytic infection of the bacterial host or during induction of a host-associated prophage. Their

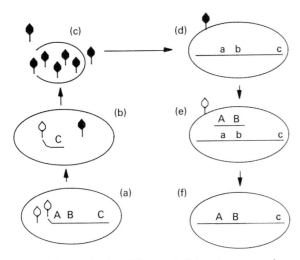

Fig. 5.6. Cotransduction of the closely linked chromosomal genetic loci A and B. During packaging of phage DNA (see Fig. 5.3) bacterial DNA is sometimes incorporated into phage capsids, in place of a phage genome, to produce a transducing particle. (a) A piece of bacterial chromosome approximately the same size as the phage genome is packaged into a capsid. As genes A and B are within this molecular length of DNA, they can be packaged into the same virion. (b) As gene C does not fall within the maximum length that can be packaged, it cannot be inserted into the same particle as genes A and B. (c) After the production of a number of virions and transducing particles, the donor cell lyses. (d) The transducing particle containing genes A and B is adsorbed to a recipient cell. (e) The chromosomal fragment containing genes A and B is injected into the recipient cytoplasm. (f) Homologous recombination allows incorporation of alleles A and B into the recipient chromosome replacing alleles a and b. As C is too far from A and B to be packaged with them, it cannot be coinherited as a single event and allele c remains unchanged.

production does not depend on homologous recombination (Ebel-Tsipis *et al.* 1972). Virulent and temperate bacteriophages have been reported to carry out generalized transduction under appropriate conditions (Susskind and Botstein 1978; Kokjohn 1989).

Despite their overall diversity, the generalized transducing phages all share certain common attributes. They replicate their genomes as multi-unit concatemers and package their DNA into phage proheads by a head-full mechanism (Sternberg and Hoess 1983; Masters 1985; Margolin 1987). DNA for encapsidation is cut by a nuclease at a particular site called *pac* and packaged progressively until the capacity of the head is reached. During this process, bacterial DNA, either chromosomal or plasmid, may be incorporated into viral particles in the place of viral DNA. Such mistakes in packaging result in transducing particles. It is thought that certain sites in the host DNA mimic the *pac* site and facilitate the process (Chelala and Margolin 1974), although homology in DNA sequence cannot be found between the *E. coli* chromosome and the *pac* site of the generalized transducing phage of *E. coli* P1 (Hanks *et al.* 1988).

Innate differences exist in the ability of various phages to transduce bacterial DNA. While phage P1 transfers all genes of the *E. coli* chromosome with essentially equal efficiency (Hanks *et al.* 1988), phage P22 transmits certain genes of the *S. typhimurium* chromosome at higher frequencies than other genes (Masters 1985). Bacteriopages normally package viral DNA much more efficiently than host DNA. The transductional frequency for a selected gene is typically in the order of 10^{-5}–10^{-6} per phage particle (Masters 1985; Saye *et al.* 1987b). However, mutants of P22 with relaxed affinity for the phage DNA *pac* site, and that package and transduce bacterial DNA with higher frequencies, have been found (Margolin 1987).

Cotransduction

In the creation of generalized transducing particles, bacterial DNA (either chromosomal or plasmid) is packaged into a viral capsid that normally contains the viral genome. The geometry and size of the capsid restrict the amount of DNA that can be packaged. Therefore, the piece of bacterial DNA that is contained in the viral capsid constitutes a contiguous piece of host DNA of approximately the same size as the phage genome (Miller *et al.* 1974;

Pemberton 1974; Miller *et al.* 1977; Masters 1985). If two genetic loci are positioned in close proximity on the host chromosome, they may be packaged into the same capsid and transferred as a single unit (Fig. 8.6); widely spaced markers cannot be transferred together in this manner. If two genes can be inherited by a recipient bacterium as the result of a single transductional event, they are said to be cotransducible. Contransduction permits the estimation of relative linkage of two genetic loci from the frequency with which they are inherited together. The greater the distance between the two genes, the lower the probability of their coinheritance (Wu 1966; Stent and Calender 1978).

Abortive transduction

To be stably inherited, chromosomal DNA introduced into a recipient cell by generalized transduction must be recombined into the chromosome of the host. This integration requires that the host cell be proficient at generalized recombination and that the injected DNA be homologous in DNA sequence to a region of the host chromosome (Sandri and Berger 1980). In addition, it appears that the decision to integrate the transduced DNA must occur soon after injection. Once the opportunity passes it does not reoccur (Hayes 1968; Masters 1985).

With so many restrictions, it is not hard to predict that most transduced chromosomal DNA is not integrated and results in the creation of an abortive transductant. The transduced DNA fragment persists in these cells but is not replicated. Genes on the transduced DNA can be expressed, leading to the formation of microcolonies under appropriate conditions (Stent 1963; Ozeki and Ikeda 1968; Hayes 1968). These colonies are the result of linear cell division in which only one of the progeny cells receives the transduced DNA. Only *one* cell in the abortively transduced clone contains the transduced DNA fragment at any one time and only this cell has the potential to express the genetic information contained in transduced DNA. This DNA fragment may be protected from degradation by an associated protein and persist with in the clone for many generations (Ikeda and Tomizawa 1965).

Specialized transduction

The production of specialized transducing particles requires replacement of one or more phage genes with bacterial DNA. Transducing particles are produced only during the induction to lytic growth of a chromosomally integrated prophage. Their origin is in an illegitimate recombination event associated with the excision of the prophage from the host genome, which replaces a portion of the phage DNA with the host DNA (Campbell 1977; Stent and Calender 1978). Packing is also by the head-full mechanism but is more stringent than packaging of generalized transducing phage DNA; it requires the presence of a DNA sequence unique to the phage genome termed *cos*. As only DNA associated with the *cos* site can be packaged, only the bacterial DNA in close proximity to the phage integration site can be incorporated into transducing particles and transferred.

The transducing particles contain a mixture of viral and host DNA and the transduction event leads to the lysogenization of the recipient cell with the transducing phage. The transductants produced contain a defective prophage, which often renders them immune to superinfection (Stent and Calender 1978).

High frequency transducing lysates

A hallmark of specialized transduction is the potential to generate high-frequency transducing (HFT) lysates. Once a transducing phage is generated it may be propagated by induction of a coresident helper phage. This helper phage is necessary to supply the functions lost from the transducing phage during the illegitimate recombination event that created it. During such propagation, the transducing phage genome will be represented in approximately half the particles present in the lysate, and such a lysate will transduce at a very high frequency.

Occurrence of transduction as mechanism of gene transfer in the various genera of bacteria

The exchange of genetic material mediated by bacteriophages has been reported for a variety of environmentally important bacterial species (Kokjohn 1989; Miller 1992). Table 5.1 summarizes the genera in which transduction has been reported and cites representative references for each system.

Table 5.1. *Taxons for which transducing phages are known*

Taxon[a]	Phage type	Transduction type	DNA source	Stress induction identified	Selected[b] references
Gram-negative bacteria					
Family: *Enterobacteriaceae*					
Genus: *Citrobacter*					
C. freundii	Temp	Spec	Chrom	Yes	43
Genus: *Enterobacter*					
E. aerogenes	Temp	Spec	Chrom	Yes	43
Genus: *Erwinia*					
E. amylovora	Temp	Spec	Chrom	Yes	43
E. chrysanthemi	Temp	Gen	Chrom	Yes	20
Genus: *Escherichia*					
E. coli	Temp	Gen	Chrom/Plas	Yes	9, 14, 37, 59, 64, 97
	Temp	Spec	Chrom	Yes	4, 5, 71
Genus: *Klebsiella*					
K. aerogenes	Temp	Gen	Chrom	Yes	9, 65
K. pnuemoniae	Temp	Gen	Chrom	nd	65
Genus: *Proteus*					
P. mirabilis	Temp	Gen	Chrom	No	27, 84, 85
	Temp	Spec	Chrom/Plas	Yes	23, 24, 25, 56
P. morganii	Temp	Gen	Chrom	Yes	21
P. rettgeri	Temp	Gen	Chrom	No	55
P. vulgaris	Temp	Gen	Chrom	No	26
	Temp	Spec	Chrom/Plas	Yes	23
Genus: *Providencia*					
P. spp.	Temp	Spec	Chrom/Plas	Yes	22, 28
Genus: *Salmonella*					
S. typhimurium	Temp	Gen	Chrom/Plas	Yes	14, 37
	Temp	Gen/Spec	Chrom	Yes	88, 97
Genus: *Serratia*					
S. marcescens	Temp	Gen	Chrom	No	51
Genus: *Shigella*					
S. dysenteriae	Temp	Gen	Chrom/Plas	No	64
Family: *Neisseriaceae*					
Genus: *Acinetobacter*					
A. spp.	Temp	Gen	Chrom	Yes	44
Family: *Pseudomonadaceae*					
Genus: *Pseudomonas*					
P. aeruginosa	Temp	Gen	Chrom/Plas	Yes	1, 46, 47, 49, 52, 54, 57, 69, 80
	Temp	Spec	Chrom	Yes	18
	Vir	Gen	Chrom	N/A	70
P. cepacia	Temp	Gen	Chrom	Yes	68
P. putida	Temp	Gen	Chrom	No	19, 48
Family: *Rhizobiaceae*					
Genus: *Rhizobium*					
R. leguminosarum	Temp	Gen	Chrom	nd	12
R. meliloti	Temp	Gen	Chrom	No	17, 73, 83

Table 5.1. *(cont.)*

Taxon[a]	Phage type	Transduction type	DNA source	Stress induction identified	Selected[b] references
Genera of unknown affiliation					
Genus: *Achromobacter* (neé *Alcaligenes?*)					
A. spp. 2	Temp	Gen	Chrom	Yes	95
Genus: *Acidilphilium*					
A. spp. 2	Temp	nd	nd	No	78
Genus: *Caulobacter*					
C. crescentus	Vir	Gen	Chrom	N/A	8, 39
Genus: *Desulfovibrio*					
D. desulfricans	Temp	Gen	Chrom	No	76
Genus: *Myxococcus*					
M. xanthus	Temp	Gen	Chrom	No	67
	Vir	Gen	Chrom	No	15, 42, 67
Family: *Vibrionaceae*					
Genus: *Vibrio*					
V. cholerae	Temp	Gen	Chrom	Yes	74
Gram-positive bacteria					
Family: *Bacillaceae*					
Genus: *Bacillus*					
B. alesti	Temp	Gen	Chrom	No	91
B. amyloliquifaciens	Temp	Gen	Chrom	nd	89
B. anthracis	Temp	Gen	Chrom/Plas	No	79
B. cereus	Temp	Gen	Chrom/Plas	No	79, 90, 91, 96
B. entomociduslimassol	Temp	Gen	Chrom	No	91
B. galleriae	Temp	Gen	Chrom	No	91
B. licheniformis	Temp	Gen	Chrom/Plas	nd	89
B. megaterium	Temp	Gen	Chrom	No	93
B. pumilus	Temp	Gen	Chrom/Plas	No	10, 61, 62, 63
	Temp	nd	Plas	Yes	11
B. sotto	Temp	Gen	Chrom	No	53
B. subtilis	Temp	Gen	Chrom/Plas	Yes	6, 41, 53, 60, 77, 81, 99
	Temp	nd	Plas	No	53, 66, 90
	Vir	Gen	Chrom	N/A	3, 16, 33, 34, 35, 36, 40
	nd	Gen	Chrom/Plas	No	75
B. thompsoni	Temp	Gen	Chrom/Plas	No	91
B. thuringiensis	Temp	Gen	Chrom/Plas	No	7, 58, 79, 91
Family: *Micrococcaceae*					
Genus: *Staphylococcus*					
S. aureus	Temp	Gen	Chrom/Plas	No	38, 72
S. epidermidis					
	Temp	Gen	Chrom/Plas	nd	98
Genus: *Streptococcus*					
Group A	Temp	Gen	Chrom	Yes	26, 29, 30, 31, 81
	Vir	Gen	Chrom/Plas	N/A	83, 92
Group C	Temp	Gen	Chrom	Yes	29, 30, 31, 94
Group G	Temp	Gen	Chrom	Yes	29, 31

Table 5.1. *(cont.)*

Taxon[a]	Phage type	Transduction type	DNA source	Stress induction identified	Selected[b] references
Family: *Streptomycetaceae*					
Genus: *Streptomyces*					
S. hydroscopicus	Temp	Gen	Chrom	No	87
S. olivaceus	Temp	Gen	Chrom	No	2
S. venezuelae		Gen	Chrom	No	86
Vir					
Genera of unknown affiliation					
Genus: *Corynebacterium*					
C. diphtheriae	Temp	Gen	Chrom	No	13
C. renale	Temp	Spec	Chrom	No	45
Genus: *Rhodococcus*					
R. erythropolis	Temp	Gen	Chrom	nd	32

Chrom, chromosomal DNA transduced; Gen, generalized transducing phage; N/A, not applicable; nd, not able to determine from reading of the references; Plas, plasmid DNA transduced; Spec, specialized transducing phage; Temp, temperate; Vir, virulent.
[a] Taxa according to Krieg and Holt (1984) and Sneath *et al.* (1986).
[b] References:
1. Akhverdian *et al.* 1985; 2, Alikhanian, Iljina and Lomovskaya 1960; 3, Alonso, Lüder and Trautner 1986; 4, Arber 1983; 5, Arber *et al.* 1983; 6, Armentrout and Rutberg 1971; 7, Barsomina *et al.* 1984; 8, Bender 1981; 9, Bender and Sambucitti 1983; 10, Bramucci and Lovett 1976; 11, Bramucci and Lovett 1977; 12. Buchanan-Wollaston 1979; 13, Buck and Groman 1981; 14, Bukhari 1976; 15, Campos *et al.* 1978; 16, Canosi *et al.* 1982; 17, Casadesús and Olivares 1979; 18, Cavenagh and Miller 1986; 19, Chakrabarty and Gunsalus 1969; 20, Chatterjee and Brown 1980; 21, Coetzee 1966; 22, Coetzee 1975a; 23, Coetzee 1975b; 24, Coetzee 1976; 25, Coetzee *et al.* 1977; 26, Coetzee *et al.* 1967; 27, Coetzee and Smit 1969; 28, Coetzee *et al.* 1966; 29 Colón *et al.* 1970; 30, Colón *et al.* 1971; 31, Colón *et al.* 1972; 32, Dabbs 1987; 33, Deichelbohrer *et al.* 1985; 34, de Lencastre and Archer 1979; 35, de Lencastre and Archer 1980; 36, de Lencastre and Archer 1981; 37, Dénarié *et al.* 1977; 38, Dyer *et al.* 1985; 39, Ely and Johnson 1977; 40, Ferrari *et al.* 1978; 41, Fink and Zahler 1982; 42, Geisselsoder *et al.* 1978; 43, Goldberg and Mekalanos 1986; 44, Herman and Juni 1974; 45, Hirai and Yanagawa 1970; 46, Holloway *et al.* 1961; 47, Holloway and Monk 1959; 48, Holloway and Van de Putte 1968a; 49 Holloway and Van de Putte 1968b; 50, Hyder and Streitfeld 1978; 51, Kaplan and Brendel 1969; 52, Kilbane and Miller 1988; 53, Kopec *et al.* 1985; 54, Krishnapillai 1971; 55, Krizsanovich *et al.* 1969; 56, Krizsanovich-Williams 1975; 57, Krylov *et al.* 1985; 58, Lecadet *et al.* 1980; 59, Lennox 1955; 60, Lipsky *et al.* 1981; 61, Lovett 1972; 62, Lovett *et al.* 1974; 63, Lovett and Young 1971; 64, Luria *et al.* 1960; 65, MacPhee *et al.* 1969; 66, Marrero *et al.* 1981; 67, Martin *et al.* 1978; 68, Matsumoto *et al.* 1986; 69, Miller *et al.* 1977; 70, Morgan 1979; 71, Morse 1954; 72, Novick *et al.* 1986; 73, Novikova and Simarov 1983; 74, Ogg *et al.* 1981; 75, Prozorov *et al.* 1980; 76, Rapp and Wall 1987; 77, Rosenthal *et al.* 1979; 78, Rowland *et al.* 1989; 79, Ruhfel *et al.* 1984; 80, Saye *et al.* 1987b; 81, Shapiro *et al.* 1974; 82, Sik *et al.* 1980; 83, Skjold *et al.* 1982; 84, Stäber and Böhme 1971a; 85, Stäber and Böhme 1971b; 86, Stuttard 1979; 87, Süss and Klaus 1981; 88, Susskind and Botstein 1978; 89, Taylor and Thorne 1963; 90, Thorne 1968; 91, Thorne 1978; 92, Totolyan *et al.* 1981; 93, Vary *et al.* 1982; 94, Wannamaker *et al.* 1973; 95, Woods and Thompson 1975; 96, Yelton and Thorne 1970; 97, Yoshikawa and Hirota 1971; 98, Yu and Baldwin 1971; 99, Zanler *et al.* 1977.

Evidence for transduction as a mechanism of horizontal gene transfer among bacterial populations in the environment

Because of the relatively restricted host range of many bacteriophages and the fact that transduction is mediated by an external factor (the transducing phage), it has been generally assumed that transduction would be a less potent form of gene transfer than conjugation in the environment. While few studies have addressed the frequency of transduction *in situ*, transductional transmission of both chromosomal and plasmid DNA has been demonstrated under the conditions found in natural environments.

Abundance of bacteria and bacteriophages in various environments and the probability of interaction

Transduction depends on the interaction of, first, a transducing phage and a host cell that will act as the donor of the transferred DNA and second, the interaction of the transducing particle produced in the donor with an appropriate recipient. As the initial interaction between host and virus is thought to be passive (see p. 56), transduction would be expected to obey a mass action model and be very sensitive to the concentration of donor and recipient organisms, as well as to the concentration of transducing phages. The concentration of the various biological components of the gene transfer system in a given ecosystem is potentially the most important factor in determining whether or not genetic transfer by any mechanism, and certainly by transduction, will take place. Studies conducted *in situ* have shown that the frequency of transfer by either conjugation (O'Morchoe *et al.* 1988) or transduction (Saye *et al.* 1987b) is significantly influenced by both the absolute numbers of cells capable of donating genetic information and their relative concentration with respect to potential recipients of that genetic determinant.

The available data suggest that in many environments the concentration of bacteria (Reanney *et al.* 1982) as well as bacteriophages (Primrose and Day 1977; Goyal *et al.* 1987; Saye and Miller 1989a) is more than sufficient to ensure contact of the various components of a transductional system. The proportion of these populations capable of participating in horizontal gene transfer is more difficult to evaluate.

Significant numbers of bacteria are found in natural environments. Natural populations of bacteria in clay soils often reach levels of 10^6–10^7 bacteria per gram (Alexander 1977; Devanas and Stotzky 1988; Zeph *et al.* 1988; Stotzky 1989). Bacterial counts of 10^9 colony-forming units (CFU)/g of forest soil have been reported (Kauri 1978) while bacterial counts often reach 10^4–10^6 CFU/g of desert soil (Skujins 1984). Variation in bacterial concentration in aquatic habitats is striking. Culturable cells in freshwater lakes and streams from various parts of the world have been estimated to range from 10^3 to 10^4 (CFU)/ml (Brayton *et al.* 1987; Saye *et al.* 1987b; O'Morchoe *et al.* 1988) while total cell counts using the acridine orange direct count method have ranged from 10^6

to 10^7 cells/ml (Saye and Miller 1989a). Higher concentrations of culturable bacteria are observed in coastal (10^4–10^5 CFU/ml and estuarine (10^6 CFU/ml) waters (Servais *et al.* 1985; Iriberri *et al.* 1987). Bacterial counts in river mud have been reported to reach 10^6 CFU/g (Fjerdugstad 1975). Bacterial counts in waste waters and wet sludge solids have been estimated to range from 10^6 to as high as 10^9 CFU/ml (Altherr and Kasweck 1982; Blackburn *et al.* 1987; Gealt 1992).

Numerous studies have reported significant titres of bacteriophage from various environmental sources including soils (Campos *et al.* 1978; Williams *et al.* 1987), marine (Moebus 1987) and freshwater (Primrose and Day 1977; Primrose *et al.* 1982; Farrah 1987; Ogunseitan *et al.* 1990) environments, waste treatment facilities (Holloway and Van de Putte 1968a,b; Kelln and Warren 1971; Bitton 1987) and hospital settings (Wannamaker *et al.* 1973; Miller and Renta Rubero 1984). The concentrations of phages from freshwater habitats may reach as high as 10^3–10^4 plaque-forming units (PFU)/ml (Farrah 1987; J. Replicon and R.V. Miller, unpublished observation). Titres in waste water and sludge have been reported to be as high as 10^8 PFU/ml (Bitton 1987) and as high as 10^{10} particles/ml of sewage have been observed by electron microscopy (Ewert and Paynter 1980). Phage titres in marine waters have been reported to range from 10^5 to 10^{11} PFU/ml (Moebus 1987; Berg *et al.* 1989; Bratbak *et al.* 1990; Proctor and Fehrman 1990). Significant seasonal variation in the concentration of bacteriophages has been observed in both marine (Moebus 1987) and freshwater (J. Replicon and R.V. Miller, unpublished observation) habitats.

Lysogeny is a common phenomenon in nature. Dhillon and Dhillon (1981) found that approximately 5 per cent of all *E. coli* isolates from sewage and faeces were lysogenic. Kelly and Reanny (1978) surveyed soil bacilli and determined that the majority of isolates were lysogenic for temperate or defective prophages. These investigators observed a high incidence of *pseudolysogeny* among *Bacillus* species, as had earlier been noted by Romig and Brodetsky (1961). Isolates of many bacterial species associated with plants have been shown to be lysogenic (Farrand 1989). Holloway (1969) estimated that 100 per cent of all natural isolates of *P. aeruginosa* were lysogenic for at least one bacteriophage and that many contained several prophages. The majority of strains of lactic acid streptococci used in cheese making throughout the world are lysogenic for temperate bacteriophages (Jarvis *et al.* 1978).

Certainly, the concentrations of bacteria and bacteriophages found in nature are sufficient to predict that interactions leading to transduction could take place (Miller and Sayler, 1992). Several studies have demonstrated that bacteriophage–bacterial interactions take place at densities of 10^4 host cells/ml and lower (Wiggins and Alexander 1985; Kokjohn *et al.* 1989). Transduction of plasmid DNA has been observed in a freshwater lake at cell concentrations of 10^4 CFU/ml and bacteriophage concentrations of 10^2–10^3 PFU/ml (Saye *et al.* 1987b).

Transduction in the soil

Although numerous transducing phages of soil organisms have been identified (see p. 62), transduction in terrestrial environments has been most extensively studied using the generalized transducing phage P1, which mediates gene transfer in the enteric species *E. coli*. Transduction of both amino acid and antibiotic resistance markers has been shown to take place *in situ* in terrestrial environments.

Germida and Khachatourians (1988) investigated generalized transduction in a sandy soil and in silty clay loam. They determined that the frequency of P1-mediated transduction to a recipient strain of *E. coli*, which they had introduced into the soil sample, was 10^{-6} per recipient organism. Similar transduction frequencies were observed for this system *in vitro*, suggesting that conditions in non-sterile soils are conducive to transduction.

Zeph *et al.* (1988) studied transduction of antibiotic resistance mediated by wild type P1 and by a temperature-sensitive, clear-plaque mutant of that phage, which produced a repressor of lytic growth that was inactive at elevated temperatures (42°C). These investigators added P1 phages and an *E. coli* recipient to their soils. In their studies, higher frequencies of transduction (10^8 transductants/g soil) were obtained when sterilized soil samples were used than when their transduction test system was added to non-sterile soil (10^6 transductants/g soil). The frequency of transduction was not affected by amendment of the soil with nutrients (Zeph and Stotzky 1988). Thus, it appears that the relatively low nutrient conditions often found in soil do not significantly reduce the potential for transduction in terrestrial environments.

In other experiments, Zeph *et al.* (1988) substituted a P1 lysogen as a source of transducing particles. Transductants of the introduced recipient were detected at a concentration of 10^5 transductants/g of sterile soil and 10^2 transductants/g when the soil was not sterilized. Although several hundred isolates of indigenous soil bacteria that exhibited the correct antibiotic-resistance phenotype were examined for the presence of transduced marker DNA (Zeph and Stotzky 1989), no transduction to the autochthonous microbiota was detected.

Transfer of plasmid DNA mediated by the generalized *P. aeruginosa* transducing phage F116L has recently been studied (Zeph and Stotzky 1988; Stotzky 1989). Using lysates of F116L grown on *P. aeruginosa* PAO containing RP4, transduction to a non-lysogenic *P. aeruginosa* recipient was detected at approximately 10^3 transductants/g of sterile soil.

Transductional transmission of DNA has been demonstrated in Gram-negative bacteria in both sterile and non-sterile soils. To date, no demonstration of transduction among the Gram-positive organisms has been made *in situ*. Whether or not transduction occurs in soil populations of Gram-positive micro-organisms must be investigated. Such studies are particularly crucial because of the abundance of Gram-positive organisms in the soil and the high frequency of lysogenization observed in soil isolates of the genus *Bacillus* (Kelly and Reanney 1978). Transducing phages have been isolated for a number of species of *Bacillus* and several have been shown to carry out both intra- and interspecies transduction (Miller 1992).

Transduction in freshwater environments

A number of studies *in situ* have addressed the potential for transduction in freshwater environments. Both chromosomal (Morrison *et al.* 1978; Saye and Miller 1989b) and plasmid DNA (Saye *et al.* 1987b) have been shown to be transferred in a freshwater lake and transduction of chromosomal DNA has been demonstrated in a freshwater river environment (Amin and Day 1988).

Morrison *et al.* (1978) investigated transduction of streptomycin resistance mediated by the generalized transducing phage F116 to a streptomycin-sensitive strain of *P. aeruginosa* in flow-through environmental chambers suspended in a freshwater lake. Both lysates of F116 prepared on a streptomycin-resistant strain of *P. aeruginosa* and an F116 lysogen were used as sources of the DNA encoding streptomycin resistance in separate chambers. Transductants were recovered over a 10-day period of incubation. Transduction frequencies reached as high as 10^{-1} transductants per surviving recipient after 10 days. Similar frequencies of F116-mediated chromosomal transduction were obtained by Amin and Day (1988) for strains of *P. aeruginosa* introduced into the epilithon of river stones submerged in a Welsh river. Saye and co-workers (Saye and Miller 1989a,b; Saye *et al.* 1990) measured the cotransduction frequencies of chromosomal markers between two *P. aeruginosa* lysogens of the generalized transducing phage F116L incubated in microcosms and in a freshwater lake. They demonstrated transduction of both auxotrophic and antibiotic-resistance markers. Gene transfer was observed when either the donor or recipient was the lysogen and when both the donor and recipient parents were lysogens.

Plasmid transduction has been extensively studied by Saye *et al.* (1987b) in a freshwater lake. These investigators measured the transduction of the conjugation-deficient plasmid Rms149 mediated by an independently isolated variant of F116 originally named øDS1 (Saye *et al.* 1987a). Three different paradigms for the source of transducing phage were evaluated *in situ*. First, lysates of phages grown on an Rms149-containing strain served as the source of plasmid DNA and were inoculated into chambers containing a suitable recipient strain. Second, a lysogenic plasmid donor was inoculated with a non-lysogenic recipient strain. Third, a non-lysogenic plasmid-donor was inoculated with a lysogenic recipient strain. The greatest number of transductants were recovered from chambers containing the third variation, where a lysogen served both as the source of the transducing phage and as the recipient of the transduced DNA.

When this paradigm was investigated further it was observed that the frequency of transductants was significantly influenced by the donor-to-recipient ratio in the test chamber (Saye *et al.* 1987b). When chambers containing sterilized lake water were compared with others containing the same introduced donor and recipient strain in the presence of the natural microbial community, the frequency of observed transduction was greatly reduced. Whether the actual rate of transduction

was reduced when indigenous micro-organisms were present or whether the ability to recover transductants was reduced due to increased simple competition between the introduced and native micro-organisms could not be determined in this study.

The observation made in all of these studies (Saye *et al.* 1987b; Amin and Day 1988; Saye and Miller 1989a,b), that the most likely reservoir of transducing phage in aquatic environments is natural lysogens, might be expected. That the most frequently recovered transductant was a transduced lysogen (Saye *et al.* 1987b; Saye and Miller 1989a,b), may be more surprising. However, one should remember that while lysogens are immune to superinfection by the resident phage, in many cases (including F116 lysogens of *P. aeruginosa*) the absorption of DNA contained in phage capsids is not inhibited. The rate of absorption of transduced DNA by lysogens and non-lysogens is essentially the same. The lysogen, however, has a distinct survival advantage over the non-lysogen because of its resistance to killing by the parental virus. The frequency of transduction of lysogens and non-lysogens in freshwater environments is likely to be similar, but the new transduced genotype is more likely to survive and be stabilized within the resident populations when the transduced cell is lysogenic for the mediating phage.

Transduction in the waste treatment facility

The water treatment facility is potentially an ideal environment for gene transmission between micro-organisms. The waste treatment process encourages the growth of degradative micro-organisms and the conditions favourable for efficient waste treatment are also those that would be predicted to be conducive to microbial interactions leading to horizontal gene exchange. Genetically engineered micro-organisms have been developed with the potential for increased degradative capacities in the hope that they can be used in waste treatment (Ghosal *et al.* 1988).

Waste treatment facilities provide a nutrient-rich and quiescent environment (Gealt 1992). The concentrations of bacteria in waste water are usually 10^6 CFU/ml or higher (Gealt 1992) and in activated sludge their numbers often reach 10^9 CFU/ml (Blackburn *et al.* 1987). In addition, titres of bacteriophages in sewage are certainly within the range predicted to lead to phage–host interaction

(Wiggins and Alexander 1985; Kokjohn *et al.* 1989). Kennedy *et al.* (1985) observed titres of coliphages of 10^3–10^4 PFU/ml of waste water. Dhillon and Dhillon (1981) discovered that 4 per cent of isolates of *E. coli* from sewage were lysogenic and that lysogens were present in at least half the sewage samples they tested.

Horizontal gene transfer by mechanisms other than conjugation has only recently been investigated in the waste treatment environment. Osmond and Gealt (1988) used a temperate coliphage isolated from a waste-water treatment facility to demonstrate transduction of both plasmid and chromosomal DNA in synthetic waste water. Both laboratory strains of *E. coli* and isolates from the sewage treatment plant were used as recipients and frequencies of transduction of 10^{-4}–10^{-5} per recipient were obtained.

Transduction in bacteria associated with animals

Transduction has been demonstrated in oysters, which were allowed to take up a streptomycin-resistant strain of *Vibrio pharhaemolyticus*, which could not degrade agar. The oysters were then transferred to a marine aquarium containing sterilized sea water and a cell-free lysate of the *Vibrio* bacteriophage P4, which had been prepared on a streptomycin-sensitive, agarase-proficient strain of *V. parahaemolyticus* was added (Baross *et al.* 1978). Streptomycin-resistant, agar-degrading *Vibrio* were isolated from the oysters after 154 h of incubation in the aquarium.

Novick and Morse (1967) investigated transduction of drug resistance between strains of *Staphylococcus aureus* in the kidneys of mice. They injected a lysogenic drug-resistant donor and a non-lysogenic recipient strain intravenously into mice. Several days later the mice were killed and *S. aureus* was isolated from the renal infections. Transduced organisms were observed in one-third to one-half of the mice, with a frequency of transduction of 1×10^{-4}/CFU. The investigators demonstrated that the transducing particles were produced in the animal and were not the result of pre-existing particles introduced with the inoculum. These experiments extended the work of Jarolmen *et al.* (1965), who had demonstrated transduction in mice by injecting large numbers (5×10^{10} PFU) into mice in which renal infections of *S. aureus* had been established.

Verification of the molecular mechanism of gene transfer in the environmental setting

Many of the studies conducted in animals (Roberts 1989; Tauxe *et al.* 1989), as well as in natural environments (Saye and Miller 1989a), have investigated the phenomenological transfer of genes, without identifying the underlying molecular mechanism of that transfer. In many cases it was assumed that this mechanism was conjugation. Several recent findings demonstrate that, even when these genes are shown to be present on extrachromosomal replicons, it cannot be assumed that transfer is mediated by conjugation. Rms149, a transfer-deficient plasmid identified in a clinical isolate of *P. aeruginosa*, was originally thought to be able to mediate its own transfer via conjugation. Only later was it discovered that transfer was actually being mediated by a coresident prophage present in the original bacterial isolate (Hedges and Jacoby 1980).

Several recent reports have demonstrated the transduction of genetic elements by phage or phage-like particles that do not produce the traditional symptoms of phage infection in their hosts. Rapp and Wall (1987) demonstrated transduction of antibiotic resistance in *Desulfovibrio desulfuricans* by cell-free lysates prepared on certain drug-resistant strains. The transducing phage did not produce plaques on recipient strains, nor could they be induced by DNA-damaging agents from the lysogenic donor strains. Lysogens could serve as recipients of transduced DNA. Electron microscopy of transducing cell-free lysates revealed the presence of complex phage particles. Starich and Zissler (1989) identified strains of *Myxococcus xanthus* that produced Mx alpha particles, which were shown to transduce specific regions of the *M. xanthus* chromosome. Again, no host able to sustain lytic growth of Mx alpha particles has been identified. Wild type isolates of several *Myxcoccus* species contained DNA sequences with homology to Mx alpha particles indicating that Mx alpha is widespread in nature. The systems identified in *D. desulfuricans* and *M. xanthus*, as well as the misinterpretation of the potential for conjugal transfer of Rms149 in *P. aeruginosa*, demonstrate the necessity for identifying the mechanism of gene transfer in environmental settings.

Environmental factors that may influence the probability of transduction

A significant number of conditions encountered in the environment may pose potential barriers to gene transmission between bacteria by any mechanism (Miller 1988). Several of these may, theoretically at least, have a profound effect on transduction in various habitats. They may alter the probability of productive interaction among the various components of the gene transfer system, the stability of the transferred DNA once transferred to the recipient cell, or the ability of the recipient to express the genetic determinants once the DNA is stably established. These factors may have either a positive or negative effect on transduction.

Temperature

We have already discussed the effect that environmental concentrations of bacteria and bacteriophages may have on host–phage interactions if such interactions are assumed to obey a mass-action model (see p. 56). Temperature may also affect any mass action process significantly. While environmental temperatures are often lower than those used in the laboratory, they are not too low to eliminate the potential for transduction. Transduction has been observed at ambient temperatures in aquatic habitats. Saye *et al.* (1987a) described transduction in a freshwater lake at 20°C, while Amin and Day (1988) detected transduction at 5°C between *P. aeruginosa* strains attached to river stones suspended 30 cm below the surface of a river in South Wales. The laboratory *P. aeruginosa* phage F116 was used in both of these studies.

Bacteriophages often show a temperature optimum for both absorption and replication (Seeley and Primrose 1980; Primrose *et al.* 1982). This temperature often reflects the ecological origin of the bacteriophage more closely than the growth optimum of the host bacterium. Therefore, phages that transduce with highest efficiency under laboratory conditions may be less effective in mediating transduction under environmental conditions than bacteriophages normally present in the habitat under study.

Ionic composition

The mineral content of a specific habitat may have a profound influence on the potential for gene transfer by transduction. Some bacteriophages require cations for adsorption to their hosts and will not infect when suspended in distilled water. Requirements are often for specific divalent cations and the ionic concentration for optimal adsorption is characteristic of both the phage and the ion. Too low or too high a concentration of the ion may lead to reduced adsorption (Tolmach 1957). The requirements for absorption of transducing particles parallels those for the parental phage. We have demonstrated that river water provided an adequate ionic environment to allow several different *P. aeruginosa* phages to infect their host efficiently (Kokjohn *et al.*, in preparation).

Solid surfaces

Whether soil or aquatic, all environments in which free-living micro-organisms find themselves are composed of a mixture of liquid and solid phases. The surfaces of particulate matter in each of these environments may have profound effects on the frequency of genetic transduction. These effects may be either positive or negative.

Soil has a high solids-to-water ratio and is unique among microbial habitats in being essentially a structured environment (Stotzky 1989). As bacteria are essentially aquatic organisms, their distribution is primarily restricted to the liquid phase. In soil, individual microhabitats are isolated from one another by the large regions made up of solids and movement of both bacteria and bacteriophages between these microhabitats is limited to locations where water bridges are present.

While the high solid content of soil may reduce the probability of gene transfer by restricting the free movement of the members of the transfer system, solids have the potential for stimulating gene transfer in those areas where they do come in contact by increasing the half life of transducing particles and by stabilizing potential interactions between cells and viruses. Both bacteria and bacteriophages tend to adhere to clay. The adhesion of bacteriophages to clays has been shown to protect them from inactivation and prolong the half life of transducing phages in soils (Bystricky *et al.* 1975; Stotzky 1986). Adhesion of bacteria to soils may, on the other hand, reduce the number of available phage attachment sites (Stotzky 1989). Roper and Marshall (1978) demonstrated that while sorption of *E. coli* to small diameter ($<0.6\,\mu m$) clay particles inhibits bacterium–bacteriophage interaction, particles of larger spherical diameter ($>2.6\,\mu m$) appear to stimulate their interaction.

Aquatic environments have a low solids-to-water ratio, and the major problem in these environments is the dilution of the components of the transducing system to levels where interaction is improbable (Kokjohn 1989). In these environments particulate matter may provide surfaces on which the various components of the gene transfer system are concentrated (O'Morchoe *et al.* 1988). In the water column of various aquatic environments the majority of the microbial biomass, as well as the majority of metabolic activity, occurs on the surfaces of suspended particulate matter (Iriberri *et al.* 1987).

Sediments have an intermediate soil-to-water ratio (Stotzky 1989). They contain enough water that movement of bacteria and bacteriophages is not limited, at the same time they contain significant concentrations of particulates, which can provide surfaces to stimulate microcolony formation and host–phage interaction. They may, therefore, be the most fertile environment for gene exchange. Unfortunately, transduction has not, at this time, been investigated in sediments of natural aquatic systems.

When solid particles increase or decrease the probability of transduction is unknown. In any case, phage-mediated genetic transmission has been demonstrated in soils (Zeph *et al.* 1988; Zeph and Stotzky 1989), in water columns (Morrison *et al.* 1978; Saye *et al.* 1987b), on surfaces of submerged rocks (Amin and Day 1988), and in the sediment-rich environment of the waste treatment plant (Osmond and Gealt 1988).

Solar radiation and transduction

While phages free in the environment have a finite half life (Wiggins and Alexander 1985; Farrah 1987; Saye *et al.* 1987b; Williams *et al.* 1987), lysogens present in these ecosystems may continuously replenish the pool of transducing particles. Indeed, several studies have implicated naturally-occurring lysogens as the most likely reservoir of transducing phages in the environment (Morrison *et al.* 1978; Saye *et al.* 1987b; Amin and Day 1988; Zeph *et al.* 1988). As can be seen from Table 5.1, many temperate prophages of environmentally important bacteria are induced to lytic growth by exposure to DNA-damaging stresses.

One of the most prevalent of these stresses, and one with which most free-living micro-organisms come in contact on a daily basis, is solar UV radiation. The problem is most acute for aquatic populations because solar radiation penetrates to considerable depths (Fletcher 1979).

Solar UV radiation contains a wide band of wavelengths extending from 290 nm to the lower 400 nm range of visible light (Peak and Peak 1989). No radiation in the 254 nm (far or UV-C) range penetrates the ozone layer and solar radiation contains only low fluences of light in the UV-B wavelengths (290 to 320 nm). However, the flux rapidly increases as the wavelength increases into the UV-A (<320 nm) region. The primary damage caused to DNA by UV-C is the formation of pyrimidine dimers, which are also produced by exposure of DNA to UV-B, although in significantly lower yields; exposure to UV-B also causes the formation of single-stranded breaks in DNA. Pyrimidine dimers are only produced in detectable levels by the shortest wavelengths of UV-A radiation, and then only in concentrations that are orders of magnitude lower than those produced by UV-B or UV-C irradiation. The types of DNA damage most commonly produced by exposure to UV-A are strand breaks, alkali-labile sites, and DNA-to-protein cross-links. DNA is probably not the primary chromophore for UV-A. The damage to DNA is an indirect result of the interaction of UV-A radiation with other compounds, which results in the formation of DNA-damaging photoproducts. Because UV-C wavelengths are absorbed by DNA at the highest efficiency, most research on prophage induction has been carried out with UV-C wavelengths. The ability of UV-A (and for that matter UV-B) to induce prophages present in the autochonthonous flora is essentially unknown.

Consideration of laboratory-generated data for some species suggests that induction of prophages in response to exposure to solar UV will occur only in physiologically active cells (Jagger 1985; Kokjohn and Miller 1987). The physiological state of the bacterial population is a critical variable affecting prophage induction. The ability to induce, termed aptitude by Andre Lwoff (1953), is dependent on the cell's growth medium. In the case of *B. megaterium*, cells growing in a rich medium were capable of supporting prophage induction, while cells cultured in minimal medium failed to induce even after exposure to a DNA-damaging stress (Lwoff 1953). Therefore, cells present in natural ecosystems may be competent to induce prophages only transiently or episodically.

Saye and Miller (1989a) initiated studies to elucidate the characteristics of induction of the *P. aeruginosa* prophage F116 *in situ*. Prophage induction frequencies as a function of the number of bacteria present were compared in microcosms housed in UV-shielding glass chambers and in UV-translucent test chambers incubated in a freshwater lake. The microcosms and test chambers were inoculated with a lysogen and monitored over a 96-h period. In laboratory-incubated microcosms containing water from the lake, the number of phages per lysogenic bacterium remained constant at a level characteristic of the spontaneous level of induction. When test chambers were incubated *in situ*, dramatic increases in the phage-to-bacterium ratio were observed. These data suggest that environmental conditions are capable of supporting stress-stimulated prophage induction. Whether prophage induction *in situ* is the direct effect of absorption of solar UV radiation by the lysogen or is a secondary effect caused by the creation of DNA-damaging photochemicals by solar UV remains to be determined.

Molecular size of transduced DNA

Transducing particles contain a single fragment of host DNA of approximately the same size as the parental phage genome. There are maximal and minimal size requirements for efficient encapsidation of DNA molecules. This characteristic of the creation of transducing particles is particularly important when the potential for transduction of plasmid DNA is considered. Saye *et al.* (1987a) found that transduction of plasmids by various isolates of the *P. aeruginosa* phage F116 was more efficient if the molecular size of the plasmid was similar to that of the phage genome. Smaller plasmids were transduced at lower frequencies. Novick *et al.* (1986) found that small *S. aureus* plasmids could be efficiently transduced by ø11 because these plasmids formed linear concatemers of larger molecular size in the host cell.

Barriers to the entry of DNA into the recipient cell

The nature of transduction as a mechanism of horizontal gene transfer among members of a natural population of bacteria leads to unique restraints on the entry of exogenous genetic material into the potential recipient cell. These entry barriers may limit the free flow of DNA between bacterial species and even among members of the same species.

Transduction is ultimately limited by the host range of the transducing phage. The initial adsorption of the transducing particle to the prospective recipient requires specific interaction of the phage-encoded capsid proteins with unique receptors on the surface of the potential host cell (see p. 56). The requirement that a specific receptor be present on the potential host often limits phages to infecting one or a small number of closely related species. In addition, genetic mutation within the host cell population may lead to phage-resistant members of sensitive species.

However, numerous examples of phages with host ranges that cross species and generic barriers can be found in the literature (Thorne 1968, 1978; Kelln and Warren 1971; Yu and Baldwin 1971; Colón *et al.* 1972; Goldberg *et al.* 1974; Coetzee 1975b; Kaiser and Dworkin 1975; Dénarié *et al.* 1977; Ruhfel *et al.* 1984; Green and Goldberg 1985). Moreover, phages of extraordinarily broad host specificity have been described. For example, phages PRR1 and PRD1 will infect any Gram-negative bacterium containing an IncP-1 plasmid, such as RP1 (Olson and Shipley 1973; Olsen *et al.* 1974). As this plasmid can be transferred and maintained in an enormous variety of Gram-negative bacterial species, the host range of PRR1 and PRD1 is extensive. It should be emphasized that in this system phage sensitivity is due to the interaction of the phage with a specific receptor coded for by the IncP-1 plasmid.

Several broad-host-range phages have been shown to be transducing (Colón *et al.* 1972; Goldberg *et al.* 1974; Kaiser and Dworkin 1975). For example, the generalized transducing bacteriophage Mu, which replicates by transposition, is capable of infecting several genera of the family Enterobacteriaceae (Dénarié *et al.* 1977). Thorne (1968, 1978) observed interspecies transduction by phage CP-54 between a variety of *Bacillus* spp. and Yu and Baldwin (1971) demonstrated interspecies transduction in the staphylococci. These data suggest that transduction in environmental communities of mixed species of bacteria may proceed not only among strains of the same species but also across species lines.

Even species that have traditionally been considered resistant to a specific phage may have susceptible subpopulations. Using a positive selection technique, Goldberg *et al.* (1974) were able to find phage P1-sensitive clones of cells from species described as resistant to P1. The extension of the effective host range of this generalized transducing phage suggests that in a suitably large population of cells, interspecies transduction events may occur at low levels.

Restriction and modification of foreign DNA

Restriction–modification systems consist of two enzymes (Roberts 1982). First, an endonuclease, which upon recognition of a unique sequence of bases in a DNA molecule, cleaves the DNA and renders it biologically and genetically inactive. Second, a methylase, which modifies the corresponding recognition sequences in the host DNA, providing protection from cleavage by the restriction endonuclease. Essentially, self-DNA is protected and foreign DNA is destroyed. Any restriction–modification system active in the potential host cell may therefore reduce heritable establishment of foreign DNA within the recipient. As these systems may not only vary between species but between strains of bacteria, they have the potential for significant alteration of the frequency of effective horizontal gene transmission among natural populations of micro-organisms.

However, the barriers imposed by restriction–modification systems may be circumvented by some phages. Antirestriction mechanisms have evolved in some species of phages (Krishnapillai 1971; Spoerel *et al.* 1979; Kruger and Bickle 1983; Kilbane and Miller 1988). In the case of F116 of *P. aeruginosa*, phage DNA as well as transduced bacterial DNA is resistant to restriction by recipient cells (Holloway and Krishnapillai 1979). Several phages have been shown to inactivate the host-cell's restriction–modification system upon infection (Spoerel *et al.* 1979; Kilbane and Miller 1988; Moffatt and Studier 1988).

Environmental factors may also act to limit restriction–modification barriers (Sanders 1987). Both elevated temperature (Rolfe and Holloway 1966) and exposure to DNA-damaging agents such as UV radiation (Walker 1984) have been shown to inactivate the restriction–modification systems of bacteria. In addition, restriction of foreign DNA is seldom complete and a few molecules usually escape (Reanney *et al.* 1982).

Superinfection immunity

As described above (see p. 57), the presence of a prophage in a lysogen will inhibit both the expression of the lytic life cycle and the establishment of lysogeny of a similar or closely related phage in the same cell. In some host–phage systems this phenomenon, known as superinfection immunity, may be accompanied by lysogenic conversion. In some cases the lysogen is modified in such a way that absorption of DNA from the supernumerary phage or transducing particle is inhibited. The cell envelope of lysogens of the specialized transducing *P. aeruginosa* phage D3 is altered so that adsorption of not only D3 but the generalized transducing phage E79 is eliminated (Kuzio and Kropinski 1983). Such lysogens would be removed from the pool of potential recipients for transduced DNA mediated by these phages. Other phages, including F116 of *P. aeruginosa*, do not elicit this type of conversion and absorption by lysogens of DNA present in F116 capsids is not reduced (Caruso and Shapiro 1982). Transduction of F116 lysogens is therefore not limited, but the transduced lysogen is protected from subsequent killing by infection with a viable phage virion. In systems of this type, superinfection immunity may actually serve to increase the observed frequency of successful transductional events in the environment. Thus Saye *et al.* (1987b, 1990; Saye and Miller 1989a,b) found that the highest frequency of transduction was observed *in situ* when a lysogens of F116 was used as the recipient.

Implications for the environmental release of genetically engineered micro-organisms

Does transduction occur at significant levels in the natural environment? To date few studies have addressed this question. Those reports that have examined transduction *in situ* present compelling evidence that gene exchange by transduction is occurring in several environmental settings (Table 5.2). In addition, transduction has been identified as a mechanism for horizontal gene transmission in a large number of environmentally significant species (see Table 5.1) and bacteriophages appear to be ubiquitous in natural environments (Goyal *et al.* 1987). The conclusion that a great potential exists for gene transmission mediated by bacterial viruses

Table 5.2. *Ecosystems in which transduction has been demonstrated*

Environment	Selected references
Soil	Germida and Khachatourians 1988; Zeph *et al.* 1988
Aquatic environments	
Lake water	Morrison *et al.* 1978; Saye *et al.* 1987b
River stones	Amin and Day 1988
Waste-treatment facility	Osmond and Gealt 1988
Animals	
Shellfish	Baross *et al.* 1978
Mammals	Novick and Morse 1967

in nature is certainly warranted. Novick *et al.* (1986) have, in fact, concluded that it may be the major mechanism for the natural dissemination of both chromosomal and plasmid DNA, at least in some bacterial species.

In 1975, the Asilomar Conference (Fox and Miller 1985) on the dangers of accidental release of genetically engineered organisms into the environment focused attention on conjugation as the means of intercellular transmission of genetic information. Ever since this conference, efforts have been made to restrict the movement of genetically engineered DNA sequences by conjugation. As we have learned more about the potential hazards and benefits of genetically engineered micro-organisms (GEMs), their potential for solving environmental problems has become increasingly exciting. With the ever-increasing likelihood of deliberate environmental release of GEMs, renewed interest in the dynamics of genetic diversity in natural microbial populations has developed. Many mechanisms, such as transduction, which produces increased genomic variability, which were once ignored or regarded as merely laboratory phenomena are now being appreciated as ecologically significant. The recent studies described in this chapter indicate that transduction is as potent, and is potentially a more potent mechanism of environmental gene exchange then conjugation.

Despite the potential for gene exchange in the environment, *E. coli*, certainly the most studied of all micro-organisms, may be described as a clonal species consisting of distinct and quasiseparate lineages (Kreiber and Rose 1986). While *E. coli* is an ecologically highly specialized micro-organism, which inhabits a unique ecological niche, these data may be interpreted to suggest that horizontal gene-

tic transmission occurs only at low levels. Alternatively, natural selection may be acting quickly to bring genetically isolated populations to equilibrium, thus reducing the chance of establishment of novel phenotypes (and genotypes) in already well-adapted environmental communities.

As our knowledge of the population genetics and ecology of micro-organisms increases, it is apparent that the genetic engineer faces new challenges of considerable significance in achieving the goal of manipulating microbes safely and reducing (or at least measuring) the probability of gene transfer from the released GEM to other organisms present in the ecosystem. Assumptions of the probability of persistence, mobility and expression of genetically engineered sequences derived from laboratory observations will need to be validated in light of these new findings.

Acknowledgements

This work was supported in part by cooperative agreements No. CR815234 to R.V.M. and CR815282 to T.A.K. from the Gulf Breeze Laboratory of the US Environmental Protection Agency. The contents of this report do not necessarily reflect the views of the Environmental Protection Agency, nor does mention of trade names or commercial products constitute endorsement or recommendation for use.

References

Akhverdian, V.Z., Khrenova, E.A., Revlets, M.A., Gerasimova, T.V., and Krylov, V.N. (1985). The properties of transposable phages of *Pseudomonas aeruginosa* belonging to two groups distinguished by DNA-DNA homology. *Genetika*, **21**, 735–47.

Alexander, M. (1977). *Introduction to soil microbiology*, (2nd edn). John Wiley and Sons, New York.

Alikhanian, S.K., Ilhina, T.S., and Lomovskaya, N.D. (1960). Transduction in actinomycetes. *Nature*, **188**, 245–6.

Alonso, J.C., Lüder, G., and Trautner, T.A. (1986). Requirements for the formation of plasmid-transducing particles of *Bacillus subtilis* bacteriophage SPP1. *EMBO Journal*, **5**, 3723–8.

Altherr, M.R., and Kasweck, K.L. (1982). *In situ* studies with membrane diffusion chambers of antibiotic resistance transfer in *Escherichia coli*. *Applied and Environmental Microbiology*, **44**, 838–43.

Amin M.K., and Day, M.J. (1988). Donor and recipient

effects on transduction frequency *in situ*. In *REGEM 1 Program*, p. 11. REGEM, Cardiff, Wales.

Arber, W. (1983). A beginner's guide to lambda biology. In *Lambda II*, (Ed. R.W. Hendrix, J.W. Roberts, F.W. Stahl, and R.A. Weisberg), pp. 381–94. Cold Spring Harbor Laboratory, Cold Spring Harbor, New York.

Arber, W., Enquist, L., Hohn, B., Murray, N.E., and Murray, K. (1983). Experimental methods for use with lambda. In *Lamda II*, (Ed. R.W. Hendrix, J.W. Roberts, F.W. Stahl, and R.A. Weisberg), pp. 433–58. Cold Spring Harbor Laboratory, Cold Spring Harbor, New York.

Armentrout, T.W., and Rutberg, L. (1971). Heat induction of prophage ø105 in *Bacillus subtilis*: replication of the bacterial and bacteriophage genomes. *Journal of Virology*, **8**, 455–68.

Bailone, A., Levine, A., and Devoret, R. (1979). Inactivation of prophage lambda repressor *in vivo*. *Journal of Molecular Biology*, **131**, 553–72.

Baross, J.A., Liston, J., and Morita, R.Y. (1978). Incidence of *Vibrio parahaemolyticus* bacteriophages and other *Vibrio* bacteriophages in marine samples. *Applied and Environmental Microbiology*, **36**, 492–9.

Barsomina, G.D., Borillar, N.J., and Thorne, C.B. (1984). Chromosomal mapping of *Bacillus thuringiensis* by transduction. *Journal of Bacteriology*, **157**, 746–50.

Bender, R.A. (1981). Improved generalized transducing bacteriophage of *Caulobacter crescentus*. *Journal of Bacteriology*, **148**, 734–5.

Bender, R.A., and Sambucitti, L.C. (1983). Recombination-induced suppression of cell division following P1-mediated generalized transduction in *Klebsiella aerogenes*. *Molecular and General Genetics*, **189**, 263–8.

Berg, Ø., Børsheim, K.Y., Bratbak, G., and Heldal, M. (1989). High abundance of viruses found in aquatic environments. *Nature*, **340**, 467–8.

Better, M., and Helinski, D.R. (1983). Isolation and characterization of the *recA* gene of *Rhizobium meliloti*. *Journal of Bacteriology*, **155**, 311–16.

Bitton, G. (1987). Fate of bacteriophages in water and wastewater treatment plants. In *Phage ecology*, (Ed. S.M. Goyal, C.P. Gerba, and G. Bitton), pp. 181–95. John Wiley and Sons, New York.

Blackburn, J.W., Jain, R.K., and Sayler, G.S. (1987). Molecular microbial ecology of a naphthalene-degrading genotype in activated sludge. *Environmental Science and Technology*, **21**, 884–90.

Bramucci, M.G., and Lovett, P.S. (1976). Low-frequency, PBS1-mediated plasmid transduction in *Bacillus pumilis*. *Journal of Bacteriology*, **127**, 829–31.

Bramucci, M.G., and Lovett, P.S. (1977). Selective plasmid transduction in *Bacillus pumilus*. *Journal of Bacteriology*, **131**, 1029–32.

Bratback, B., Heldal, M., Norland, S., and Thingstad, T.F. (1990). Viruses as partners in spring bloom microbial trophodynamics. *Applied and Environmental Microbiology*, **56**, 1400–5.

Brayton, P.R., Tamplin, M.L., Huq, A., and Colwell, R.R. (1987). Enumeration of *Vibrio cholerae* 01 in Bangladesh waters by fluorescent-antibody direct viable count. *Applied and Environmental Microbiology*, **53**, 2862–5.

Buchanan-Wollaston, V. (1979). Generalized transduction in *Rhizobium leguminosarum*. *Journal of General Microbiology*, **112**, 135–42.

Buck, G.A., and Groman, N.B. (1981). Genetic elements novel for *Corynebacterium diphtheriae*: specialized transducing elements and transposons. *Journal of Bacteriology*, **148**, 143–52.

Bukhari, A.I. (1976). Bacteriophage mu as a transposition element. *Annual Review of Genetics*, **10**, 389–412.

Bystricky, V., Stotzky, G., and Schiffenbauer, M. (1975). Electron microscopy of T_1-bacteriophage adsorbed to clay minerals: application of the critical point drying method. *Canadian Journal of Microbiology*, **21**, 1278–82.

Campbell, A. (1977). Defective bacteriophages and incomplete prophages. In *Comprehensive virology*, Vol. 8, (Ed. H. Fraenkel-Conrat and R.R. Wagner), pp. 259–328. Plenum Press, New York.

Campos, J.M., Geisselsoder, J., and Zusman, D.R. (1978). Isolation of bacteriophage MX4, a generalized transducing phage of *Myxococcus xanthus*. *Journal of Molecular Biology*, **119**, 167–78.

Canosi, U., Lüder, G., and Trautner, T.A. (1982). SPP1-mediated plasmid transduction. *Journal of Virology*, **44**, 431–6.

Caruso, M., and Shapiro, J.A. (1982). Interactions of Tn7 and temperate phage F116L of *Pseudomonas aeruginosa*. *Molecular and General Genetics*, **188**, 292–8.

Casadesús, J., and Olivares, J. (1979). General transduction in *Rhizobium meliloti* by a thermosensitive mutant of bacteriophage DF2. *Journal of Bacteriology*, **139**, 316–17.

Cavenagh, M., and Miller, R.V. (1986). Specialized transduction of *Pseudomonas aeruginosa* PAO by bacteriophage D3. *Journal of Bacteriology*, **165**, 448–52.

Chakrabarty, A.M., and Gunsalus, I.C. (1969). Autonomous replication of a defective transducing phage in *Pseudomonas putida*. *Virology*, **38**, 92–104.

Chatterjee, A.K., and Brown, M.A. (1980). Generalized transduction in the enterobacterial phytopathogen *Erwinia chrysanthemi*. *Journal of Bacteriology*, **143**, 1444–9.

Chelala, C., and Margolin, P. (1974). Effects of deletions on cotransduction linkage in *Salmonella typhimurium*. *Molecular and General Genetics*, **131**, 97–112.

Clark, A.J. (1973). Recombination deficient mutants of *Escherichia coli* and other bacteria. *Annual Review of Genetics*, **7**, 67–86.

Coetzee, J.N. (1966). Transduction in *Proteus morganii.* *Nature*, **210**, 220.

Coetzee, J.N. (1975a). Specialized transduction of kanamycin resistance in a providence strain. *The Journal of General Microbiology*, **88**, 307–16.

Coetzee, J.N. (1975b). Transduction of a *Proteus vulgaris* strain by a *Proteus mirabilis* bacteriophage. *The Journal of General Microbiology*, **89**, 299–309.

Coetzee, J.N. (1976). Intra-species transduction with *Proteus mirabilis* high frequency transducing phages. *Journal of General Microbiology*, **93**, 153–65.

Coetzee, J.N., and Smit, J.A. (1969). Restriction of a transducing bacteriophage in a strain of *Proteus mirabilis. The Journal of General Virology*, **4**, 593–607.

Coetzee, J.N., Smit, J.A., and Prozesky, O.W. (1966). Properties of providence and *Proteus morganii* transducing phages. *Journal of General Microbiology*, **44**, 167–76.

Coetzee, J.N., de Klerk, H.C., and Smit, J.A. (1967). A transducing bacteriophage for *Proteus vulgaris. The Journal of General Virology*, **1**, 561–4.

Coetzee, J.N., Krizsanovich-Williams, K., and Williams, J.A. (1977). Cotransduction of morganocinogenic plasmid 174 and R factor R772. *Journal of General Microbiology*, **100**: 299–308.

Colón, A.E., Cole, R.M., and Leonard, C.G. (1970). Transduction in group A streptococci by ultraviolet-irradiated bacteriophages. *Canadian Journal of Microbiology*, **16**, 201–2.

Colón, A.E., Cole, R.M., and Leonard, C.G. (1971). Lysis and lysogenization of group A, C, and G streptococci by a transducing bacteriophage induced from group G *Streptococcus. Journal of Virology*, **8**, 103–10.

Colón, A.E., Cole, R.M., and Leonard, C.G. (1972). Intergroup lysis and transduction by streptococcal bacteriophages. *Journal of Virology*, **9**, 551–3.

Dabbs, E.R. (1987). A generalized transducing bacteriophage for *Rhodococcus erythropolis. Molecular and General Genetics*, **206**, 116–20.

Dallmann, G., Papp, P., and Oroxz, L. (1987). Related repressor specificity of unrelated phages. *Nature*, **330**, 398–401.

de Lancastre, H., and Archer, L.J. (1979). Transducing activity of bacteriophage SPP1. *Biochemical and Biophysical Research Communications*, **86**, 915–19.

de Lancastre, H., and Archer, L.J. (1980). Characterization of bacteriophage SPP1 transducing particles. *Journal of General Microbiology*, **117**, 347–55.

de Lancastre, H., and Archer, L.J. (1981). Molecular origin of transducing DNA in bacteriophage SPP1. *Journal of General Microbiology*, **122**, 345–9.

Deichelbohrer, I., Alonao, J.C., Lüder, G., and Trautner, T.A. (1985). Plasmid transduction by *Bacillus subtilis* bacteriophage SPP1: effects of DNA homology between plasmid and bacteriophage. *Journal of Bacteriology*, **162**, 1238–43.

Dénarié, J., Rosenberg, C., Bergeron, B., Boucher, C., Michel, M., and Barate de Bertalmio, M. (1977). Potential of RP4::mu plasmids for *in vivo* genetic engineering of gram-negative bacteria. In *DNA insertion elements, plasmids, and episomes*, (Ed. A.I. Bukhari, J.A. Shapiro, and S. Adhya), pp. 507–20. Cold Spring Harbor Laboratory, Cold Spring Harbor, New York.

Devanas, M.A., and Stotzky, G. (1988). Survival of genetically engineered microbes in the environment: effect of host/vector relationship. In *Developments in industrial microbiology*, Vol. 29, (Ed. G. Pierce), pp. 287–96. Elsevier Science Publishers, Amsterdam.

Dhillon, T.S., and Dhillon, E.K.S. (1981). Incidence of lysogeny, colicinogeny, and drug resistance in enterobacteria isolated from sewage and from rectum of humans and some domesticated species. *Applied and Environmental Microbiology*, **41**, 894–902.

Doermann, A.H. (1952). Intracellular growth of bacteriophages. I. liberation of intracellular bacteriophage T4 by premature lysis with another phage or cyanide. *Journal of General Physiology*, **35**, 645–56.

Downs, D.M., and Roth, J.R. (1987). A novel P22 prophage in *Salmonella typhimurium. Genetics*, **117**, 367–80.

Duckworth, D.H. (1976). Who discovered bacteriophage? *Bacteriological Reviews*, **40**, 793–802.

Duckworth, D.H. (1987). History and basic properties of bacterial viruses. In *Phage ecology*, (Ed. S.M. Goyal, C.P. Gerba, and G. Bitton), pp. 1–44. John Wiley and Sons, New York.

Dyer, D.W., Rock, M.I., Lee, C.Y., and Iandolo, J.J. (1985). Generation of transducing particles in *Staphylococcus aureus. Journal of Bacteriology*, **161**, 91–5.

Ebel-Tsipis, J., Botstein, D., and Fox, M. (1972). Generalized transduction by phage P22 in *Salmonella typhimurium*. I. molecular origin of transducing DNA. *Journal of Molecular Biology*, **185**, 433–48.

Echols, H., and Green, L. (1971). Establishment and maintenance of repression by bacteriophage lambda: the role of the *cI, cII,* and *cIII* proteins. *Proceedings of the National Academy of Sciences* (USA), **68**, 2190–4.

Ellis, E.L., and Delbruck, M. (1939). The growth of bacteriophage. *Journal of General Physiology*, **22**, 365–84.

Ely, B., and Johnson, R.C. (1977). Generalized transduction in *Caulobacter crescentus. Genetics*, **87**, 391–9.

Ewert, D.L., and Paynter, M.J.B. (1980). Enumeration of bacteriophages and host bacteria in sewage and the activated sludge treatment process. *Applied and Environmental Microbiology*, **39**, 576–83.

Farrah, S.R. (1987). Ecology of phage in freshwater environments. In *Phage ecology*, (Ed. S.M. Goyal, C.P. Gerba and G. Bitton), pp. 125–36. John Wiley and Sons, New York.

Farrand, S.K. (1989). Conjugal transfer of bacterial genes on plants. In *Gene transfer in the environment*, (Ed. S.B. Levy and R.V. Miller), pp. 261–85. McGraw-Hill, New York.

Ferrari, E., Canosi, U., Galizzi, A., and Mazza, G. (1978). Studies on transduction process by SPP1. *Journal of General Virology*, **41**, 563–72.

Fink, P.S., and Zahler, S.A. (1982). Specialized transduction of the *ilvD-thyB-ilvA* region mediated by *Bacillus subtilis* bacteriophage SP. *Journal of Bacteriology*, **150**, 1274–9.

Fjerdugstad, E. (1975). Bacteria and fungi. In *River ecology*, (Ed. B.A. Whitton), pp. 129–40. University of California Press, Berkeley.

Fletcher, M. (1979). The aquatic environment. In *Microbial ecology: a conceptual approach*, (Ed. J.M. Lynch and N.J. Poole), pp. 92–114. John Wiley and Sons, New York.

Fox, J.L., and Miller, J.A. (1985). *ASM News*, **51**, 225–6.

Freifelder, D. (1983). *Molecular biology. A comprehensive introduction to prokaryotes and eukaryotes*. Science Books International, Portola Valley, California.

Gealt, M.A. (1992). Gene transfer in waste treatment. In *Microbial ecology: principles, methods, and application in environmental biotechnology*, (Ed. M. Levin, R. Seidler, and M. Rogul), pp. 327–44. McGraw-Hill Publishing Co., New York.

Geisselsoder, J., Campos, J.M., and Zusman, D.R. (1978). Physical characterization of bacteriophage MX4, a generalized transducing phage for *Mysococcus xanthus*. *Journal of Molecular Biology*, **119**, 179–89.

Germida, J.J., and Khachatourians, G.G. (1988). Transduction of *Escherichia coli* in soil. *Canadian Journal of Microbiology*, **34**, 190–3.

Ghosal, D., You, I.S., Chattergee, D.K., and Chakrabarty, A.M. (1988). Capacity of aquatic bacteria to act as recipients of plasmid DNA. *Applied and Environmental Microbiology*, **54**, 115–17.

Goldberg, I., and Mekalanos, J.J. (1986). Cloning of the *Vibrio cholerae recA* gene and construction of a *Vibrio cholerae recA* mutant. *Journal of Bacteriology*, **165**, 715–22.

Goldberg, R.B., Bender, R.A., and Streicher, S.L. (1974). Direct selection of P1-sensitive mutants of enteric bacteria. *Journal of Bacteriology*, **118**, 810–914.

Goyal, S.M., Gerba, C.P., and Bitton, G. (1987). *Phage ecology*. John Wiley and Sons, New York.

Green, J., and Goldberg, R.B. (1985). Isolation and preliminary characterization of lytic and lysogenic phages with wide host range within the *Streptomycetes*. *Journal of General Microbiology*, **131**, 2459–65.

Hanks, M.C., Newman, B., Oliver, I.R., and Masters, M. (1988). Packaging of transducing DNA by bacteriophage P1. *Molecular and General Genetics*, **214**, 523–32.

Hayes, W. (1968). *The genetics of bacteria and their viruses*, (2nd edn). John Wiley and Sons, New York.

Hedges, R.W., and Jacoby, G.A. (1980). Compatibility and molecular properties of plasmid Rms 149 in *Pseudomonas aeruginosa* and *Escherichia coli*. *Plasmid*, **3**, 1–6.

Herman, N.J., and Juni, E. (1974). Isolation and characterization of generalized transducing bacteriophage for *Acinetobacter*. *Journal of Virology*, **13**, 46–52.

Herskowitz, I., and Hagen, D. (1980). The lysis–lysogeny decision of phage lambda: explicit programming and responsiveness. *Annual Review of Genetics*, **14**, 399–445.

Hirai, K., and Yanagawa, R. (1970). Generalized transduction of *Corynebacterium renale*. *Journal of Bacteriology*, **101**, 1086–7.

Holloway, B.W. (1969). Genetics of *Pseudomonas*. *Bacteriological Reviews*, **33**, 419–43.

Holloway, B.W., and Krishnapillai, V. (1979). Bacteriophages and bacteriocins. In *Genetics and biochemistry of Pseudomonas*, (Ed. P.H. Clarke and M.H. Richmond), pp. 99–132. John Wiley and Sons, New York.

Holloway, B.W., and Monk, M. (1959). Transduction in *Pseudomonas aeruginosa*. *Nature*, **184**, 1426–7.

Holloway, B.W., and Van de Putte, P. (1968a). Transducing phage for *Pseudomonas putida*. *Nature*, **217**, 459–60.

Holloway, B.W., and Van de Putte, P. (1968b). Lysogeny and bacterial recombination. In *Replication and recombination of genetic material*, (Ed. W.J. Peacock and R.D. Brook), pp. 175–83. Australian Academy of Science, Canberra.

Holloway, B.W., Egan, J.B., and Monk, M. (1961). Lysogeny in *Pseudomonas aeruginosa*. *Australian Journal of Experimental Biology*, **38**, 321–9.

Hyder, S.L., and Streitfeld, M.M. (1978). Transfer of erythromycin resistance from clinically isolated lysogenic strains of *Streptococcus pyogenes* via their endogenous phage. *Journal of Infectious Disease*, **138**, 281–6.

Ikeda, H., and Tomizawa, J.-T. (1965). Transducing fragments in generalized transduction by phage P1. II. Association of DNA and protein in the fragments. *Journal of Molecular Biology*, **14**, 110–19.

Iriberri, J., Unanue, M., Barcina, I., and Egea, L. (1987). Seasonal variation in population density and heterotrophic activity of attached and free-living bacteria in coastal waters. *Applied and Environmental Microbiology*, **53**, 2308–14.

Jagger, J. (1985). *Solar-UV actions on living cells*. Praeger Publishers, New York.

Jarolmen, H., Bondi, A., and Crowell, R.L. (1965). Transduction of *Staphyloccus aureus* to tetracycline resistance *in vivo*. *Journal of Bacteriology*, **89**, 1286–90.

Jarvis, A.W., Heap, H.A., and Lawrence, R.C. (1978).

The origin of bacteriophages in cheese factories. In *Microbial ecology*, (Ed. M.W. Loutit and J.A.R. Miles), pp. 438–42. Springer-Verlag, Berlin.

Kaiser, D., and Dworkin, M. (1975). Gene transfer to a mycobacterium by *E. coli* phage P1. *Science*, **187**, 653–4.

Kaplan, R.W., and Brendel, M. (1969). Formation of prototrophs in mixtures of two auxotrophic mutants of *Serratia marcescens* HY by a transducing bacteriophage produced by some auxotrophs. *Molecular and General Genetics*, **104**, 27–39.

Kauri, T. (1978). Aerobic bacteria in the horizons of a beech forest soil. In *Microbial ecology*, (Ed. M.W. Loutit and J.A.R. Miles), pp. 110–12. Springer-Verlag, Berlin.

Kelln, R.A., and Warren, R.A.J. (1971). Isolation and properties of a bacteriophage lytic for a wide range of psuedomonads. *Canadian Journal of Microbiology*, **17**, 677–82.

Kelly, W.J., and D.C. Reanney (1978). Virulent, temperate and defective bacteriophages from Gram-positive soil bacilli. In *Microbial ecology*, (Ed. M.W. Loutit and J.A.R. Miles), pp. 113–20. Springer-Verlag, Berlin.

Kennedy, J.E., Jr, Bitton, G., and Oblinger, J.L. (1985). Comparison of selective media for assay of coliphages in sewage effluent and lake water. *Applied and Environmental Microbiology*, **49**, 33–6.

Kilbane, J.J., and Miller, R.V. (1988). Molecular characterization of *Pseudomonas aeruginosa* bacteriophages: identification and characterization of the novel virus B86. *Virology*, **164**, 193–200.

Kokjohn, T.A. (1989). Transduction: mechanism and potential for gene transfer in the environment. In *Gene transfer in the environment*, (Ed. S.B. Levy and R.V. Miller), pp. 73–97. McGraw-Hill, New York.

Kokjohn, T.A., and Miller, R.V. (1987). Characterization of the *Pseudomonas aeruginosa recA* analog and its protein product: *rec-102* is a mutant allele of the *P. aeruginosa* PAO *recA* gene. *Journal of Bacteriology*, **169**, 1499–508.

Kokjohn, T.A., and Miller, R.V. (1988). Characterization of the *Pseudomonas aeruginosa recA* gene: the Les⁻ phenotype. *Journal of Bacteriology*, **170**, 578–82.

Kokjohn, T.A., Sayler, G.S., and Miller, R.V. (1989). Replication of *Pseudomonas aeruginosa* bacteriophages: lack of dependency on host-cell density. In *Abstracts of the annual meetings of the American Society for Microbiology 1989*, p. 367. American Society for Microbiology, Washington DC.

Kopec, L.K., Yasbin, R.E., and Marrero, R. (1985). Bacteriophage SPO2-mediated plasmid transduction in transpositional mutagenesis within the genus *Bacillus. Journal of Bacteriology*, **164**, 1283–7.

Kourilsky, P. (1973). Lysogenization by bacteriophage lambda. I. Multiple infection and the lysogenic response. *Molecular and General Genetics*, **122**, 183–95.

Kreiber, M., and Rose, M.R. (1986). Molecular aspects of the species barrier. *Annual Review of Ecology and Systematics*, **17**, 465–86.

Krieg, N.R., and Holt, J.G. (1984). *Bergey's Manual of Systematic Bacteriology*, Vol. 1. Williams and Wilkins, Baltimore.

Krishnapillai, V. (1971). A novel transducing phage: its role in recognition of a possible new host-controlled modification system in *Pseudomonas aeruginosa. Molecular and General Genetics*, **114**, 134–43.

Krizsanovich, K., de Klerk, H.S., and Smit, J.A. (1969). A transducing bacteriophage for *Proteus rettgeri. The Journal of General Virology*, **4**, 437–9.

Kirzsanovich-Williams, K. (1975). Specialized transduction of a leucine marker by *Proteus mirabilis* phage 5006M. *Journal of General Microbiology*, **91**, 213–16.

Kruger, D.H., and Bickle, T.A. (1983). Bacteriophage survival: multiple mechanisms for avoiding deoxyribonucleic acid restriction systems of the hosts. *Microbiological Reviews*, **47**, 345–60.

Krylov, V.N., Akhverdian, V.A., Bogush, V.G., Khrenova, E.A., and Revlets, M. (1985). The modular structure of transposable phage genomes of *Pseudomonas aeruginosa. Genetika*, **21**, 724–34.

Kuzio, J., and Kropinski, A.M. (1983). O-Antigen conversion in *Pseudomonas aeruginosa* strain PAO by bacteriophage D3. *Journal of Bacteriology*, **155**, 203–12.

Lecadet, M.-M., Blondel, M.-O., and Ribier, J. (1980). Generalized transduction in *Bacillus thuringiensis* var. *berliner* 1715 using bacteriophage CP-54Ber. *The Journal of General Microbiology*, **121**, 203–12.

Lederberg, J., and Tatum, E.L. (1946a). Novel genotypes in mixed cultures of biochemical mutants of bacteria. *Cold Spring Harbor Symposium on Quantitative Biology*, **11**, 113–14.

Lederberg, J., and Tatum, E.L. (1946b). Gene recombination in *E. coli. Nature*, **158**, 558.

Lederberg, J., Lederberg, E.M., Zinder, N.D., and Lively, E.R. (1951). Recombination analysis of bacterial heredity. *Cold Spring Harbor Symposium on Quantitative Biology*, **16**, 413–43.

Lennox, E.S. (1955). Transduction of linked genetic characters of the host by bacteriophage P1. *Virology*, **1**, 190–206.

Levin, D.B., and DuBow, M.S. (1987). Cloning and localization of the repressor gene (*c*) of the mu-like transposable phage D108. *FEBS Letters*, **222**, 199–203.

Levine, M. (1957). Mutations in the temperate phage P22 and lysogeny in *Salmonella. Virology*, **3**, 22–41.

Levine, A., Bailone, A., and Devoret, R. (1979). Cellular levels of the prophage and 434 repressors. *Journal of Molecular Biology*, **131**, 655–61.

Lipsky, R.H., Rosenthal, R., and Zanler, S.A. (1981).

Defective specialized SPβ transducing bacteriophages of *Bacillus subtilis* that carry the *sup-3* and *sup-4* gene. *Journal of Bacteriology*, **148**, 1012–15.

Lovett, P.S. (1972). PBS1: a flagella specific bacteriophage mediating transduction in *Bacillus pumilis*. *Virology*, **47**, 743–52.

Lovett, P.S., and Young, F.E. (1971). Linkage groups in *Bacillus pumilus* determined by bacteriophage PBS1-mediated transduction. *Journal of Bacteriology*, **106**, 697–9.

Lovett, P.S., Bramucci, K.D., Bramucci, M.G., and Burdick, B.D. (1974). Some properties of the PBP1 transduction system in *Bacillus pumilus*. *Journal of Virology*, **13**, 81–4.

Lovett, C.M., Love, P.E., Yashbin, R.E., and Roberts, J.W. (1988). SOS-like induction in *Bacillus subtilis*: induction of the RecA protein analog and a damage-inducible operon by DNA damage in Rec$^+$ and DNA repair-deficient strains. *Journal of Bacteriology*, **170**, 1467–74.

Low, K.B., and Porter, R.D. (1978). Modes of gene transfer and recombination in bacteria. *Annual Review of Genetics*, **12**, 249–87.

Luria, S.K., Adams, J.N., and Ting, R.C. (1960). Transduction of lactose-utilizing ability among strains of *E. coli* and *S. dysenteriae* and the properties of the transducing phage particles. *Virology*, **12**, 348–90.

Lwoff, A. (1953). Lysogeny. *Bacteriological Reviews*, **17**, 269–337.

MacPhee, D.G., Sutherland, I.W., and Wilkinson, J.F. (1969). Transduction in *Klebsiella*. *Nature*, **221**, 475–6.

Margolin, P. (1987). Generalized transduction. In *Escherichia coli and Salmonella typhimurium. Cellular and molecular biology*, Vol. 2, (Ed. F.C. Neidhardt), pp. 1154–68. American Society for Microbiology, Washington DC.

Marrero, T., Chiafari, F.A., and Lovett, P.S. (1981). SPO2 particles mediating transduction of a plasmid containing spo2 cohesive ends. *Journal of Bacteriology*, **147**, 1–8.

Martin, S., Sodergren, E., Masuda, T., and Kaiser, D. (1978). Systematic isolation of transducing phages for *Mysococcus xanthus*. *Virology*, **88**, 44–53.

Masters, M. (1985). Generalized transduction. In *Genetics of bacteria*, (Ed. J. Scaife, D. Leach, and A.Galizzi), pp. 197–216. Academic Press, New York.

Matsumoto, H., Yoshifumi, I., Ohta, S., and Terawaki, Y. (1986). A generalised transducing phage of *Pseudomonas cepacia*. *Journal of General Microbiology*, **132**, 2583–6.

Melechen, N.E., and Go, G. (1980). Induction of lamboid prophages by amino acid deprivation: differential inducibility; role of *recA*. *Molecular and General Genetics*, **180**, 147–55.

Miller, R.V. (1988). Potential for transfer and establishment of engineered genetic sequences. In *Planned release of genetically engineered organisms (Trends in Biotechnology and Trends in Ecology and Evolution special publication)*, (Ed. J. Hodgson and A.M. Sugden), pp. S23–S27. Elsevier Publications, Cambridge.

Miller, R.V. (1992). Methods for evaluating transduction: an overview with environmental considerations. In *Microbial ecology: principles, methods, and application in environmental biotechnology*, (Ed. M. Levin, R. Seidler, and P.H. Pritchard), pp. 229–51. McGraw-Hill Publishing Company, New York.

Miller, R.V., and Kokjohn, T.A. (1987). Cloning and characterization of the *cI* repressor of *Pseudomonas aeruginosa* bacteriophage D3; a functional analog of phage lambda *cI* protein, *Journal of Bacteriology*, **169**, 1847–52.

Miller, R.V., and Kokjohn, T.A. (1990). General microbiology of *recA*: environmental and evolutionary significance. *Annual Review of Microbiology*, **44**, 365–94.

Miller, R.V., and Renta Rubero, V.J. (1984). Mucoid conversion by phages of *Pseudomonas aeruginosa* strains from patients with cystic fibrosis. *Journal of Clinical Microbiology*, **19**, 717–19.

Miller, R.V., and Sayler, G.S. (1992). Bacteriophage–host interactions in aquatic systems. In *Genetic interactions among microorganisms in the natural environment*, (Ed. E.M. Wellington and J.D. Van Elsas), pp. 176–93. Pergamon Press, Oxford.

Miller, R.V., Pemberton, J.M., and Richards, K.E. (1974). F116, D3 and G101: temperate bacteriophages of *Pseudomonas aeruginosa*. *Virology*, **59**, 566–9.

Miller, R.V., Pemberton, J.M., and Clark, A.J. (1977). Prophage F116: evidence for extrachromosomal location in *Pseudomonas aeruginosa* PAO. *Journal of Virology*, **22**, 844–7.

Moebus, K. (1987). Ecology of marine bacteriophages. In *Phage ecology*, (Ed. S.M. Goyal, C.P. Gerba, and G. Bitton), pp. 137–56. John Wiley and Sons, New York.

Moffatt, B.A., and Studier, F.W. (1988). Entry of bacteriophage T7 DNA into the cell and escape from host restriction. *Journal of Bacteriology*, **170**, 2095–105.

Morgan, A.F. (1979). Transduction of *Pseudomonas aeruginosa* with a mutant of bacteriophage E79. *Journal of Bacteriology*, **139**, 137–40.

Morrison, W.D., Miller, R.V., and Sayler, G.S. (1978). Frequency of F116 mediated transduction of *Pseudomonas aeruginosa*. *Applied and Environmental Microbiology*, **36**, 724–30.

Morse, M.L. (1954). Transduction of certain loci in *Escherichia coli* K-12. *Genetics*, **39**, 984–5.

Novick, R.P., and Morse, S.I. (1967). In vivo transmission of drug resistance factors between strains of *Staphylococcus aureus*. *Journal of Experimental Medicine*, **125**, 45–59.

Novick, R.P., Edelman, I., and Lofdahl, S. (1986).

Small *Staphylococcus aureus* plasmids are transduced as linear multimers that are formed and resolved by replicative processes. *Journal of Molecular Biology*, **192**, 209–20.

Novikova, N.I., and Simarov, B.V. (1983). Isolation of transducing phages of *Rhizobium meliloti*. *Genetika*, **19**, 331–2.

Ogg, J.E., Timmo, T.L., and Alemohammad, M.M. (1981). General transduction in *Vibrio cholerae*. *Infection and Immunity*, **31**, 737–41.

Ogunseitan, O.A., Sayler, G.S., and Miller, R.V. (1990). Dynamic interaction of *Pseudomonas aeruginosa* and bacteriophages in lake water. *Microbial Ecology*, **19**, 171–85.

Olsen, R.H., and Shipley, P. (1973). Host range and properties of the *Pseudomonas aeruginosa* R Factor R1822. *Journal of Bacteriology*, **113**, 772–80.

Olsen, R.H., Siak, J., and Gray, R.H. (1974). Characteristics of PRD1, a plasmid-dependent broad host range DNA bacteriophage. *Journal of Virology*, **14**, 689–99.

O'Morchoe, S., Ogunseitan, O., Sayler, G.S., and Miller, R.V. (1988). Conjugal transfer of R68.45 and FP5 between *Pseudomonas aeruginosa* in a natural freshwater environment. *Applied and Environmental Microbiology*, **54**, 1923–9.

Osmond, M.A., and Gealt, M.A. (1988). Wastewater bacteriophages transduce genes from the chromosome and a recombinant plasmid. In *Abstracts of the annual meetings of the American Society for Microbiology 1988*, p. 254. American Society for Microbiology, Washington DC.

Ozeki, H., and Ikeda, H. (1968). Transduction mechanisms. *Annual Review of Genetics*, **2**, 245–78.

Peak, M.J., and Peak, J.G. (1989). Solar-ultraviolet-induced damage to DNA. *Photodermatology*, **6**, 1–15.

Pemberton, J.M. (1974). Size of the chromosome of *Pseudomonas aeruginosa* PAO. *Journal of Bacteriology*, **119**, 748–52.

Primrose, S.B. and Day, M. (1977). Rapid concentration of bacteriophage from aquatic habitats. *Journal of Applied Bacteriology*, **42**, 417–21.

Primrose, S.G., Seeley, N.D., Logan, K.B., and Nicolson, J.W. (1982). Methods for studying aquatic bacteriophage ecology. *Applied and Environmental Microbiology*, **43**, 694–701.

Proctor, L.M., and Fuhrman, J.A. (1990). Viral mortality of marine bacteria and cyanobacteria. *Nature*, **343**, 60–2.

Prozorov, A.A., Belova, T.S., and Surikov, N.N. (1980). Transformation and transduction of *Bacillus subtilis* strains with the Bsu R restriction-modification system by means of modified and unmodified DNA of pUB110 plasmid. *Molecular and General Genetics*, **180**, 135–8.

Ptashne, M. (1967). Isolation of the lambda phage repressor. *Proceedings of the National Academy of Sciences* (USA), **57**, 306–13.

Ptashne, M. (1986). *A genetic switch: gene control and phage* λ. Cell Press and Blackwell Scientific Publications, Palo Alto, California.

Quillardet, P., Moreay, P.L., Ginsberg, H., Mount, D.W., and Devoret, R. (1982). Cell survival, UV reactivation and induction of prophage lambda in *Escherichia coli* K-12 overproducing *recA* protein. *Molecular and General Genetics*, **188**, 37–43.

Rapp, B.J., and Wall, J.D. (1987). Genetic transfer in *Desulfovibrio desulfuricans*. *Proceedings of the National Academy of Sciences* (USA), **84**, 9128–30.

Reanney, D.C. (1976). Extrachromosomal elements as possible agents of adaptation and development. *Bacteriological Reviews*, **40**, 552–90.

Reanney, D.C., Roberts, W.P., and W.J. Kelly (1982). Genetic interactions among microbial communities. In *Microbial interactions and communities*, Vol. 1, (Ed. A.T. Bull and J.H. Slater), pp. 287–322. Academic Press, New York.

Roberts, R. (1982). Restriction and modification enzymes and their recognition sequences. *Nucleic Acids Research*, **10**, r117–r144.

Roberts, M.C. (1989). Gene transfer in the urogenital and respiratory tract. In *Gene transfer in the environment*, (Ed. S.B. Levy and R.V. Miller), pp. 347–76. McGraw-Hill, New York.

Roberts, J., and Devoret, R. (1983). Lysogenic induction. In *Lambda II*, (Ed. R.W. Dendriz, J.W. Roberts, F.W. Stahl, and R.A. Weisberg), pp. 123–44. Cold Spring Harbor Laboratory, Cold Spring Harbor, New York.

Roberts, J.W., and Roberts, C.W. (1975). Proteolytic cleavage of bacteriophage lambda repressor in induction. *Proceedings of the National Academy of Sciences* (USA), **75**, 4714–18.

Rolfe, B., and Holloway, B.W. (1966). Alterations in host specificity of bacterial deoxyribonucleic acid after an increase in growth temperature of *Pseudomonas aeruginosa*. *Journal of Bacteriology*, **92**, 43–8.

Romig, W.R., and Brodetsky, A.M. (1961). Isolation and preliminary characterization of bacteriophages for *Bacillus subtilis*. *Journal of Bacteriology*, **82**, 135–41.

Roper, M.M., and Marshall, K.C. (1978). Effect of clay particle size on clay-*Escherichia coli*-bacteriophage interactions. *Journal of General Microbiology*, **106**, 187–9.

Rosenthal, R., Toye, P.A., Korman, R.Z., and Zahler, S.A. (1979). The prophage of SPβc2dcitK$_1$, a defective specialized transducing phage of *Bacillus subtilis*. *Genetics*, **92**, 721–39.

Rowland, M.L., Watkins, C.S., Bruhn, D.F., and Ward, T.E. (1989). Studies on a bacteriophage which infects members of the genus *Acidiphilium*. In *Abstracts of the annual meetings of the American Society for Microbiology 1989*, p. 279. American Society for Microbiology, Washington DC.

Ruhfel, R.E., Robillard, N.J., and Thorne, C.B. (1984). Interspecies transduction of plasmids among *Bacillus anthracis*, *B. cereus*, and *B. thuringiensis*. *Journal of Bacteriology*, **157**, 708–11.

Sanders, M.E. (1987). Bacteriophages of industrial importance. In *Phage ecology*, (Ed. S.M. Goual, C.P. Gerba, and G. Billon), pp. 211–44. John Wiley and Sons, New York.

Sandri, R.M., and Berger, H. (1980). Bacteriophage P1-mediated generalized transduction in *Escherichia coli*: fate of transduced DNA in Rec$^+$ and RecA$^-$ recipients. *Virology*, **106**, 14–29.

Sauer, R.T., Ross, M.J., and Ptashne, M. (1982). Cleavage of the λ and P22 repressors by *recA* protein. *Journal of Biological Chemistry*, **257**, 4458–62.

Saye, D.J., and Miller, R.V. (1989a). The aquatic environment: consideration of horizontal gene transmission in a diversified habitat. In *Gene transfer in the environment*, (Ed. S.B. Levy and R.V. Miller), pp. 223–59. McGraw-Hill, New York.

Saye, D.J., and Miller, R.V. (1989b). F116L-mediated transduction of the *Pseudomonas aeruginosa* chromosome in a freshwater environment. In *Abstracts of the annual meetings of the Americal Society for Microbiology 1989*, p. 354. American Society for Microbiology, Washington DC.

Saye, D.J., Kokjohn, T.A., and Miller, R.V. (1987a). *Pseudomonas aeruginosa* phage DS1 suppresses the Les$^-$ phenotype. In *Abstracts of the annual meetings of the American Society for Microbiology 1987*, p. 238. American Society for Microbiology, Washington DC.

Saye, D.J., Ogunseitan, O., Sayler. G.S., and Miller, R.V. (1987b). Potential for transduction of plasmids in a natural freshwater environment: effect of plasmid donor concentration and a natural microbial community on transduction in *Pseudomonas aeruginosa*. *Applied and Environmental Microbiology*, **53**, 987–95.

Saye, D.J., Ogunseitan, O.A. Sayler, G.S., and Miller, R.V. (1990). Transduction of linked chromosomal genes between *Pseudomonas aeruginosa* during incubation *in situ* in a freshwater habitat. *Applied and Environmental Microbiology*, **56**, 140–5.

Seeley, N.D., and Primrose, S.B. (1980). The effect of temperature on the ecology of aquatic bacteriophages. *Journal of General Virology*, **46**, 87–95.

Servais, P., Billen, G., and Rego, V.J. (1985). Rate of bacterial mortality in aquatic environments. *Applied and Environmental Microbiology*, **49**, 1448–54.

Shapiro, J.A., Dean, D.H., and Halvorson, H.O. (1974). Low-frequency specialized transduction with *Bacillus subtilis* bacteriophage ø105. *Virology*, **62**, 393–403.

Sik, T., Horváth, J., and Chatterjee, S. (1980). Generalized transduction in *Rhizobium meliloti*. *Molecular and General Genetics*, **178**, 511–16.

Skjold, S.A., Maxted, W.R., and Wannamaker, L.W.

(1982). Transduction of the genetic determinant for streptolysin S in group A streptococci. *Infection and Immunity*, **38**, 183–8.

Skujins, J. (1984). Microbial ecology of desert soils. In *Advances in microbial ecology*, Vol. 7, (Ed. K.C. Marshall), pp. 49–91. Plenum Press, New York.

Sneath, P.H.A., Mair, N.S., Sharpe, M.E., and Holt, J.G. (1986). *Bergey's manual of systematic bacteriology*, Vol. 2. Williams and Wilkins, Baltimore.

Spoerel, N., Herrlich, P., and Bickle, T.A. (1979). A novel bacteriophage defence mechanism: the anti-restriction protein. *Nature*, **276**, 30–4.

Stäber, H., and Böhme, H. (1971a). Transdukton mit dem temperenten *Proteus mirabilis*-phagen π$_1$. I. Charakterisierung des transduktionssystems. *Zeitung für Allgemeine Mikrobiologie*, **11**, 221–30.

Stäber, H., and Böhme (1971b). Transdukton mit dem temperenten *Proteus mirabilis*-phagen π$_1$. II. Einfluss der vermehrungsmiltiplizität auf die transduktionsfahigkeit des phagen. *Zeitung für Allgemeine Mikrobiologie*, **11**, 231–6.

Starich, T., and Zissler, J. (1989). Movement of multiple DNA units between *Myxococcus xanthus* cells. *Journal of Bacteriology*, **171**, 2323–36.

Stent, G.S. (1963). *Molecular biology of bacterial viruses*. W.H. Freeman, San Francisco.

Stent, G.S., and Calendar, R. (1978). *Molecular genetics: an introductory narrative*, (2nd edn). W.H. Freeman and Company, San Francisco.

Sternberg, N., and Hoess, R. (1983). The molecular genetics of bacteriophage P1. *Annual Review of Genetics*, **17**, 123–54.

Stotzky, G. (1986). Influence of soil mineral colloids on metabolic processes, growth, adhesion, and ecology of microbes and viruses. In *Interactions of soil minerals with natural organics and microbes*, (Ed. P.M. Huang and M. Schnitzer), pp. 305–428. Soil Science Society of America, Madison, Wisconsin.

Stotzky, G. (1989). Gene transfer in soil. In *Gene transfer in the environment*, (Ed. S.B. Levy and R.V. Miller), pp. 165–222. McGraw-Hill, New York.

Stuttard, C. (1979). Transduction of auxotrophic markers in a chloramphenicol-producing strain of *Streptomyces*. *Journal of General Microbiology*, **110**, 479–82.

Süss, F., and Klaus, M. (1981). Transduction in *Streptomyces hydroscopicus* mediated by the temperate bacteriophage SH10. *Molecular and General Genetics*, **181**, 552–5.

Susskind, M.M., and Botstein, D. (1978). Molecular genetics of bacteriophage P22. *Microbiological Reviews*, **42**, 385–413.

Tauxe, R.V., Holmberg, S.D., and Cohen, M.L. (1989). The epidemiology of gene transfer in the environment. In *Gene transfer in the environment*, (Ed. S.B. Levy and R.V. Miller), pp. 377–404. McGraw-Hill, New York.

Taylor, M.J., and Thorne, C.B. (1963). Transduction of

Bacillus licheniformis and *B. subtilis* by each of two phages. *Journal of Bacteriology*, **86**, 452–61.

Thorne, C.B. (1968). Transducing bacteriophage of *Bacillus cereus*. *Journal of Virology*, **2**, 657–62.

Thorne, C.B. (1978). Transduction in *Bacillus thuringiensis*. *Applied and Environmental Microbiology*, **35**, 1109–15.

Tolmach, L.J. (1957). Attachment and penetration of cells by viruses. *Advances in Virus Research*, **4**, 63–110.

Totolyan, A.A., Boitsov, A.S., Kol', K.L., and Golubkov, V.I. (1981). Comparative characteristics of the transducing virulent streptococcal phages A25 and CA1. *Molecular Biology (Moscow)*, **15**, 894–900.

Vary, P.S., Garge, J.C., Franzen, M., and Frampton, E.W. (1982). MP13, a generalized transducing bacteriophage for *Bacillus megaterium*. *Journal of Bacteriology*, **149**, 1112–19.

Walker, G.C. (1984). Mutagenesis and inducible responses to deoxyribonucleic acid damage in *Escherichia coli*. *Microbiological Reviews*, **48**, 60–93.

Wannamaker, L.W., Almquist, S., and Skjold, S. (1973). Intergroup phage reactions and transduction between group C and group A streptococci. *Journal of Experimental Medicine*, **137**, 1338–53.

Weisberg, R.A. (1987). Specialized transduction. In *Escherichia coli and Salmonella typhimurium. Cellular and molecular biology*, Vol. 2, (Ed. F.C. Neidhardt), pp. 1169–76. American Society for Microbiology, Washington DC.

Wiggins, B.A., and Alexander, M. (1985). Minimum bacterial density for bacteriophage replication: implications for significance of bacteriophages in natural ecosystems. *Applied and Environmental Microbiology*, **49**, 19–23.

Williams, S.T., Mortimer, A.M., and Manchester, L. (1987). Ecology of soil bacteriophages. In *Phage ecology*, (Ed. S.M. Goyal, C.P. Gerba, and G. Bitton), pp. 157–79. John Wiley and Sons, New York.

Woods, D.R., and Thompson, J.A. (1975). Unstable generalized transduction in *Achromobacter*. *Journal of General Microbiology*, **88**, 86–92.

Wu, T.T. (1966). A model for three-point analysis of random general transduction. *Genetics*, **54**, 405–10.

Yamagishi, H., and Takahashi, I. (1968). Transducing particles of PBS1. *Virology*, **36**, 639–45.

Yelton, D.B., and Thorne, C.B. (1970), Transduction in *Bacillus cereus* by each of two bacteriophages. *Journal of Bacteriology*, **102**, 573–9.

Yoshikawa, M., and Hirota, Y. (1971). Impaired transduction of R213 and its recovery by a homologous resident R factor. *Journal of Bacteriology*, **106**, 523–8.

Yu, L., and Baldwin, J.N. (1971). Intraspecific transduction in *Staphylococcus epidermidis* and interspecific transduction between *Staphylococcus aureus* and *Staphylococcus epidermidis*. *Canadian Journal of Microbiology*, **17**, 767–73.

Zanler, S.A., Korman, R.Z., Rosenthal, R., and Kemphill, H.E. (1977). *Bacillus subtilis* bacteriophage SPβ: localization of the prophage attachment site, and specialized transduction. *Journal of Bacteriology*, **129**, 556–8.

Zeph, L.R., and Stotzky, G. (1988). Transduction by bacteriophage P1 in soil. In *Abstracts of the annual meetings of the American Society for Microbiology 1988*, p. 298. American Society for Microbiology, Washington DC.

Zeph, L.R., and Stotzky, G. (1989). Use of a biotinylated DNA probe to detect bacteria transduced by bacteriophage P1 in soil. *Applied and Environmental Microbiology*, **55**, 661–5.

Zeph, L.R., Onaga, M.A., and Stotzky, G. (1988). Transduction of *Escherichia coli* by bacteriophage P1 in soil. *Applied and Environmental Microbiology*, **54**, 1731–7.

Zinder, N.D., and Lederberg, J. (1952). Genetic exchange in *Salmonella*. *Journal of Bacteriology*, **64**, 679–99.

6

Gene transfer in the environment: transformation

Gregory J. Stewart

Introduction

OTHER CHAPTERS IN THIS BOOK have discussed gene transfer in the environment by conjugation and transduction. Recent work has clearly shown that gene transfer occurs by these means in the environment or in environmental simulations (Gealt *et al*. 1985; Saye *et al*. 1987; Trevors and Starodub 1987; Trevors *et al*. 1987; Bale *et al*. 1988; Genthner *et al*. 1988; Zeph *et al*. 1988; Zeph and Stotzky 1989). Both of these processes utilize a protective system for transfer of DNA from donors to recipients (Low and Porter 1978). As a consequence, the probability for genetic exchange in the field is reasonably high for plasmid- or phage-mediated gene transfer. The process of transformation, as taken in its traditional context, involves an intermediate form of DNA that is unprotected from the environment (Avery *et al*. 1944). Consequently, when one considers the potential for horizontal gene transfer by transformation, one must consider not only the environmental factors that allow organisms to take up and express DNA, but also the stability of DNA as a component, although perhaps a transient one, of the environment. Thus, factors that influence DNA stability also influence the capacity of the environment to support transformation.

Another important consideration is that our perception of gene transfer has changed in recent years. The traditional trichotomy of conjugation, transduction and transformation, as clearly defined and distinguishable processes (Low and Porter 1978), can no longer serve to encompass horizontal

gene transfer. Our view of conjugation as a plasmid-associated process has been challenged by the discovery and characterization of conjugal elements and transposons (Clewell and Gawron-Burke 1986). The gene transfer agent of *Rhodobacter capsulatus* shares many properties with transduction yet no viable phage has been shown to be intrinsic to the process (Solioz *et al*. 1975). The report of transformation between intact cells of *Pseudomonas* (Stewart *et al*. 1983), *Haemophilus* (Albritton *et al*. 1982) and *Bacillus* (Ephrati-Elizur 1968) challenge the notion that transformation is solely a recipient function and that DNA release is an accidental or incidental process. With studies underway to measure gene transfer as an environmentally significant process, we must consider that the traditional view of gene transfer events may limit our ability to detect and measure horizontal transfer events.

We are also limited by the fact that bacterial genetic exchange has been, until very recently, a process only studied in the laboratory. While it is a property of a wide range of bacteria (Stewart 1989) our understanding of the mechanism of natural transformation results from studies in only a few model systems. Genetic exchange was viewed as a tool to the geneticist and conditions that optimized the process were employed. One would predict that environmental gene transfer would be suboptimal compared to controlled laboratory processes. Thus, as we consider the potential for environments to support genetic exchange, we must also consider those factors that reduce the probability of the process. In this chapter we will deal specifically with the process of transformation and, more specifically, natural transformation. Processes such as $CaCl_2$ precipitation, polyethylene glycol/protoplast fusion and electroporation will not be addressed. Instead, we will review our understanding of the physiology of natural transformation in bacteria, of the factors that influence DNA stability and the potential for environments to support natural transformation.

The physiology of natural transformation

The ability of certain bacteria to take up and express exogenous DNA is a complex process. The phenotype of gene conversion that is most commonly employed to detect transformation is the consequence of the activity of a number of genes (Morrison *et al.* 1982). Several reviews have thoroughly addressed the physiology of natural transformation (Venema 1979; Smith *et al.* 1981; Goodgal 1982; Stewart and Carlson 1986). In this discussion we will provide a basic overview of transformation, but with a particular focus on those aspects of the process that are particularly relevant to environmental manipulation or influence. The process of natural transformation is generally categorized in four steps. Competence development is the physiological shift in the cells that allows them to bind, take up and express exogenous DNA. Once competence is induced, exogenous DNA binds to the surface of the competent cell. The bound DNA is taken into the cell and, finally, the acquired sequence replaces a homologous region in the endogenote of the cell. Inhibition of any of these processes is sufficient to prevent gene conversion.

Competence development

Competence development is controlled differently in various transforming bacteria, but the net effect is the same. In *Bacillus subtilis* (Dooley *et al.* 1971), *Streptococcus pneumoniae* (Tomasz 1966) and certain other Gram-positive bacteria, competence is induced when the aqueous environment surrounding a cell accumulates a small molecular-weight protein in sufficient concentration. This protein, referred to as competence factor, is synthesized constitutively by the population of cells in which competence is to be induced (Tomasz 1970). While competence is induced transiently in *S. pneumoniae*

during the exponential phase, in many Gram-positive bacteria it typically induces at relatively high cell concentrations; those often found as cells shift to unbalanced growth at the exponential phase/stationary phase transition. Competence in Gram-negative bacteria, including *Haemophilus influenzae* (Spencer and Herriott 1965), *Pseudomonas stutzeri* (Carlson *et al.* 1983) and *Azotobacter vinelandii* (Page 1982) is not induced by an exogenous soluble competence factor but is internally regulated. Again, induction of competence seems to be associated with a reduction in growth rate or a shift to unbalanced growth (Stewart and Carlson 1986). Finally, competence appears to be constitutively expressed in at least one organism, *Neisseria gonorrhoeae*, but only certain strains of this bacterium are capable of natural transformation (Sparling 1966).

DNA binding

A number of other transformation-related genes are expressed as a consequence of competence development. Among these are genes for membrane-associated proteins that are capable of binding DNA. The specificity of these DNA-binding proteins varies. In the case of *Haemophilus* and *Neisseria* species, binding proteins appear to have a strong preference for particular and distinct DNA sequences that are conserved and repeated on the genomes of these two bacteria (Scocca *et al.* 1974; Graves *et al.* 1982). In other cases, *Streptococcus*, *Pseudomonas* and *Bacillus*, for example, sequence specificity has not been detected and non-homologous DNA has been shown to compete effectively with DNA homologous with the endogenote for DNA binding sites (Stewart and Carlson 1986). In all cases studied thus far, DNA must be in the double-stranded form for the effective transformation of competent cells (Morrison 1977). Single-stranded DNA and RNA usually compete poorly for binding sites (Smith *et al.* 1981).

DNA binding is usually described in loose and tight binding stages. Loose binding almost always precedes tight binding and is characterized as an association of double-stranded DNA (dsDNA) with the cell in a form that can be dissociated by salt washes (Ceglowski *et al.* 1980). Tight binding, then, is binding of DNA to the surface of competent cells in a form that is resistant to removal by washing but that remains sensitive to enzymatic digestion by DNase I (Seto and Tomasz 1974).

Binding appears to be the result of specific outer surface-associated DNA binding proteins that are, in many cases, competence-specific (Scocca *et al.* 1974; Concino and Goodgal 1981). Mutants that fail to transform have been isolated and at least some classes of these mutants appear blocked at DNA binding steps (Lacks *et al.* 1974). In some cases a correlation of loss of DNA binding and the loss of competence-specific membrane proteins has been shown (Kahn *et al.* 1979; Kooistra *et al.* 1980).

DNA uptake

Once tightly bound, transforming DNA must penetrate the outer surface of the cell to allow integration into the genome of the recipient. The process of DNA uptake differs among transforming bacteria, but certain generalizations can be made. First, while dsDNA is required for binding, DNA entering the cytoplasm of the bacterium is single-stranded (Stewart and Carlson 1986). Second, this double-strand to single-strand conversion is mediated by one or more nucleases, which are often tightly associated with the DNA binding protein (Lacks 1977). Third, strand specificity for uptake has not been detected. This is to say that either strand seems to have an equal probability for penetration, while the other is usually degraded (Raina and Ravin 1980). Finally, steps are taken to ensure that the penetrating strand is protected from degradation by single-strand-specific nucleases in the cell. This may involve protection by single-stranded DNA binding proteins in the cytoplasm, or it may be the result of coordinate recombination and entry (Morrison 1977; Kahn and Smith 1984).

The means by which these events are accomplished can differ. Upon binding of exogenous DNA to the surface of competent cells, at least one strand of DNA is nicked, providing a site for processing to single-stranded DNA. Uptake of one strand is coincidental with the release of trichloroacetic acid-soluble components from the transforming DNA, resulting from the degradation of the second (non-penetrating) strand. In *S. pneumoniae* and *B. subtilis*, degradation of the second strand results in release of nucleotides to the exogenous environment (Lacks 1977). In *H. influenzae* and *H. parainfluenzae* the DNA binding protein is associated with membrane structures called transformasomes (Kahn *et al.* 1983). These transformasomes are the site of binding for exogenous DNA

and seem to allow the dsDNA to sequester within the transformasome, protecting the transforming DNA from both exogenous and endogenous nucleases. The single-strand conversion process occurs within the transformasome and entry occurs, presumably through a pore to the cytoplasm. Entry of the transforming strand allows immediate formation of a recombination complex for strand exchange (Kahn and Smith 1984).

Recombination

The recombination events in natural transformation involve generalized recombination functions for the most part (Dubnau *et al.* 1973). This integration event utilizes a *recA*-like homologous recombination system where the transported strand replaces its homologous region in the endogenote (de Vos and Venema 1983). While natural transformation has been most extensively studied for chromosomal markers, transformation of plasmid DNA has been demonstrated in cases where sufficient homology exists between the exogenous plasmid and the endogenote of the recipient cell. In the case of *P. stutzeri* the frequency of transformation is a function of the amount of the plasmid that is homologous to the endogenote, and linearization of the plasmid prior to uptake must occur in the homologous region to ensure stable establishment of the plasmid in the recipient cell (Carlson *et al.* 1984). Homologous sequences in the endogenote may exist on other extrachromosomal elements or they may be chromosomally linked. Thus, natural transformation may provide a means for horizontal transfer of both chromosomal and extrachromosomal elements in the environment.

Environmental factors that may influence natural transformation

When one considers the physiology of transformation, and especially of competence development, one can see that a number of factors could influence the transformation process in the field. The first two steps of the process – competence development and DNA binding – are particularly susceptible to environmental influence.

Concentration

For the competence factor-induced systems, the need to accumulate competence factor protein to a sufficient concentration to trigger competence development implies that both dilution effects (in reduction of competence factor below the critical concentration), and exogenous proteases (that could degrade competence factor) would prevent initiation of competence development and consequently would prevent transformation.

Environmental influences

Environmental influences could also affect competence development in those bacteria that regulate this developmental shift internally. The fact that changes in growth conditions, especially in nutrient concentration, can trigger the competence cycle has been shown for a number of bacteria. Competence is induced for *Haemophilus* species by shifting cells from a rich medium to a non-growth medium (Spencer and Herriott 1965). Competence is initiated in *Pseudomonas* (Carlson *et al.* 1983) and *Acinetobacter* (Juni 1972) by growing cultures to stationary phase. Competence development has been shown to be controlled by a catabolite repression-like phenomenon (Wise *et al.* 1973) and by starvation for iron (Page 1982). Thus, changes in the environment that influence nutrient concentration, that alter growth rates for bacteria or that result in flux of specific growth factors, could potentially induce competence. Another consideration is that induction of competence has only been characterized under laboratory growth conditions; conditions where most cultures can achieve and maintain exponential rates of growth for extended periods of time. Competence is usually repressed under exponential growth conditions (Stewart and Carlson 1986) and 'induction of competence' may therefore be an artificially-induced laboratory artifact. One presumes that rates of growth under most environmental conditions would more closely approximate a laboratory stationary phase (Brock and Madigan 1988) – the phase during which competence is normally expressed for many bacteria. Thus, competence may be a normal physiological state for certain bacteria in their native environments.

One would predict that DNA binding and initiation of uptake of DNA would also be susceptible to environmental influence. In virtually all transformable species, the ability to bind DNA can be prevented or reversed (at least early in the binding process) by high salt washes (Seto and Tomasz 1974). Thus, the ionic strength of the environment may influence proficiency for transformation. Additionally, concentrations of dissolved DNA and of RNA may affect transformation efficiency; these molecules have been shown to serve as competitive inhibitors for binding of homologous DNA to cell surfaces (Smith *et al.* 1981), although to varying degrees, depending on the system. Thus, the concentration of salts and nucleic acids in waters, sediments, soils and other environmental constituents, may play an important role in determining the potential for environmental gene transfer by natural transformation.

There is also compelling evidence that divalent cations, particularly magnesium ions, may be critical to the initiation of DNA uptake in transformation. One can prevent DNA uptake, but not DNA binding, by the addition of EDTA to competent cells (Seto and Tomasz 1976). This ability to block DNA uptake is a consequence of a required magnesium cofactor for one or more of the nucleases for the double-strand to single-strand conversion of DNA (Lacks and Neuberger 1975). Thus, total concentration of magnesium ions and the presence of natural chelating agents are sources for environmental influence in natural transformation.

Sources of DNA in the environment

The ability to transform is of little use to an organism if there is no exogenous DNA available for uptake. Natural environments contain significant amounts of DNA. Paul and colleagues have shown significant concentrations of dissolved DNA in both marine (DeFlaun *et al.* 1987) and freshwater environments (DeFlaun *et al.* 1986). Significant amounts of DNA have also been isolated from soils and sediments (Ogram *et al.* 1987; Steffen *et al.* 1988). The question of the source of this DNA has not been resolved, but at least reports in marine systems indicate that a substantial portion of the dissolved DNA is bacterial in origin (Paul *et al.* 1988).

The mechanism or mechanisms for release of DNA from cells is also unclear. For many years it was assumed that exogenous DNA available to cells for transformation was probably a result of random lysis events. However, several reports have indicated a controlled, perhaps intentional, release of DNA by some fraction of the bacterial population. In *B. subtilis*, cells appear to be 'prone to lysis' at the onset of competence development (Sinha and

Iyer 1971). This ensures that DNA release is coordinately controlled with competence development. A number of bacteria are also known to express genes for autolytic enzymes as they shift their growth to stationary phase (Brock and Madigan 1988) – the same physiological shift that induces competence in many bacteria.

The issue of DNA release by cell lysis is especially important as we consider the application of genetically-engineered micro-organisms (GEMs) to environmental tasks. This debate has focused its attention thus far on the release of GEMs and on the survival and persistence of these organisms in the environment. Concerns on the impact of such releases have been discussed (Curtiss 1988). A number of methods have been proposed to 'kill' GEMs once their task was completed. These include inhibition of energy production by the organism through phased expression of a transmembrane protein (Gerdes et al. 1986a,b). This, and other approaches that result in loss of viability, would probably lead to eventual lysis of the cell and release of the recombinant DNA to the environment. As a consequence, while the GEM has been controlled, the recombinant DNA still persists, at least transiently, and is available for natural transformation.

It is also possible that DNA may be released from cells without lysis, constituting a 'donor' function for natural transformation. In B. subtilis, ordered release of genetic markers has been detected in synchronous cultures. Transfer initiates near the origin of replication and progresses unidirectionally around the chromosome of the bacterium (Raina et al. 1979). This DNA served as a source of exogenous genetic material for competent recipient cells. In H. influenzae a 'cell-to-cell' genetic exchange event has been reported that is partially sensitive to DNase I, suggesting that the DNA is exogenous at some point in the process (Albritton et al. 1982). Gene transfer has also been reported within a colony of H. influenzae without the addition of exogenous DNA (Stuy 1985); donor functions have also been reported in P. stutzeri. In this case, frequencies of transformation obtained by mixing genetically distinct strains on a filter were orders of magnitude higher than frequencies obtained when exogenous DNA was provided to a recipient strain (Stewart et al. 1983). This process is not strain-specific, that is, all strains tested were capable of both donor and recipient function. Donor activity could not be correlated to the presence of any extrachromosomal element, was DNase I-sensitive and was inhibited by the additon of nalidixic acid, an inhibitor of DNA synthesis. It was not inhibited by the addition of the RNA synthesis inhibitor, rifampicin, nor by streptomycin, an inhibitor of protein synthesis (Stewart et al. 1983).

The capacity of intact cells to serve as reservoirs for genes and to allow those genes to disseminate through a population by cell contact transformation brings to task the question of the importance of exogenous nucleases in inhibiting or reducing transformation frequencies in the environment. Even if transformation were an infrequent event due to low levels of competence or by effective scavaging by nucleases, the rare occurrence of exogenous DNA being taken up by a competent cell and incorporated into the genome of that recipient ensures stable replication of that element. The genetic sequence could then be transferred, as is any chromosomal marker, by transduction or conjugation, or it could transfer to other recipients by a cell contact transformation mechanism.

The stability of DNA in the environment

If one wishes to predict potential for transformation in the environment it is not enough to consider only the uptake and expression of that DNA, one must also consider the fact that DNA is susceptible to fates other than transformation. The frequency of turnover by nucleases and the capacity of other materials in the environment to bind DNA are examples of alternative fates for exogenous DNA.

Nuclease activity

Evidence for nuclease activity in natural environments is extensive. Greaves and Wilson (1970) have shown that DNA and RNA are turned-over rapidly when added to non-sterile soils. DNA turnover rates are also high in marine environments and Paul et al. (1987) have shown that DNA is degraded very rapidly in marine environments. This information might suggest that the potential for transformation would be minimal in the field, but there are other factors to be considered. DNA is apparently protected from degradation when attached to sand (Lorenz and Wackernagel 1987) and when bound to marine sediments (Aardema et al. 1983), yet this DNA can contribute to horizontal gene transfer by natural transformation (Lorenz *et al.* 1988). Thus, binding of DNA to sediments, sands or other surfaces may protect the nucleic acid from turnover until it is acquired by competent cells in the environment. This ability to slow the rate of turnover of exogenous DNA by surface binding would become especially important if the state of competence were to fluctuate in transformable bacteria in their environments. Oscillation of competence is a distinct possibility based on laboratory studies on competence induction, given the many and varied factors that influence competence development.

The role of solid surfaces

A second factor to consider is the role of solid surfaces in facilitating the natural transformation process. *Pseudomonas stutzeri* shows higher frequencies of transformation for exogenous DNA when recipient cells and DNA are immbolized on membrane filters and placed on agar plates than when liquid culture is used (Carlson *et al.* 1983). This difference in transformation frequency is even more striking for cell contact transformation between intact donor and recipient cells (Stewart *et al.* 1983). In field simulations using *P. stutzeri* strain ZoBell as a model, frequencies of transformation in water column samples were below the limit of detection, while frequencies were easily detected when recipient cells were mixed with exogenous DNA in sterile and non-sterile sediments (Stewart *et al.* 1991). These studies also indicated that transformed cells were tightly associated with sediments. Transformants were not detected in column flow through or in high salt washes, but were measured when the sediments were extracted from the column and vortex-mixed. Thus, we may find that surface zones may be especially productive for horizontal movement of exogenous DNA.

Direct evidence for natural transformation in the field

There are a limited number of reports of transformation in the environment or in microenvironment simulations.

Soil

Graham and Istock (1978) reported transformation of *B. subtilis* in sterile soils. Frequencies of trans-

formation were lower than those obtained in liquid culture but the frequency of cotransfer of markers was considerably higher, suggesting that the sediments might stabilize the DNA/recipient complex, thus preventing interrupted transformation. DNase did not reduce genetic exchange frequencies. It was suggested that either adsorption of the enzyme to sediment components, or binding of the DNA to proteins or membranes prevented degradation by the nuclease. They also reported that calf thymus DNA failed to compete with *B. subtilis* DNA for transformation. Later reports (Graham and Istock 1979, 1981) discussed the role of transformation in natural selection and population genetics. It was suggested that the sediment transformation experiments revealed the potential for significant changes in population dynamics, especially under conditions of selection or counterselection. However, as discussed by Stotzky (1989), the value of sterile sediments as models for field transformation must be questioned. Rates of degradation of exogenous DNA are much greater in non-sterile soils than in γ-irradiated soil (Greaves and Wilson 1970), indicating that DNase from viable bacterial cells is probably more important than exogenous enzymes in DNA turnover in sediments. Evidence for transformation in non-sterile soils is only now forthcoming. There have been reports of transformation of amino acid biosynthetic genes in *B. subtilis* in sterile and non-sterile soils (Lee and Stotzky 1989). Additionally, recent experiments indicate that antibiotic resistance genes can also transfer in non-sterile soils (G. Stotzky, personal communication).

Aquatic environments

There is also limited evidence of transformation in aquatic environments. Again, using the Gram-positive bacterium *B. subtilis* as a model, transformation has been detected in sterile marine sediments (Aardema *et al.* 1983) and sterile sand (Lorenz and Wackernagel 1987). Addition of sand to marine sediments reduced the frequency of transformation in sediments in the absence of DNase, but significantly increased transformation frequencies in the presence of exogenous nuclease (Aardema *et al.* 1983). Furthermore, the presence of DNA adsorbed to sand seems to influence the binding of competent cells, and therefore the frequency of transformation. Approximately ten times as many competent cells were bound to sea sand pretreated with DNA than to sand that was not pretreated. The cells were apparently released following transformation (Lorenz *et al.* 1988), strongly suggesting that the interaction between the recipient cell and the sand grains was mediated by exogenous DNA.

Recent experiments in this laboratory using the Gram-negative marine isolate *P. stutzeri* strain ZoBell (Dohler *et al.* 1987) indicate that transformation occurs in both sterile and non-sterile marine sediments (Stewart and Sinigalliano 1990b). The composition of the sediment seems to influence efficiency of transformation, with those natural sediments containing higher organic content better supporting transformation. The role of sediment composition was further explored by the construction of synthetic sediments. Again, as the organic content of the sediment increased so did transformation frequency (Stewart and Sinigalliano 1990b). Transformation frequencies were reduced in sediments by the addition of DNase I, and transformation frequencies were significantly lower in non-sterile sediments than they were in sterile sediment systems. Frequencies of transformation were not detected using exogenous DNA or cell contact transformation when sterile or non-sterile marine waters were used for transformation assays. Thus, surface association appears to be important in facilitating transformation in marine and freshwater environments.

The likelihood of transformation in the environment

The lack of extensive investigations into natural transformation as an environmentally-significant process makes it difficult to predict the likelihood of horizontal movement of genetic information by natural transformation. It is possible to speculate on the significance the process in the field by considering what we know about the regulation of transformation in pure culture, controlled laboratory conditions and by extrapolation of the limited studies done thus far on DNA uptake and turnover in field simulations.

We know that transformable bacteria must undergo a developmental shift from the non-competent to the competent state (Smith *et al.* 1981; Stewart and Carlson 1986). The exception to this rule seems to be *Neisseria gonorrhoeae*, where competence is constitutive (Sparling 1966).

Although the physiology and genetics of this developmental shift have only been studied in a limited number of species, two general approaches to regulation of competence development appear to have evolved. In the case of Gram-positive organisms (at least for species of *Bacillus* and *Streptococcus*) a small protein is excreted and must accumulate to a sufficient concentration to initialize the development shift. One can postulate, then, that a process that is dependent on the accumulation of extracellular effector would be unlikely in an aqueous environment with a large volume and with free exchange of solutes. However, the sediment microenvironment in freshwater or marine habitats (Atlas and Bartha 1987) and the hydration zones that surround clay cutans (particle complexes), silt and sand in soils (Stotzky 1989) provide microhabitats with restricted exchange of solutes relative to water column environments. These restricted environments could favour the accumulation of a soluble competence factor, allowing induction or maintenance of competent cells. The alternative approach to induction of the competent state, common in a number of Gram-negative transformable species (Stewart and Carlson 1986) involves internal regulation of competence. While the trigger mechanisms can be quite different for various bacteria, most reflect some type of nutrient limitation, shift in growth rate (to slower growth) or other change that can induce, or is characteristic of, unbalanced growth. Nutrient concentrations are often quite limited in nature (relative to laboratory conditions) and bacterial growth rates are usually much slower in nature (Brock and Madigan 1988). These conditions would favour the types of growth changes associated with competence induction in Gram-negative organisms. Thus, the conditions found in natural environments would tend to support competence development. In fact, the non-competent state in many Gram-negative organisms may be a consequence of laboratory growth conditions – conditions that are highly unlikely in the field. While it is impossible to conclude that competent organisms exist in natural environments, our understanding of the control of competence in Gram-negative, transformable bacteria would suggest that the competent state was a distinct possibility for naturally-occurring organisms.

We can conclude that the process of transformation is only one alternative fate for exogenous DNA. The relatively high rates of DNA turnover detected in the environment would suggest that the residence time for dissolved DNA would be quite short. However, previously cited evidence indicates that binding of DNA to surfaces protects the nucleic acid from degradative enzymes and still allows DNA to be taken up and expressed by bacteria. As surfaces seem to facilitate the natural transformation process, at least in some organisms (Stewart *et al.* 1983), transformable cells bound to, or in association with, sediments or soils may be especially important in taking up and stabilizing exogenous DNA sequences.

The complication of cell contact transformation, or subsequent movement of transformed sequences by transduction, conjugation, or some other exchange process must also be considered. The combination of relatively low transformation frequencies in environmental simulations (Stewart and Sinigalliano 1990a,b; Stewart *et al.* 1991) and the activity of exogenous nucleases probably ensures that transformation of a marker will be a very low frequency event. However, should a sequence succeed in entry and establishment within a cell in the natural population, the ability of that bacterium to replicate its DNA, including the transformed sequence, allows a means for amplification of the sequence. Once established in the endogenote of a recipient cell, this sequence could then transfer by any means into other cells in the population. As transduction, conjugation and related processes are DNase-resistant, extracellular nucleases are ineffective in inhibiting horizontal gene flow. Even transmission by cell contact transformation might gain from an initial low-frequency transformation event, and reports of partial DNase resistance have been reported (Albritton *et al.* 1982). Additional factors such as endonuclease restriction barriers, species segregation and fluctuation in growth rates would also become less important in restricting horizontal gene transfer into a native bacterium.

Another important consideration in determining the stabilization of transformed sequences is the impact of expression of any genes encoded by those sequences. If a transformed marker provided a selective advantage for the transformant relative to non-transformed cells in the population, the transformed cells could increase their numbers relative to the total population of bacteria, thus amplifying the sequence. Examples of sequences that could provide a selective advantage might be genes that extend the nutritional diversity of the bacterium or that encode resistance to inhibitors present in the environment. Likewise, if a transformed gene re-

duced the fitness of the bacterium, by slowing the growth rate of the transformed cell or by increasing the susceptibility of the cell to death, the transformed species would reduce its number relative to the total bacterial population. Thus, the phenotype conferred by expression of a transformed sequence may play an important role in the persistence of the sequence. Population changes would influence frequency for a given sequence in the total genome of the community, thus influencing the probability of subsequence transfer events.

The composition of the microenvironment would be expected to influence the probability of transformation. The degree of binding of nucleic acids to sediments and other surfaces influences both survival of the sequence (Lorenz *et al.* 1981; Aardema *et al.* 1983;) and its availability for uptake by competent cells (Lorenz *et al.* 1988). The concentration of cations in the environment may influence initial binding of DNA to transformable cells (Seto and Tomasz 1974) and availability of divalent cations influences the ability of a cell to take up bound DNA (Seto and Tomasz 1976). Thus, changes in the environment, including pH, ionic strength, volume to surface ratios, etc. may influence the potential for horizontal transfer of DNA by natural transformation.

Concluding remarks

Although transformation has been studied as a laboratory phenomenon for more than 50 years, we still do not understand its importance, if any, as a means for horizontal movement of genetic sequences in the environment. A number of areas of information must be filled before we can make intelligent predictions of the potential for natural transformation in the field. First, we must broaden the number of model systems where the physiology and genetics of natural transformation are extensively studied. Our general models for the process of natural transformation are based on a very limited pool of organisms. A more extensive study should provide a better basis for predicting the influence of environmental factors on the transformation process. Second, further research is needed on the turnover of DNA and the role of environmental factors in enhancing turnover and in DNA stabilization. Additionally, the role of surfaces and sediments in facilitating gene transfer should be explored. Finally, more extensive environmental simulations of both aquatic and soil environments are needed to monitor acquisition and expression of exogenous DNA in both introduced species and native organisms.

The limited information currently available to us allows certain predictions. First, environmental conditions probably support competence development and uptake of exogenous DNA by transformable bacteria. Second, certain environments, sediments and soils for example, are more likely to support natural transformation than other environments, i.e. water column. Thus, sediments, soils and perhaps surface interfaces are potential microenvironments for natural transformation. Third, binding of DNA to sediments, sands, etc. probably increases the stability of the nucleic acid in the environment, therefore increasing the probability of its eventual uptake. Finally, even low-frequency transfer of exogenous DNA may be sufficient to impact the natural community because cell contact transformation, conjugation, transduction and other gene transfer events could occur from transformed cells as subsequent events.

At the moment it is impossible to state whether or not transformation occurs in the environment at a significant rate, or to predict its relative importance in horizontal gene transfer. However, one can conclude that environmental conditions are certainly capable of supporting natural transformation and that there is significant potential for natural transformation to play an important role in the horizontal movement of genes in the environment.

Acknowledgements

The author would like to thank Drs J.H. Paul, and W.H. Jeffrey, G. Stotzky, M. Day and J. Fry for discussions that contributed to the content of this paper.

References

Aardema, B.W., Lorenz, M.G., and Krumbein, W.E. (1983). Protection of sediment-absorbed transforming DNA against enzymatic inactivation. *Applied and Environmental Microbiology*, **46**, 417–20.

Albritton, W.L., Setlow, J.K., and Slaney, L. (1982). Transfer of *Haemophilus influenzae* chromosomal genes by cell-to-cell contact. *Journal of Bacteriology*, **152**, 1066–70.

Atlas, R.M. and Bartha, R. (1987). *Microbial ecology: fundamentals and applications*. Benjamin/Cummings Publishing Company, Menlo Park, California.

Avery, A.T.C., MacLeod, M., and McCarty, M. (1944). Studies on the chemical nature of the substance inducing transformation in pneumococcal types. *Journal of Experimental Medicine*, **79**, 137–59.

Bale, M.J., Fry, J.C., and Day, M.J. (1988). Transfer and occurrence of large mercury resistance plasmids in river epilithon. *Applied and Environmental Microbiology*, **54**, 972–8.

Brock, T.D., and Madigan, M.T. (1988). *Biology of microorganisms*. Prentice Hall, Englewood Cliffs, New Jersey.

Carlson, C.A., Pierson, L.S., Rosen, J.J., and Ingraham, J.L. (1983). *Pseudomonas stutzeri* and related species undergo natural transformation. *Journal of Bacteriology*, **153**, 93–9.

Carlson, C.A., Steenbergen, S.M., and Ingraham, J.L. (1984). Natural transformation of *Pseudomonas stutzeri* by plasmids that contain cloned fragments of chromosomal DNA. *Archives of Microbiology*, **140**, 134–8.

Ceglowski, P., Kawczynski, M., and Dobrzanki, W.T. (1980). Purification and properties of deoxyribonucleic acid binding factor isolated from the surface of *Streptococcus sanguis* cells. *Journal of Bacteriology*, **141**, 1005–14.

Clewell, D.B., and Gawron-Burke, C. (1986). Conjugative transposons and the dissemination of antibiotic resistance in streptococci. *Annual Review of Microbiology*, **40**, 635–59.

Concino, M.F., and Goodgal, S.H. (1981). *Haemophilus influenzae* polypeptide involved in DNA uptake detected at a cellular surface protein iodination. *Journal of Bacteriology*, **148**, 220–31.

Curtiss, R. (1988). Engineering organisms for safety: What is necessary? In *The release of genetically-engineered micro-organisms*, (Ed. M. Sussman, C.H. Collins, F.A. Skinner, and D.E. Stewart-Tull), pp. 7–20. Academic Press, London.

DeFlaun, M.E., Paul, J.H., and Davis, D. (1986). Simplified method for dissolved DNA determination in aquatic environments. *Applied and Environmental Microbiology*, **52**, 115–17.

DeFlaun, M.E., Paul, J.H., and Jeffrey, W.H. (1987). Distribution and molecular weight of dissolved DNA in subtropical estuarine and oceanic environments. *Marine Ecology Progress Series*, **38**, 65–73.

de Vos, W.M., and Venema, G. (1983). Transformation of *Bacillus subtilis* competent cells: identification of the *recE* gene product. *Molecular and General Genetics*, **181**, 424–33.

Dohler, K., Huss, V.A.R., and Zumft, W.G. (1987). Transfer of *Pseudomonas perfectomarina* Baumann, Bowditch, Baumann and Beaman 1983 to *Pseudomonas stutzeri* (Lehmann and Neuman, 1896) Sijderius 1946.

International Journal of Systematic Bacteriology, **37**, 1–3.

Dooley, D.C., Hadden, C.T., and Nester, E.W. (1971). Macromolecular synthesis in *Bacillus subtilis* during development of the competent state. *Journal of Bacteriology*, **108**, 668–79.

Dubnau, D., Davidoff-Abelson, R., Scher, B., and Cirigiano, C. (1973). Fate of transforming deoxyribonucleic acid after uptake by competent *Bacillus subtilis*: phenotypic characterization of radiation-sensitive recombination-deficient mutants. *Journal of Bacteriology*, **114**, 273–86.

Ephrati-Elizur, E. (1968). Spontaneous transformation in *Bacillus subtilis*. *Genetic Research*, **11**, 83–96.

Gealt, M.A., and Chai, M.D., Alpert, K.B., and Boyer, J.C. (1985). Transfer of plasmids pBR322 and pBR325 in wastewater from laboratory strains of *Escherichia coli* to bacteria indigenous to the waste disposal system. *Applied and Environmental Microbiology*, **49**, 836–41.

Genthner, F.J., Chatteerjee, P., Barkay, T., and Bourquin, A.W. (1988). Capacity of aquatic bacteria to act as recipients of plasmid DNA. *Applied and Environmental Microbiology*, **54**, 115–17.

Gerdes, K., Bech, F.W., Jorgensen, S.T., Lobner-Olesen, A., Rasnussen, P.B., Atlung, T., Boe, L., Karlstrom, O., Molin, S., and Von Meyenburg, K. (1986a). Mechanism of postsegregational killing by the *hok* gene product of the *parB* system of plasmid R1 and its homology with the *relF* gene product of the *E. coli relB* operon. *EMBO Journal*, **5**, 2023–9.

Gerdes, K., Rasmussen, P.B., and Molin, S. (1986b). Unique types of plasmids maintenance function: Postsegregation killing of plasmid-free cells. *Proceedings of the National Academy of Sciences (USA)*, **83**, 3116–20.

Goodgal, S.H. (1982). DNA uptake in *Haemophilus* transformation. *Annual Review of Genetics*, **16**, 169–92.

Graham, J.B., and Istock, C.A. (1978). Genetic exchange in *Bacillus subtilis* in soil. *Molecular and General Genetics*, **166**, 287–90.

Graham, J.B., and Istock, C.A. (1979). Gene exchange and natural selection cause *Bacillus subtilis* to evolve in soil cultures. *Science*, **204**, 637–9.

Graham, J.B., and Istock, C.A. (1981). Parasexuality and microevolution in experimental populations of *Bacillus subtilis*. *Evolution*, **36**, 954–63.

Graves, J.F., Biswas, G.D., and Sparling, P.F. (1982). Sequence-specific DNA uptake in transformation of *Neisseria gonorrhoeae*. *Journal of Bacteriology*, **152**, 1071–7.

Greaves, M.P., and Wilson, M.J. (1970). The degradation of nucleic acids and montmorillonite-nucleic acid complexes by soil microorganisms. *Soil Biology and Biochemistry*, **2**, 257–68.

Juni, E. (1972). Interspecies transformation of *Acinetobacter*: genetic evidence for a ubiquitous genus. *Journal of Bacteriology*, **112**, 917–31.

Kahn, M.E., and Smith, H.O. (1984). Transformation in *Haemophilus*: a problem in membrane biology. *Journal of Membrane Biology*, **81**, 89–103.

Kahn, M., Concino, M., Gromkova, R., and Goodgal, S.H. (1979). DNA binding activity of vesicles produced by competence-deficient mutants of *Haemophilus*. *Biochemical and Biophysical Research Communications*, **87**, 764–72.

Kahn, M.E., Barany, F., and Smith, H.O. (1983). Transformasomes: specialized membrane structures that protect DNA during *Haemophilus* transformation. *Proceedings of the National Academy of Sciences (USA)*, **80**, 6927–31.

Kooistra, J., van Boxel, T., and Venema, G. (1980). Deoxyribonucleic acid binding properties and membrane protein composition of a competence-deficient mutant of *Haemophilus influenzae*. *Journal of Bacteriology*, **144**, 22–7.

Lacks, S.A. (1977). Binding and entry of DNA in bacterial transformation. In *Microbial interactions, receptors, and recognition*, (Ed. J.L. Reissig), pp. 179–232. Chapman and Hall, London.

Lacks, S.A., and Neuberger, M. (1975). Membrane location of a deoxyribonuclease implicated in the genetic transformation of *Diplococcus pneumoniae*. *Journal of Bacteriology*, **124**, 1321–9.

Lacks, S., Greenberg, B., and Neuberger, M. (1974). Role of deoxyribonuclease in the genetic transformation of *Diplococcus pneumoniae*. *Proceedings of the National Academy of Sciences USA*, **71**, 2305–9.

Lee, G.H., and Stotzky, G. (1989). Transformation is a mechanism of gene transfer in soil. (Abstract.) *Wind river conference on genetic exchange*, Estes Park, Colorado.

Lorenz, M.G., and Wackernagel, W. (1987). Adsorption of DNA to sand and variable degradation rates of adsorbed DNA. *Applied and Environmental Microbiology*, **53**, 2948–52.

Lorenz, M.G., Aardema, B.W., and Krumbein, W.E. (1981). Interaction of marine sediment with DNA and DNA availability to nucleases. *Marine Biology*, (Berlin) **64**, 225–30.

Lorenz, M.G., Aardema, B.W., and Krumbein, W.E. (1988). Highly efficient genetic transformation of *Bacillus subtilis* attached to sand grains. *Journal of General Microbiology*, **134**, 107–12.

Low, K.B., and Porter, D.D. (1978). Modes of gene transfer and recombination in bacteria. *Annual Review of Genetics*, **12**, 249–87.

Morrison, D.A. (1977). Transformation in pneumococcus: existence and properties of a complex involving donor deoxyribonucleate single strands in eclipse. *Journal of Bacteriology*, **132**, 576–83.

Morrison, D.A., Mannerelli, B., and Vijayakumar, M.N. (1982). Competence for transformation in

Streptococcus pneumoniae: an inducible high capacity system for genetic exchange. In *Microbiology – 1982*, (Ed. D. Schlessinger), pp. 136–8. American Society for Microbiology, Washington DC.

Ogram, A., Sayler, G.S., and Barkay, T. (1987). The extraction and purification of microbial DNA from sediments. *Journal of Microbial Methods*, 7, 57–66.

Page, W.J. (1982). Optimal conditions for competence development in nitrogen-fixing *Azotobacter vinelandii*. *Canadian Journal of Microbiology*, 28, 389–97.

Paul, J.H., Jeffrey, W.H., and DeFlaun, M.E. (1987). Dynamics of extracellular DNA in the marine environment. *Applied and Environmental Microbiology*, 53, 170–9.

Paul, J.H., DeFlaun, M.F., and Jeffrey, W.H. (1988). Mechanisms of DNA utilization by estuarine microbiol populations. *Applied and Environmental Microbiology*, 54, 1682–8.

Raina, J.L., and Ravin, A.W. (1980). Switches in macromolecular synthesis during induction of competence for transformation of *Streptococcus sanguis*. *Proceedings of the National Academy of Sciences (USA)*, 77, 6062–6.

Raina, J.L., Metzer, E., and Ravin, A.W. (1979). Translocation of the presynaptic complex formed upon DNA uptake by *Streptococcus sanguis* and its inhibition by ethidium bromide. *Molecular and General Genetics*, 170, 249–59.

Saye, D.J., Ogunseitan, O., Salyer, G.S., and Miller, R.V. (1987). Potential for transduction of plasmids in a natural freshwater environment: effect of plasmid donor concentration and a natural community on transduction in *Pseudomonas aeruginosa*. *Applied and Environmental Microbiology*, 53, 987–95.

Scocca, J.J., Poland, R.L., and Zoon, K.C. (1974). Specificity in deoxyribonucleic acid uptake by transformable *Haemophilus influenzae*. *Journal of Bacteriology*, 118, 369–73.

Seto, H., and Tomasz, A. (1974). Early stages in DNA binding and uptake during genetic transformation of pneumococci. *Proceedings of the National Academy of Sciences (USA)*, 71, 1493–8.

Seto, H., and Tomasz, A. (1976). Calcium-requiring step in the uptake of deoxyribonucleic acid molecules through the surface of competent pneumococci. *Journal of Bacteriology*, 126, 1113–18.

Sinha, R.A., and Iyer, V.N. (1971). Competence for genetic transformation and the release of DNA from *Bacillus subtilis*. *Biochimica et Biophysica Acta*, 232, 61–71.

Smith, H.O., Danner, D.B., and Deitch, R.A. (1981). Genetic transformation. *Annual Review of Biochemistry*, 50, 41–68.

Solioz, M., Yen, H.C., and Marrs, B. (1975). Release and uptake of gene transfer agent by *Rhodopseudomonas capsulata*. *Journal of Bacteriology*, 123, 651–6.

Sparling, P.E. (1966). Genetic transformation of *Neisseria gonorrhoeae* to streptomycin resistance. *Journal of Bacteriology*, 92, 1364–71.

Spencer, H.T., and Herriott, R.M. (1965). Development of competence of *Haemophilus influenzae*. *Journal of Bacteriology*, 90, 911–20.

Steffen R.J., Goksøyr, J., Bej, A.K., and Atlas, R.M. (1988). Recovery of DNA from soils and sediments. *Applied and Environmental Microbiology*, 54, 2908–15.

Stewart, G.J. (1989). The mechanism of natural transformation. In *Gene transfer in the environment*, (Ed. S.B. Levy and R.V. Miller), pp. 139–64. McGraw-Hill Publishing Company, New York.

Stewart, G.J., and Carlson, C.A. (1986). The biology of natural transformation. *Annual Review of Microbiology*, 40, 211–35.

Stewart, G.J., and Sinigalliano, C.D. (1990a). Detection and characterization of natural transformation in the marine bacterium *Pseudomonas stutzeri* strain ZoBell. *Archives of Microbiology*, 152, 520–6.

Stewart, G.J., and Sinigalliano, C.D. (1990b). Detection of horizontal gene transfer by natural transformation in native and introduced species of bacteria in marine and synthetic sediments. *Applied and Environmental Microbiology*, 56, 1818–24.

Stewart, G.J., Carlson, C.A., and Ingraham, J.L. (1983). Evidence for an active role of donor cells in natural transformation in *Pseudomonas stutzeri*. *Journal of Bacteriology*, 156, 30–5.

Stewart, G.J., Sinigalliano, C.D., and Garko, K.A. (1991). Binding of exogenous DNA to marine sediments and the effect of DNA/sediment binding on natural transformation of *Pseudomonas stutzeri* strain ZoBell in sediment columns. *FEMS Microbiology Ecology*, 85, 1–8.

Stotzky, G. (1989). Gene transfer among bacteria in soil. In *Gene transfer in the environment*, (Ed. S.B. Levy and R.V. Miller), pp. 165–222. McGraw-Hill Publishing Company, New York.

Stuy, J.H. (1985). Transfer of genetic information within a colony of *Haemophilus influenzae*. *Journal of Bacteriology*, 162, 1–4.

Trevors, J.T., and Starodub, M.E. (1987). R-plasmid transfer in non-sterile agricultural soil. *Systematic and Applied Microbiology*, 9, 312–15.

Trevors, J.T., Barkay, T., and Bourquin, A.W. (1987). Gene transfer among bacteria in soil and aquatic environments: a review. *Canadian Journal of Microbiology*, 33, 191–8.

Tomasz, A. (1966). Model for the mechanism controlling the expression of competent state in *Pneumococcus* cultures. *Journal of Bacteriology*, 94, 562–70.

Tomasz, A. (1970). Cellular metabolism in genetic transformation in pneumococci: Requirement for protein synthesis during induction of competence. *Journal of Bacteriology*, 101, 860–71.

Venema, G. (1979). Bacterial transformation. *Advances*

in Microbial Physiology, **19**, 245–331.

Wise, E.M., Alexander, S.P., and Powers, M. (1973). Adenosine 3′, 5′-cyclic monophosphate as a regulator of bacterial transformation. *Proceedings of the National Academy of Sciences (USA)*, **70**, 471–5.

Zeph, L.R., and Stotzky, G. (1989). Use of biotinylated DNA probe to detect bacteria transduced by

bacteriophage P1 in soil. *Applied and Environmental Microbiology*, **55**, 661–5.

Zeph, L.R., Onaga, M.A., and Stotzky, G. (1988). Transduction of *Escherichia coli* by bacteriophage P1 in soil. *Applied and Environmental Microbiology*, **54**, 1731–7.

7

Lethal genes in biological containment of released micro-organisms

Stephen M. Cuskey

Introduction

RECENT ADVANCES in *in vitro* DNA manipulation have encouraged industrial and academic desires for the release of microbial strains to the environment. This, in turn, has generated concern, within both the scientific community and the general public, about the unintended consequences of such releases. These concerns centre around survival of released organisms and their supposed potential ability to disrupt ecological processes, as has been evident with the introduction of macrobiological species into new habitats (Obukowicz *et al.* 1986). Additionally, released micro-organisms may be able to disseminate their 'genetically engineered' DNA to the indigenous population, which again may alter microbially-mediated ecological processes. These debates mirror those that occurred when we first began to 'engineer' bacteria and other organisms through *in vitro* DNA manipulation. At that time, considerable effort was made to construct organisms and cloning systems that were considered 'safe' (Armstrong *et al.* 1977; Blattner *et al.* 1977; Curtiss *et al.* 1977; Levine *et al.* 1983), and that would ensure that altered strains would not survive accidental environmental release.

Safeguards

Safeguards used to ensure the impossibility of accidental environmental release and survival bacteria and cloning vectors incapable of replication and/or transfer except under well-defined laboratory conditions.

Such solutions have been adapted in several cases for organisms deliberately introduced into the environment. For example, removal of the outer protein coat from a strain of the baculovirus, *Autographa californica* reduced the long-term viability of virus particles released as biological pest control agents (Bishop, 1988; see Chapter 11). The potential for conjugal plasmid transfer to the indigenous population has also been reduced through the deletion of genes required for self-transfer (Jones *et al.* 1988); the use of small, non-self-transmissible, poorly-mobilizable plasmids as cloning vectors; and by placement of the novel genetic information on the bacterial chromosome (Obukowicz *et al.* 1986).

The above-mentioned strategies may be useful for the reduction of gene transfer to the native populations but the use of debilitated hosts does not adequately address the problem of survival of organisms intentionally released to the environment. Obviously, intentionally released strains must be able to survive and compete successfully with indigenous populations, at least for a while, to perform their desired function. Concerns about the long-term survival of these organisms may be assuaged, however, if they contain a conditionally-lethal phenotype, which kills the released cells at a time chosen by the investigator. This could theoretically satisfy the requirement for both the survival under defined environmental conditions and for containment of released strains. These phenotypes may be constructed by placing lethal genes under the control of promoters, which are activated under certain environmental conditions and would trigger cell death when those conditions are met. Such systems may be considered superior to traditional conditional lethal mutations (e.g. temperature sensitivity) because they can be quickly transferred *in vitro* between different bacterial hosts (Cuskey 1990).

Lethal genes

Many bacterial genes are potentially lethal if expressed at altered levels or times in the cell life-cycle or without expression of concomitant 'protecting' genes. These include, but are not limited to, genes encoding bacteriocins, restriction endonucleases and the numerous lethal genes found on large, low-copy-number plasmids (Figurski *et al.* 1982; Ogura and Hiraga 1983; Winans and Walker 1985; Gerdes *et al.* 1986). These last genes have been proposed most often for use in biological containment schemes and will be the focus of the remainder of this chapter. The function of these plasmid-encoded lethal genes, which cannot be cloned unless the cell contains a concomitant 'protecting' gene, is not known but they may be associated with plasmid replication and maintenance. For example, the *hok* (*host killing*) gene located on plasmid R1 encodes a protein of 52 amino acids, which disrupts the cell membrane electrical potential and results in cell death. Plasmid R1 also contains the protecting *sok* (*suppression of killing*) gene, which produces antisense RNA that binds to *hok* messenger RNA preventing protein synthesis (Gerdes *et al.* 1986). Due to the greater stability of *hok* RNA, daughter cells not receiving a copy of R1 at cell division will die as a result of *hok* expression (Gerdes *et al.* 1988). The *hok* gene has been shown to be lethal to a wide variety of bacterial cells (Molin *et al.* 1987) and, thus, is an attractive candidate for use in biological containment strategies. The selection of a particular lethal gene for containment, however, is less important than the selection of the regulatory DNA that controls its expression upon environmental release. By analogy, the interesting feature of the use of *lac* gene fusions in exploring regulation of gene expression is not the *lac* reporter gene but the regulatory sequences placed upstream of it that control β-galactosidase expression. Consequently, the next sections of the chapter will detail schemes involving alternative control of lethal gene expression and their application to biological containment of released micro-organisms.

Biological containment with the R1 *hok* gene

As stated above, the small size and broad host range lethality of the R1 *hok* gene make it an attractive candidate in biological containment schemes. Several groups have constructed, and tested in the laboratory, model containment plasmids based on alternate control of *hok* expression. For example, Molin *et al.* (1987), at the Technical University of Denmark, have constructed plasmid pKG345, which contains the *hok* gene downstream from the lambda P_R promoter, and the *c*1857 temperature-sensitive repressor gene. Cells containing this plasmid grow at normal rates at 30°C but show decreased viability at 42°C, when the repressor is inactivated. As environmental temperatures rarely reach 42°C, the main use of this construct is to show the ability of *hok* to control cell populations *in vitro*. Another construct, pPKL100 contains the *hok* gene downstream from the *Escherichia coli fim*A promoter (Molin *et al.* 1987). This invertible promoter (Abraham *et al.* 1985), responds to the ratio of two regulatory proteins encoded by the *fim*E and *fim*B genes (Klemm 1986) and stochastically activates *hok* gene expression through inversion of the promoter site (Molin *et al.* 1987). Cells with this plasmid, therefore, show increased doubling times compared with control strains because *hok* is randomly activated in individual cells. Doubling times may be increased or decreased through manipulation of the expression of the two regulatory genes. This sort of strategy, based on random cell killing, may prove useful in containment of organisms released to the environment.

Ronald Atlas and colleagues, at the University of Louisville, have also used the *hok* gene for construction of a model biological containment system (Table 7.1) designed to test the ability to contain released organisms (Bej *et al.* 1988). Plasmid pBAP19*h* contains the *hok* gene downstream from the well-characterized *lac* promoter. Cells with this plasmid must also contain the *lac*I^q gene to prevent uninduced read-through from the strong *lac* promoter to the *hok* gene. Reduced viability of transformed cells is seen when the gratuitous *lac* inducer (derepressor) IPTG is added to the culture medium. This model containment strategy is based

Table 7.1. *Biological containment systems based on lethal bacterial genes*

Plasmid	Origin of replication	Promoter	Lethal gene	Regulatory gene	Lethal signal	Source[a]
pKG345	ColE1	lambda P_R	hok	cl857	temperature	1
pPKL100	ColE1	fimA	hok	fimE, fimB	random inversion of the fimA promoter	1
pEPA70	ColE1	OP2	kilA	xylS	benzoate	2
pEPA88	ColE1, IncPII	OP2	kilA	xylS	benzoate	2
pEPA95	ColE1	OP2	hok	xylS	benzoate	2
pEAP112	ColE1, IncPII	OP2	hok	xylS	benzoate	2
pBAP19h	ColE1	lac	hok	lacI	IPTG	3

[a] Plasmids are from the laboratories of: 1, Soren Molin, the Technical University of Denmark, Copenhagen (Molin *et al*. 1987); 2, Stephen Cuskey, United States Environmental Protection Agency, Environmental Research Laboratory, Gulf Breeze, Florida (unpublished results); 3, Ronald Atlas, the University of Louisville, Louisville, Kentucky (Bej *et al*. 1988). Further details are provided in the text.

on activation of a lethal gene upon addition of an innocuous chemical to the environment to kill released organisms, though it is unlikely that IPTG would be used for this purpose in an actual release.

Biological containment with the RK2 *kil*A gene

Another plasmid-borne lethal gene tested in biological containment schemes is the *kil*A gene from plasmid RK2. This gene is less well characterized than *hok* but it is lethal to *Escherichia coli* cells, which are normally protected from its action through expression of a protein encoded by the *kor*A (kill override) gene (Figurski *et al*. 1982). I have utilized the *kil*A gene in the construction of several plasmids designed for biological containment of released micro-organisms. Plasmids pEPA70 and pEPA88 (Table 7.1) both contain the *kil*A gene downstream from the OP2 operator-promoter region (Mermod *et al*. 1984; Cuskey and Sprenkle 1988) of the TOL plasmid pWWO. This catabolic plasmid carries genes responsible for the degradation of toluene and related aromatic hydrocarbons and their aromatic acid metabolic intermediates. The OP2 region activates downstream genes in the presence of inducing aromatic acids such as benzoate (Inouye *et al*. 1981; Mermod *et al*. 1987) and either the product of the plasmid *xyl*S regulatory gene or a chromosomal gene putatively involved in regulation of chromosomally-encoded benzoate metabolism (Cuskey and Sprenkle 1988). Plasmid pEPA70 replicates in enteric bacteria and pEPA88 will also replicate in pseudomonads; cells containing either are sensitive to additions of

benzoate to the growth medium. My results with *Pseudomonas aeruginosa*, however, suggest that the *kil*A gene product causes a non-lethal cessation of growth, which may be reversed if OP2 promoter activity is repressed. The containment strategy using the above plasmids is based on addition of benzoate to the environment but the proposed non-lethality of *kil*A in non-enteric hosts may alter the utility of the gene for this purpose (see below).

Problems with biological containment

The above preliminary attempts at construction of biological containment systems illustrate several of the problems in controlling organisms intentionally released to the environment. Population killing is incomplete with all of the constructs used to date (Molin *et al*. 1987; Bej *et al*. 1988; Cuskey *et al*. 1990), as survivors that have lost the containment plasmid, or that have deleted, mutated or rearranged plasmids that no longer confer a conditional lethal phenotype on the host cells are readily isolated. Genetic instability may be overcome through the use of two or more containment systems in a cell and through placement of the system(s) on the chromosome. However, instability may be less of a problem when cells are incubated under environmental conditions (Bej *et al*. 1988). Experiments with cells transformed with pBAP19*h* incubated in a soil microcosm showed that the relatively few survivors of IPTG additions contained mutations that rendered them resistant to the action of the *hok* gene, but that the survivors were also incapable of normal cell growth.

In addition to the problem of genetic instability, most of the above attempts at biological control of released organisms rely on the addition of normally innocuous chemicals to the growth medium for control. This may suffice if the released organisms are not transported beyond the release site but may result in incomplete containment of organisms not receiving the lethal signal. This can be solved through use of a strategy employing the invertible *fim*A promoter as proposed by Molin and co-workers (1987). Another approach is to utilize the 'protecting' and lethal genes in a containment strategy. In this strategy, the lethal gene would be expressed from its own promoter or from another constitutive promoter. The protecting gene would be placed under control of a promoter recognizing an environmental signal, such as the presence of a toxic compound. A potential 'suicide' plasmid based on this concept is presented in Fig. 7.1. Cells carrying this construct would survive in the presence of the toxic compound but would be unprotected from the action of the lethal gene after its degradation (Cuskey and Bourquin 1987). The use of a non-lethal gene like *kil*A in this type of construction could potentially allow toxic waste-degrading bacteria to lie dormant in the environment until a toxic chemical is encountered, but allow for a form of containment in its absence. The 'suicide' plasmid shown in Fig. 7.1 is designed to allow cell survival in the presence of toxic chemicals but could potentially be constructed to respond to a variety of environmental signals. Examples might include temperature, light or the presence of other, non-toxic chemicals such as nitrate, sulphate, etc.

Finally, the components of containment systems outlined in Table 7.1 reveal that most of them originate and are constructed and tested in *E. coli*. This is probably due to the huge preponderance of information available on this organism, but containment strategies based on *E. coli* promoters, regulatory and lethal genes may not function outside this organism. For example, *E. coli* genes and promoter sequences are not efficiently expressed in *Pseudomonas* (Jeenes *et al.* 1986). Thus, biological containment strategies developed in organisms other than *E. coli* may require investigation into conditions regulating gene expression in those organisms.

Gene transfer

There is another potential use for lethal bacterial genes in organisms designed for environmental release, i.e. for counter-selection of donors in gene transfer experiments. Most studies designed to measure gene transfer rates to the indigenous population merely illustrate the potential for transfer because laboratory-bred and genetically marked strains are added to the test system as recipients. This is done to aid donor counter-selection, which is necessary to detect the relatively small number of recipients that has acquired new genetic capabilities from the added donor. In the laboratory gene exchange experiments, donor organisms are usually counter-selected by making them auxotrophic and selecting for recipients on minimal medium. Conversely, recipients are made resistant to a particular antibiotic, which is added to the selection medium to kill donor cells. Neither strategy is satisfactory when dealing with a heterogenous recipient population. Introduction of auxotrophic mutations may alter the survival of donor strains, thus artificially lowering transfer rates, and some potential recipients may not grow on minimal medium. In addition, it is obviously impractical to make all the potential recipients antibiotic resistant. Inclusion of a conditional lethal determinant to the donor may allow for counter-selection in these experiments. Plasmids pEPA95 and pEPA112 (Table

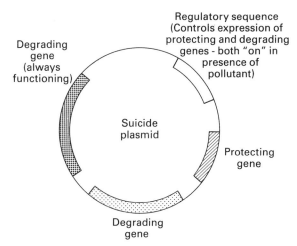

Fig. 7.1. Model of a biological containment plasmid based on constitutive expression of a lethal gene. Expression of the concomitant 'protecting' gene is dependent on the presence of a toxic chemical. Cells carrying such a construct will survive until the toxic chemical is degraded. See text for details.

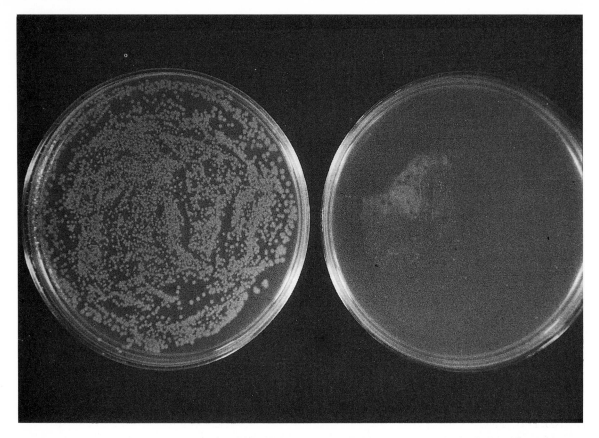

Fig. 7.2. Cells of *Escherichia coli* carrying plasmid pEPA91 (contains the *xyl*S regulatory gene on an IncQ plasmid) were transformed with plasmid pEPA95 carrying the lethal *hok* gene downstream from the TOL plasmid OP2 operator–promoter region. Cells were inoculated onto L-agar (with ampicillin for selection of pEPA95-bearing cells) with (right) and without (left) 0.05 per cent benzoate. Benzoate-dependent, *xyl*S-dependent expression of the lethal *hok* gene is seen. See text for details.

7.1) contains the *hok* gene downstream from the TOL plasmid OP2 promoter (see above for a discussion of this promoter). Cells transformed with either plasmid do not grow on L-broth with 0.05 per cent benzoate (when the *xyl*S gene is also present and thus may be useful as donors in gene transfer experiments (Fig. 7.2). The few survivors seen on the benzoate-containing medium have lost the *xyl*S regulatory plasmid. Thus, placement of this lethal contruct on the chromosome may be necessary to increase stability and further enhance its utility in donor counter-selection.

References

Abraham, J.M., Freitag, C.S., Clements, J.R., and Eisenstein, B.I. (1985). An invertible element of DNA controls phase variation of type fimbriae of *Escherichia coli*. *Proceedings of the National Academy of Sciences (USA)*, **82**, 5724–7.

Armstrong, K.A., Hershfield, V., and Helinski, D.R. (1977). Gene cloning and containment properties of plasmid ColE1 and its derivatives. *Science*, **196**, 172–4.

Bej, A.K., Perlin, M.H., and Atlas, R.M. (1988). Model suicide vector for containment of genetically engineered microorganisms. *Applied and Environmental Microbiology*, **54**, 2472–7.

Bishop, D.H.L. (1988). The release into the environment of genetically engineered viruses, vaccines and viral pesticides. *Trends in Biotechnology*, **6**, 512–15.

Blattner, F.R., Williams, B.G., Blechl, A.E., Denniston-Thompson, K., Faber, H.E., Furlong, L.-A., Grunwald, D.J., Kiefer, D.O., Moore, D.D., Schumm, J.W., Sheldon, E.L., and Smithies, O. (1977). Charon phages: safer derivatives of bacteriophage lambda for DNA cloning. *Science*, **196**, 161–9.

Curtiss, R., Inoeu, M., Pereira, D., Hsu, J.C., Alexander, L., and Rock, L. (1977). Construction and use of safer bacterial host strains for recombinant DNA research. In *Molecular cloning of recombinant DNA*, (Ed. W.A. Scott and R. Werner), pp. 99–111. Academic Press, New York.

Cuskey, S.M. (1990). Biological containment of genetically engineered microorganisms. In *Classical and molecular methods to assess environmental applications of microorganisms*, (Ed. M. Levin, R. Seidler, and P. Pritchard). American Society for Microbiology, Washington DC.

Cuskey, S.M., and Bourquin, A.W. (1987). Construction of bacteria with conditional lethal genetic determinants for the control of released microorganisms in the environment. *Abstracts of the Annual Meeting of the American Society of Microbiology*, **Q127**, 303.

Cuskey, S.M., and Sprenkle, A.B. (1988). Benzoate-dependent induction from the OP2 operator-promoter region of the TOL plasmid pWWO in the absence of known plasmid regulatory genes. *Journal of Bacteriology*, **170**, 3742–6.

Cuskey, S.M., Reagin, M., Jeffrey, W.H., and Coffin, R.E. (1990). Conditional expression of the RK2 *kilA* gene in *Pseudomonas aeruginosa* PAO1 can cause a non-lethal inhibition of growth. *Abstracts of the Annual Meeting of the American Society of Microbiology*, **H173**.

Figurski, D.H., Pohlman, R.F., Bechhofer, D.H., Prince, A.S., and Kelton, C.A. (1982). Broad host range plasmid RK2 encodes multiple *kil* genes potentially lethal to *Escherichia coli* host cells. *Proceedings of the National Academy of Sciences (USA)*, **76**, 1935–9.

Gerdes, K., Rasmussen, P.B., and Molin, S. (1986). Unique type of plasmid maintenance function: post-segregational killing of plasmid-free cells. *Proceedings of the National Academy of Sciences (USA)*, **83**, 3116–20.

Gerdes, K., Helin, K., Christensen, O.W., and Lobner-Olesen, A. (1988). Translational control and differential RNA decay are key elements regulating postsegregational expression of the killer protein encoded by the *parB* locus of plasmid R1. *Journal of Molecular Biology*, **203**, 119–29.

Inouye, S., Nakazawa, A., and Nakazawa, T. (1981). Molecular cloning of gene *xlyS* of the TOL plasmid: evidence for positive regulation of the *xylDEGF* operon by *xylS*. *Journal of Bacteriology*, **148**, 413–18.

Jeenes, D.J., Soldati, L., Baur, H., Watson, J.M., Mercenier, A., Reimmann, C., Leisinger, T., and

Hass, D. (1986). Expression of biosynthetic genes from *Pseudomonas aeruginosa* and *Escherichia coli* in the heterologous host. *Molecular and General Genetics*, **203**, 421–9.

Jones, D.A., Ryder, M.H., Clare, B.G., Farrand, S. K., and Kerr, A. (1988). Construction of a Tra⁻ deletion mutant of pAgK84 to safeguard the biological control of crown gall. *Molecular and General Genetics*, **212**, 207–14.

Klemm, P. (1986). Two regulatory *fim* genes, *fimB* and *fimE*, control the phase variation of type 1 fimbriae in *Escherichia coli*. *EMBO Journal*, **5**, 1389–96.

Levine, M.M., Kaper, J.B., Lockman, H., Black, R.E., Clements, K.L., and Falkow, S. (1983). Recombinant DNA risk assessment studies in man: efficacy of poorly mobilizable plasmids in biologic containment. *Recombinant DNA Technology Bulletin*, **6**, 89–97.

Mermod, N., Lehrbach, P.R., Reineke, W., and Timmis, K.N. (1984). Transcription of the TOL plasmid toluate catabolic pathway operon of *Pseudomonas putida* is determined by a pair of coordinately and positively regulated overlapping promoters. *EMBO Journal*, **3**, 2461–6.

Mermod, N., Ramos, J.L., Bairoch, A., and Timmis, K.N. (1987). The *xylS* gene positive regulator of TOL plasmid pWWO: identification, sequence analysis and overproduction leading to constitutive expression of the *meta* cleavage operon. *Molecular and General Genetics*, **207**, 349–54.

Molin, S., Klemm, P., Poulsen, L.K., Biehl, H., Gerdes, K., and Anderson, P. (1987). Conditional suicide system for containment of bacteria and plasmids. *Biotechnology*, **5**, 1315–18.

Obukowicz, M.G., Perlak, F.L., Kusano-Kretzmer, K., Mayer, E.J., Bolten, S.L., and Watrud, S.L. (1986). Tn5-mediated integration of the delta-endotoxin gene from *Bacillus thuringiensis* into the chromosome of root-colonizing Pseudomonads. *Journal of Bacteriology*, **168**, 982–9.

Ogura, T., and Hiraga, S. (1983). Mini-F plasmid genes that couple host cell division to plasmid proliferation. *Proceedings of the National Academy of Sciences (USA)*, **80**, 4784–8.

Simberloff, D. (1985). Predicting ecological effects of novel entities: evidence from higher organisms. In *Engineered organisms in the environment: scientific issues*, (Ed. H.O. Halvorson, D. Pramer and N. Rogul), pp.152–61. American Society for Microbiology, Washington DC.

Winans, S.C., and Walker, G.C. (1985). Identification of pKM101-encoded loci specifying potentially lethal gene products. *Journal of Bacteriology*, **161**, 417–24.

8

Survival and mortality of bacteria in natural environments

P. Servais, J. Vives-Rego and G. Billen

Introduction

IT HAS BEEN 15 YEARS since molecular biologists first spliced DNA from one microbe to another to create a recombinant bacterium. Developments in biotechnology have moved forward rapidly during these years, making it possible to construct genetically engineered micro-organisms (GEMs) that may provide useful applications. GEMs can be used in industrial processes, participating in the synthesis of products of economic or social important. Many laboratories are now working with recombinant microbes and many industrial plants will do so in the future.

The risk of accidental release of such micro-organisms or of some part of their genetic material into the natural environment is therefore obvious and probably unavoidable. In some cases, there might be reasons for concern about the survival or even growth of these GEMs in the environment.

Conversely, some applications will involve the planned introduction of GEMs into natural habitats. Other chapters in this book provide numerous examples in agriculture and waste-water treatment. In these cases, the survival of the released micro-organism is the first condition to be considered.

Whatever the purpose might be, the fate of GEMs in natural environments, either feared or desired, is currently the subject of intensive questioning. To answer these questions, it is necessary to discuss a closely related and more general topic: the control of survival and mortality in the natural environment of micro-organisms in general. Two

cases must be envisaged: (i) the case of natural, active, microbial populations, the dynamics of which depend on the equilibrium between growth and mortality rates; and (ii) the case of allochthonous bacteria, which may be unable to grow in the habitat where they have been introduced and may develop active or passive mechanisms for escaping normal mortality processes until better conditions are restored. Both cases are of interest from the point of view of the question set above and they will be discussed in parallel.

The chapter will first attempt to define the concept of mortality in microbes, summarizing the rather scarce literature dealing directly or indirectly with this important question located at the intersect of philosophy and methodology. We will then discuss the various processes leading to mortality of microbes in natural environments, including predation, autolysis, physical or chemical injury, etc.; a summary of the data available to quantify the rate of bacterial mortality in the natural environment will be presented. The various strategies developed by microbes to overcome these mortality processes will then be presented and we will conclude with a brief discussion of the relevance of this information to the question set by the accidental or intentional introduction of GEMs in natural ecosystems.

Partly because the literature is more rich in this field, and partly because of our own research interests, the emphasis of this chapter is on aquatic environments. Some data are presented to show that basically similar conclusions can be reached for the soil environment. We can feel less guilty about this aquatic bias because Chapter 11 specifically deals with the survival and dissemination of GEMs in soils.

Concepts and methods related to microbial mortality

As said with much humour by Gray and Postgate (1977), in the preface of one of the very few books dealing with the question of survival of vegetative microbes: 'a scientific visitor from another planet, studying the state of Earth's science, might well conclude that a taboo exists among microbiologists regarding the death of bacteria'. From an epistemological point of view, however, the little that has been written on this subject constitutes a very interesting example of how knowledge of a subject can evolve according to the techniques available and the practical problems involved in the subject study.

For many years following the onset of microbiology, the mortality of microbes was not as extensively studied as their growth and metabolism. The development of procedures for killing microbes was the only angle from which microbial death was approached. Beside this, microbes were considered as immortal, as they are theoretically capable of indefinite vegetative reproduction. This capability for vegetative reproduction was taken by Postgate *et al.* (1961) as the criterion for defining the first operational procedure for assessing microbial viability. Microbial death was defined as the loss of viability, where viability was the ability to grow on an optimal medium.

During the late 1970s and early 1980s, this definition was faced with several methodological and conceptual difficulties, arising from the observation that microbes that were unable to grow on standard, supposedly optimal, media could be shown to be quite active in some other important respect. A range of techniques was developed for assessing or even quantifying various activities of individual micro-organisms. This allowed Colwell and co-authors (Roszack *et al.* 1984a; Colwell *et al.* 1985; Roszak and Colwell 1987) to define the concept of viable but not culturable micro-organisms. Improvement in the techniques for seeing and enumerating micro-organisms allowed scientists to define death as the loss of cellular integrity.

The latest developments in this field, and the discovery that extracellular genetic material from one bacterial strain can be transmitted to others, have resulted in a more basic definition of death based on the destruction of the genetic material itself.

Postgate's viability concept and the difference between plate and direct bacterial counts

The usual method for estimating bacterial numbers in natural environments has been the plate count method. In principle this method records the numbers of organisms, in a known volume, able to form observable colonies after incubation on a rich medium that is supposed to be adequate for the organisms being counted. However, working with pure cultures of bacteria, Postgate *et al.* (1961; Postgate and Hunter 1962) were the first to stress that direct microscopic observations generally yield higher counts than plate count methods. This was the basis of their definition of viability: the viability of a microbial population is the proportion of its members that are capable of multiplication when provided with optimal conditions for growth (Postgate *et al.* 1961). In this sense, a micro-organism should be called dead (or non-viable) if it does not multiply in such conditions, even if it keeps its morphological integrity and most of its biological functions. As stressed by Postgate (1977), this leads us to consider as non-viable micro-organisms that are not: 'any more dead than, say, a woman post-menopause', but this is, according to him, the only pragmatic and operational criterion to be applied for micro-organisms. Moreover, it was explicitly assumed by these authors (Postgate and Hunter 1962; Postgate 1977) that loss of the ability to grow on rich media is loss of an enzyme necessary for replication, and that this results from a genetic defect.

Attempts to apply Postgate's criterion to natural assemblages of micro-organisms have revealed severe difficulties. As pointed out by Gray (1977), this criterion: 'is inappropriate for investigations on mixed cultures of natural origin since a single recovery medium does not allow all microbial cells to develop and, even if it did, differential growth rates would lead to difficulties in counting colonies, and the presence of motile cells would make colony counts inaccurate'.

Indeed, the development of direct microscopic methods – including scanning electron microscopy

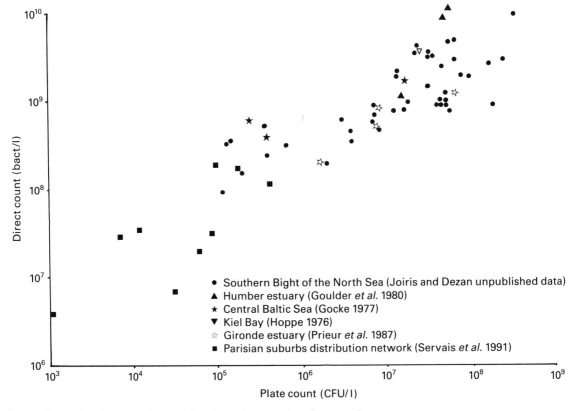

Fig. 8.1. Comparison between plate and direct bacterial counts in various aquatic ecosystems.

and epifluorescence microscopy – for enumerating bacteria in natural environments has thrown suspicion on the validity of plate count methods. Direct epifluorescence microscopic counting of bacteria, filtered on polycarbonate membranes and stained by a fluorochrome, has become the standard method for ecological microbiologists. Various fluorochromes have been proposed, including fluorescein isothiocyanate (Fliermans et al. 1975), euchrysine (Jones and Simons 1975), acridine orange (Daley and Hobbie 1975) and 4′-6 diaminido-2-phenylindole (DAPI) (Porter and Feig 1980). In aquatic environments, the classic plate count of total bacteria usually yields as low as 1 per cent of direct epifluorescence bacterial counts and sometimes less (Fig. 8.1). A similar discrepancy is observed in other environments. For soil bacteria, Bakken and Olsen (1987) have showed that only 2–4 per cent of the total microscopical count was observed as colony-forming cells.

Does this really mean, as implied by Postgate's criterion, that such a low proportion of the total

bacterial community is viable? Or rather, is it the indication that the conditions used for plate counts are not suitable for growth of a large fraction of the natural bacterial community. In support of the latter interpretation, Bakken and Olsen (1987) have shown that only 0.2 per cent of soil bacteria with diameter of less than 0.4 μm produced colonies when spread on nutrient agar, while up to 30–40 per cent of the largest bacteria (diameter ≥ 0.6 μm) were colony-forming, clearly indicating that cultivation on rich agar media selects for a particular component of the natural population. This selectivity can be due to the nutrient composition of agar media, which might be either unavailable or even inhibitory for some bacterial strains. In other cases it can be the result of a high incubation temperature (generally 20–37°C) that might be unfavourable for the development of many natural bacterial strains. A complete discussion of the inadequacy of the widely used agar-based cultivation media can be found in the reviews by Buck (1979) and Fry (1982).

The above-mentioned methods for direct microscopic enumeration of total, morphologically intact micro-organisms do not allow a selective count of a specific group of bacteria as, for example, is often required in public health microbiology. Enumeration on specific agar media therefore continues to be used routinely for that purpose. However, the recent development of immunoassays now offers a method for direct enumeration of specific microbial groups.

The basis for this assay is the presence of specific antigenic determinants in bacteria, which can be recognized by a specific antiserum (Peters *et al.* 1985). The combination of the antigen–antibody reactions and epifluorescence microscopy makes possible the direct counting of bacteria bearing the specific antigen in natural habitats (Kaspar and Tartera 1989). Polyclonal antisera (Xu *et al.* 1984) often present high cross-reactivity with particles or other micro-organisms. These problems may be partly solved by double-staining techniques (Hoff 1988), but the specificity and sensitivity of immunological assays have been enhanced by the use of monoclonal antibodies (Brayton *et al.* 1987; Obst *et al.* 1989).

Immunofluorescence techniques are now widely used for direct detection of enterobacteria in waters. They have also been used for enumeration of nitrifying bacteria in soils (Schmidt 1974) and in waters (Ward 1984). Again, in both cases it has been found that direct counts yield higher numbers than classic methods using growth on specific media (Colwell *et al.* 1985). In the case of potentially pathogenic bacteria, this discrepancy might have important implications for the reliability of the agar method currently used as a legal standard. How confident can we be of the non-pathogenicity of the bacteria that are morphologically intact but non-viable in the sense of Postgate (1977).

Viable but not culturable bacteria: methods for individual activity assessment

The preceding discussion calls for the assessment of the metabolic state of those morphologically intact micro-organisms unable to grow on routinely used bateriological media.

Several methods have been developed for assessing the metabolic activity of individual cells. These methods allowed Colwell and co-authors (Roszak *et al.* 1984a; Colwell *et al.* 1985; Roszak and Colwell 1987) to introduce the concept of viable but not

culturable organisms. For these authors, viability is defined as the ability to grow, metabolize, respire or divide. The viable but non-culturable bacteria do not undergo cell division on agar media but are morphologically intact and active. This definition of viability is obviously closely dependent on the methods used to assess bacterial activity. Microautoradiography was the first of these methods. It consists of incubating the cells in the presence of radiolabelled substrates, such as amino acids (Hoppe 1976), glucose (Meyer-Reil 1978) and thymidine (Tabor and Neihof 1984). The cells that are metabolically active and able to incorporate the radioactive substrate are detected under the microscope by the presence of silver grains in a film emulsion. The total number of bacteria can be counted in the same microscope fields. In aquatic environments the percentage of active bacteria determined by this method is in the range 20–100 per cent of the total number of bacteria determined by direct count.

The number of active cells is, however, dependent on the experimental conditions (substrates used, substrate concentration, duration and temperature of incubation). Tabor and Neihof (1982b, 1984) have shown quite large differences between three substrates used with coastal sea water samples.

Electron transport system (ETS) activity was advocated to be a more general characteristic of metabolically active micro-organisms than the uptake of particular substrate. ETS activity is linked to many metabolic pathways in virtually all bacteria and thus a sensitive measurement of ETS activity is useful in indicating the rate of respiration (not restricted to oxygen consumption) of the composite population and in the recognition of metabolically active bacteria. The methods for ETS activity assessment involve replacement of natural electron acceptors of the micro-organisms by a redox dye, which is usually a tetrazolium salt; dehydrogenase activity and respiration cause reduction of these salts to red insoluble formazan compounds. Various tetrazolium salts have been used, such as 2, 3, 5-triphenyl tetrazolium chloride (TTC) (Ryssov-Nielsen 1975), but the most commonly employed is the 2(*p*-iodophenyl)-3-(*p*-nitrophenyl)-5-phenyltetrazolium chloride (INT) (Zimmerman *et al.* 1978; Maki and Remson 1981; Tabor and Neihof 1982a, 1984; Fukui and Takii 1989).

Kogure *et al.* (1979, 1980) have proposed a method (direct viable count) based on the inhibition of DNA synthesis by nalidixic acid to distinguish active from inactive cells. Bacteria sensitive to nalidixic acid are specifially inhibited in DNA synthesis but continue to synthesize cellular components causing cells to become enlarged because cell division does not occur. Bacteria are incubated in the presence of nalidixic acid and yeast extract for some hours; after incubation the proportion of enlarged cells determined by microscopy is a measure of the active cells fraction.

Table 8.1 summarizes the results obtained by the application of the above methods of assessing individual activity of bacteria from natural aquatic environments. The percentage of active cells ranges between 0.5 and 97 per cent. Generally, this is obviously much higher than the percentage of plate count with respect to direct count (Fig. 8.1), confirming the view that lack of culturability cannot be used as a criterion of mortality. This concept of viable but not culturable bacteria is of particular importance for the study of the fate of allochthonous pathogenic micro-organisms in natural environments, because viable but not culturable cells may retain or recover virulence.

Colwell and co-workers have studied the ability of various enteric bacteria, such as *Escherichia coli* (Xu *et al.* 1982; Palmer *et al.* 1984; Grimes and Cowell 1986), *Salmonella enteritidis* and *Vibrio choleraea* (Roszak *et al.* 1984b; Colwell *et al.* 1985) to survive in sea water. When placed in sea water, enteric bacteria maintain cellular integrity for several days to weeks, but rapidly lose their ability to be cultivated by classic methods. However, when the non-culturable cells are examined by the nalidixic acid method described above, a fraction gives a positive answer. Moreover, these viable but not culturable cells have been shown to retain virulence; the strain *E. coli* H 10407 retained virulence plasmids (Grimes and Colwell 1986), *V. cholerae* produces a positive response and was again fully culturable when injected into ligated rabbit ileal loops (Colwell *et al.* 1985).

Mortality as the destruction of genetic material

The existence of significant amounts of dissolved (cell-free) DNA in aquatic environments has been demonstrated (Pillai and Ganguly 1972; Minear 1972; De Flaun *et al.* 1986). Determination of its molecular weight distribution (<0.1–>30 kilobases) indicates a size range sufficient to contain gene sequences (De Flaun *et al.* 1987). These authors have suggested that these naturally occur-

ring portions of extracellular DNA might be involved in the process of transformation after being taken up by competent bacteria and integrated into their genome or an extrachromosomal element (Stewart and Carlson 1986).

The pool of dissolved cell-free DNA may therefore mediate a flux of genetic information through aquatic microbial populations. This possibility is particularly important in the context of the introduction of pathogens of GEMs to natural environments. In this respect, the definition of mortality of a microbe should include the destruction of its DNA because, in a sense, a microbe is not entirely dead as long as its genetic information remains intact.

Interestingly enough, a method proposed recently by Servais *et al.* (1985, 1989) for determining the overall rate of bacterial mortality in natural environments meets this requirement. This method consists of labelling the DNA of a natural assemblage of bacteria using ^{3}[H]thymidine, and following the rate of disappearance of radioactivity from the DNA. As DNA is basically a conservative molecule, undergoing little turnover within the cell (if we neglect the processes of DNA repair, which are probably quantitatively insignificant), the decrease of radioactivity from the DNA constitutes a measure of bacterial mortality, whatever its cause (spontaneous lysis, grazing, etc.) may be. This method is currently the only one that allows a quantitative determination of the rate of overall bacterial mortality, even in conditions of net growth of the population. The method, initially developed for the study of the mortality of natural assemblages of aquatic environments, can also be used for studying the fate of allochthonous bacteria released in natural waters (Martinez *et al.* 1989; Garcia-Lara *et al.* 1991; Servais and Menon 1991).

A major advantage of this method is that it refers to mortality of bacteria as the loss of integrity of their genetic material: a bacterium is considered as dead when, and only when, its DNA has been hydrolysed. For example, in the case of an extreme stress, e.g. autoclaving or sonication of a natural sample, which destroys any micro-organisms and their DNAses, the method fails to show any mortality. This is because labelled DNA, liberated by the killed bacteria, remains intact under the sterile conditions established, unless the sample is reinoculated with a natural assemblage of bacteria.

Another approach to the quantification of the survival and mortality of specific genes (instead of

Table 8.1. *Percentage of active bacteria determined by three different methods with respect to total count (acridine orange direct count)*

Methods	Environments	Percentage	References
Autoradiography			
[³H]-amino acids	Marine (Kiel Bay, Germany)	41.0	Hoppe 1976
[³H]-glucose	Marine (Kiel Bay, Germany)	2.3–56.2	Meyer-Reil 1978
[³H]-amino acids	Marine (Chesapeake Bay, USA)	45.0–76.0	Tabor and Neihof, 1982b
[³H]-acetate		49.0–59.0	Tabor and Neihof 1982b
[³H]-thymidine		15.0–94.0	Tabor and Neihof 1982b
[³H]-thymidine	Marine (Southern California Bight, USA)	28.0–50.0	Fuhrman and Azam 1982
[³H]-amino acids	Marine (Chesapeake Bay, USA)	17.0–82.0	Tabor and Neihof 1984
[³H]-acetate	Surface samples	6.0–51.0	Tabor and Neihof 1984
[³H]-thymidine	Marine (North Sea coastal zone, Denmark)	20.0–80.0	Rieman *et al.* 1984
INT reduction	Marine Eckernoförde Bight, Baltic Sea)	8.0–97.0	Itturiaga, 1979
	Marine (Chesapeake Bay, USA)	61.0	Tabor and Neihof 1982a
	Marine (Chesapeake Bay, USA)	25.0–94.0	Tabor and Neihof 1984
	Bottled drinking water (France)	10.0–20.0	Oger *et al.* 1987
	Lake (Leman Lake—Epilimnion	1.4–26.0	Dufour, personal
	Hypolimnion	0.6–14.0	communication
Enlargement induced	Marine (Chesapeake Bay, USA)	7.0–15.0	Tabor and Neihof 1984
by nalidixic acid	Marine (Puerto Rico)	0.5–6.0	Kogure *et al.* 1987

individual micro-organisms) is offered by the use of recently developed DNA probes. This technique, based on the hybridization properties of complimentary DNA–DNA sequences, allows direct identification and quantification of specific genes associated with natural microbial communities (Ogram and Sayler 1988). Species-specific or function-specific nucleic acid probes have been developed. The former have been used to detect specific microbial pathogens in aquatic environment, soils or foods (Fitts *et al.* 1983; Hill *et al.* 1983a,b; Festl *et al.* 1986; Holben *et al.* 1988, Kaspar and Tartera 1989). Function-specific probes allowed an assessment of the potential of the microbial community to perform a given function, for example nitrogen fixation or xenobiotic biodegradation (Sayler *et al.* 1985).

DNA probes have also been devised for the direct detection and monitoring of specific engineered genes, and for tracking their survival and maintenance in natural environments (Chaudhry *et al.* 1989; Amy and Hiatt 1989).

Mortality processes in natural environments

The importance of bacterial production in aquatic and terrestrial ecosystems is now recognized to account for a significant fraction of primary production (Pomeroy 1974; Azam *et al.* 1983; Porter *et al.* 1985). The question of the fate of this production, and its possible transfer to higher trophic levels, is presently the object of intensive research, and is obviously closely related to the question of bacterial mortality.

Grazing by protozoa (mainly nanoflagellates and nanociliates) is generally considered as the major process of removal of bacterial biomass in aquatic environments, but several other processes may also play a significant role (McManus and Fuhrman 1988; Pace 1988). In this section, we present a review of these processes, the methods available for determining their rate and the results they yield in various natural environments.

For many years, the question of the disappearance of allochthonous bacteria, mainly of faecal bacteria, in the natural environment was approached from an independent angle, primarily concerned with the public health implications of the survival of these bacteria. There is now convincing evidence that the same processes are responsible for the elimination of allochthonous bacteria and for the mortality of autochthonous bacteria in natural environments. These two questions will therefore be discussed together in this section.

Table 8.2. *Grazing rates in aquatic environments measured by various methods*

Method	n	Grazing rate (10^{-3}/h)	Environment	Reference
Dilution technique	2	21–45	Kaneohe Bay (Hawaii)	Landry *et al.* 1984
Selective filtration	2	44	Essex Estuary	Wright and Coffin 1984
Interpretation of predator prey oscillations	4	21–52	Aarhus Bay	Andersen and Fenchel 1985
Selective inhibitors	12	7–108	New York coastal zone	Fuhrman and McManus 1984
Genetically marked minicells	1	9	Bothnian Sea	Wikner *et al.* 1986
	8	5.2–152	Baltic Sea	
	3	22–27	Mediterranean Sea	
Decrease of radioactivity after ^{3}[H]thymidine labelling		0–25	Various marine and freshwater ecosystems	(see Table 8.4)

Grazing on bacteria

Measurement of bacterial grazing

Various methods have been proposed for the estimation of the grazing rates of phagotrophic protozoa on bacteria in aquatic environments. They have been reviewed recently by McManus and Fuhrman (1988).

A first group of methods involves the comparison of bacterial growth with and without grazers. The procedure used for eliminating the predators consists of selective filtration (Wright and Coffin 1984), dilution with bacteria-free water (Landry *et al.* 1984; Ducklow and Hill 1985; Campbell and Carpenter 1986) or addition of various inhibitors to prevent the activity of grazers (Newell *et al.* 1983; Fuhrman and McManus 1984; Campbell and Carpenter 1986; Sanders and Porter 1986; Sherr *et al.* 1986).

A second group of methods measures the rate of intake by predators of 'marked' bacteria. The marked bacteria might consist in ^{3}[H]thymidine-labelled natural assemblages of bacteria (Holli-baugh *et al.* 1980; Gast 1985), genetically marked minicells of *E. coli* (Wikner *et al.* 1986), inert fluorescent particles (Borsheim 1984; McManus and Fuhrman 1986) or monodispersed fluorescently labelled bacteria (Albright *et al.* 1987; Sherr *et al.* 1987).

Another method proposed by Servais *et al.* (1985, 1989) and Becquevort *et al.* (unpublished data) does not fit into the former categories. It involves a comparison of the rate of decrease of radioactivity from ^{3}[H]thymidine-labelled bacterial DNA, with and without predators, using both

filtration and inhibitors for suppressing grazing.

Application of these methods in various aquatic environments generally show grazing rates in the range 0.0005–0.15/h (Table 8.2). Data are expressed in terms of first order grazing rates of the bacterial population. Similar methods have occasionally been applied to other environments. Thus, Montagna (1984) measured meiofaunal grazing rates on bacteria as high as 30/h in natural sediments of a salt marsh in South Carolina, polychaetes being the most active predators.

Grazing by protozoa has also been shown to play a role in the disappearance of enteric bacteria released into aquatic environments. It has been known for many years that enteric bacteria survive longer in either heat- or filter-sterilized natural water than in untreated water (Zobell 1936; Nusbaum and Garver 1955; Carlucci *et al.* 1961; McCambridge and McMeekin 1979). Use of antibiotics does not suppress the disappearance of *E. coli* resistant to those antibiotics used, while use of eukaryotic inhibitors does (Enzinger and Cooper 1976; McCambridge and McMeekin 1979), suggesting that grazing by protozoa is at least partly responsible for eliminating *E. coli*. This also explains the observation that enteric bacteria in natural water disappears at a higher rate in summer than in winter (Anderson *et al.* 1983; Barcina *et al.* 1986a,b).

Grazing rate of specific bacterial predators

Table 8.3 gathers the estimates, obtained with various methods, of the average specific grazing rate per predator. In this case, the results are expressed in terms of number of bacteria grazed by

Table 8.3. *Specific grazing rates of bacteriovorous protozoa, in monospecific cultures of nanoflagellates or in natural assemblages, estimated using various methods*

Source	Method	Protozoa or site	Protozoa numbers (10^3/ml)	Bacteria numbers (10^6/ml)	Grazing rate (bacteria/protozoan/h)
Fenchel (1982)	Disappearance of cultured bacteria	*Monosiga* sp.	0.1–1.0	5–35	27*
		Paraphysomonas sp.		10–115	254*
		Actinomonas sp.		2–85	107*
		Ochromonas sp.		5–170	190*
		Pleuromonas sp.			54*
		Pseudobodo sp.		5–50	84*
Davis and Sieburth (1984)	FDDC method	*Pseudobodo* sp. *Bodo* sp. *Monas* sp. *Paraphysomonas* sp. *Ochromonas* sp.	up to 400	0.1–10	30–200
Landry et al. (1984)	Dilution method	*Acantheopsis* Hetorotrophic microflagellates, Kaneohe Bay Hawaii	0.89–0.97	1.3–1.4	16–37
Caron et al. (1985)	Disappearance of prey density	*Paraphysomonas* sp.	–	–	23.3–117.5*
Sherr et al. (1986)	Selective inhibition	Natural assemblage, Georgia coast	1.3–2.2	3.3–11	20–80
Lucas et al. (1986)	Predator–prey oscillations	*Pseudobodo* sp.	1.11	7–9	18–19
Sherr et al. (1983)	Disappearance of cultured bacteria	*Monas* sp.	–	25–8000	10–75
McManus and Fuhrman (1986)	Uptake of fluoroscent microspheres	Natural assemblage, Chesapeake Bay	–	–	2–25
Cynar and Sieburth (1986)	Uptake of fluorescence microspheres	*Monas* sp. *Pseudobodo* sp. *Codosiga* sp.	–	3.9–18.3	0–144**
Sherr et al. (1988)	Uptake of fluorescently labelled bacteria	Mixed culture of flagellate incubated at 12 to 22°C	–	2.5–9.5	27.4–52
Becquevort et al. (unpublished data)	Decrease of ^{3}[H]-labelled DNA	Natural assemblages of nanoglagellates (North Sea)	0.17–1.55	1.8–6.8	65–108
		Natural assemblages of nanoflagellates (Scheldt estuary)	0.002–7	4.1–19.5	35–53
		Nanoflagellate culture *Bodo* sp.	0.36–2.1	3.3–4.0	36

* maximum rates; – no data.
** bacterial-sized particles.

a predator per hour. Data show grazing rates in the range of 10–254 bacteria/nanoflagellate/h.

More detailed work with monospecific cultures of nanoflagellates has allowed the determination of the dependence of grazing rate per protozoa on bacterial abundance. A Michaelis–Menten relationship generally fits the observed results, with maximum rate in the range 27–254 bacteria/nanoflagellate/h and half saturation constant in the range 1.3×10^6–30×10^6 bacteria/ml (Fenchel 1982).

Ciliates can also play a role as bacterial predators, but the species able to ingest small food particles (0.21–1 μm, the size range of most bacteria in natural environments) require high concentrations of prey. Fenchel (1980) measured the minimum concentration of bacteria required to sustain the growth of various ciliates to be in the range 4×10^6–2.5×10^7 bacteria/ml. Similarly, Turley et al. (1986) mentioned a threshold concentration of 6×10^6 cells/ml and 3×10^6 cells/ml for the ciliates *Uronema* and *Euplotes*, respectively. Such high bacterial abundance only occurs in very eutrophic waters.

Some aquatic filter-feeding metazoa can also contribute to bacterial grazing. From an analysis of the properties of their feeding appendages. Jorgensen (1984) identified *Cladocerans*, in fresh water, and *Appendicularians*, in marine waters as efficient bacterial grazers. Direct experimental evidences confirm this view (Peterson et al. 1978; King et al. 1980; Porter et al. 1983). Bacterial grazing rates by *Cladocerans* of up to 0.02/h have been calculated in Lake Ogelthorpe (Georgia, USA; Pace 1988).

Non-grazing mortality

At the present time no method exists for the direct determination of the part of bacterial mortality that is not attributable to nanozooplankton grazing. This can only be measured by difference, comparing overall mortality in an untreated sample and a sample treated to remove or inhibit grazers (Servais et al. 1985, 1989; Becquevort et al. unpublished data). However, no procedure can be considered 100 per cent efficient for the latter purpose: filtration might fail by retaining the smallest protozoa (Fuhrman and McManus 1984; Andersen and Fenchel 1985) and inhibitors may not be completely inhibitory.

Independent evidence for the existence of non-grazing mortality is therefore required. Becquevort

et al. (unpublished data) present data on three aquatic environments. They show that although the rate of bacterial mortality strongly correlated with the abundance of nanoflagellates, the regression of these two variables extrapolates to positive values for zero concentration of the predators (Fig. 8.2). The residual mortality rate, non-attributable to the enumerated grazers, ranged between 0.003 and 0.01/h.

The processes of bacteriophage-induced lysis, bacterial predation or lysis and autolysis may be involved in non-grazing mortality; these will now be discussed further.

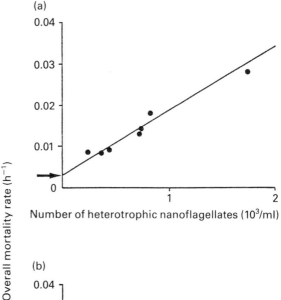

Fig. 8.2. Correlation between overall first order mortality rate and the number of heterotrophic nanoflagellates (a) in the Belgian coastal zone of the North Sea and (b) in the Scheldt estuary and its tributary the Rupel (from Bequevort et al. unpublished data).

Bacteriophage-induced lysis

Bacteriophages specific for natural populations of bacteria have been shown to be present in aquatic environments. Thus, Spencer (1957, 1960) isolated seven strains of bacteriophages active against bacteria from the same area in the North Sea. Moebus (1980) described an improved method for isolating bacteriophages from sea water and studied the specificity of phage–host systems isolated from Atlantic and North Sea waters (Moebus and Nattkemper 1981).

There is also evidence that bacteriophages can affect the population of their host bacteria in soils (Evans *et al*. 1979; Barnet 1980). Conversely, several authors concluded that bacteriophages are not responsible for the decline of enteric bacteria in natural waters (Carlucci and Pramer 1960; Pretorius 1962; Chambers and Clarke 1966), probably because the density of the host bacterial species is below the threshold of 10^4 susceptible bacteria per ml required for bacteriophage replication (Wiggins and Alexander 1985).

Recently Bergh *et al*. (1989) have developed a new method for virus enumeration in water (involving ultracentrifugation and transmission electron microscopy) by which they determined bacteriophage abundance ranging from 5×10^6 to 250×10^6/ml in fresh and marine waters. The authors presented theoretical calculations, based on the adsorption rate constant determined for the T4 phage–*E. coli* system, which showed that a rate of bacterial lysis of about 0.01/h is quite plausible. On the basis of an analysis of the turnover rate of viruses, Heldal and Bratbak (1991) estimated that phage may lyse 2–24 per cent of the bacterial population per hour. Proctor and Fuhrman (1990) calculated that 60 per cent of the total mortality of marine heterotrophic bacteria is due to viral infection.

Bacterial predation or lysis

Besides bacteriophages, predatory bacteria might also play a role in bacterial mortality. Some myxobacterial groups are bacteriolytic, lysing the host cells by the action of extracellular enzymes, including proteases, nucleases, lipsases and cell-wall-lytic enzymes. A wide range of prey can be attacked, including both Gram-negative and Gram-positive organisms (Schlegel 1986). Evaluation of the role of bacteriolytic bacteria in the general mortality processes in nature is still lacking. Casida (1980a,b) suggests that bacterial predators are involved in the disappearance of added allochthonous bacteria in soil.

Bacteria of the genera *Bdellovibrio* are obligate predators, requiring Gram-negative bacteria as prey for growth and multiplication (Shilo 1984). *Bdellovibrio* species are frequent in natural environments. Thus, Hentzschel (1980) showed the occurrence of *Bdellovibrio* in the North Sea at densities varying from 20 000–25 000 plaque-forming units per litre, from offshore to inshore waters. Studies, carried out in batch culture and in continuous culture showed that maintenance of the *Bdellovibrio* populations required a density of their specific prey of more than 10^5 cells per ml (Fenchel and Jorgensen 1977; Danso *et al*. 1975). Such high prey densities do not seem to be encountered in the bulk of most aquatic or terrestrial environments, an observation that leads to the suggestion that the reservoir of active *Bdellovibrio* could be confined to some enriched microniches (Shilo 1984).

Two other types of predatory bacteria have been described recently (Esteve *et al*. 1983; Guerrero *et al*. 1986). *Vampirococcus* is a Gram-negative anaerobe, which adheres to the surface of bacteria, degrading the cytoplasm of the prey cell without entering it. *Daptobacter* is a facultative parasitic bacterium, which is the first prokaryote described that penetrates into the cytoplasm of other bacteria; *Bdellovibrio* remains in the periplasmic space.

The significance of these two groups of bacteria in the processes of natural mortality is still unknown.

Autolysis

Autolysis of bacteria is another process that can be of significance in non-grazing mortality. Its mechanism remains to be fully elucidated. The growth of bacteria requires the interaction of biosynthesis and degradation of various structural polymers, including peptidoglycans. The enzymes responsible for cell wall polymer hydrolysis are autolysins, and they have a double action: constructive and destructive (Rogers 1979). They control the construction of the new cell wall during daughter cell separation, turnover and expansion of the cell wall and morphological differentiation. When the control of autolysin activity breaks down, the destructive role of autolysins is manifested by cell wall disruption leading to the destruction of the cell (Rogers and Forsberg 1971; Leduc *et al*. 1982). Very little attention has been paid to the destructive potential of autolysins. The

Table 8.4. *Overall mortality rates, grazing rates and non-grazing mortality rates measured* in situ *in various aquatic environments by Servais et al. (1985)'s method*

	n	Temperature (°C)	k_d (10^{-3}/h)	k_{dg} (10^{-3}/h)	k_{dr} (10^{-3}/h)
Rivers					
Meuse (Belgium)	14	8–22	6.1–23.7	1.6–6.8	2–19.4
Ebro (Spain)	4	18–20	12–22	5–9	6–11
Escaut (Belgium)	15	4.1–15	5.4–18.5	1–14	1.9–7.3
Rupel (Belgium)	7	2–16	7.5–22	0.6–19	4.3–12
Lakes					
Leman (France)	2	18	13–20	10–14	3–6
Bois Chambre (B)	1	20	33	6	27
Banyoles (Spain)	1	20	20	12	8
Ciso (Spain)	1	20	16	15	1
Vilar (Spain)	1	20	17	14	3
Sea water					
North Sea					
Belgian coastal zone	14	9–13	4.3–21	0–21	0.3–6
Holland coastal zone	1	14	17	6	11
English Channel	5	12	10–22	5–13	4–9
Mediterranean coastal zone					
Barcelona (Spain)	10	18–20	12–60	0–25	11–45
Costa Brava (Spain)	1	18–20	18	8	10
Ebro delta (Spain)	1	18–20	26	16	10
Antarctica					
Scotia Sea	2	1.7–3.2	3.6–5.3	2–2.4	1.6–2.9
Confluence	2	(−0.5)–0.5	5.8–6.5	3.7–3.9	1.9–2.8
Marginal ice zone	2	(−1.7)–0.9	4.9–5	3.1–3.9	1.1–1.8

n number of measurements; k_d, overall mortality rate; k_{dg}, grazing mortality rate; k_{dr}, non-grazing (residual) mortality rate.

destructive activity of autolysins is a general phenomenon in micro-organisms and can be expected when growth is stopped under conditions of starvation or chemical or physical stress. Location of the autolysins and the factor that triggers their destructive action are not yet known.

Overall rate of mortality

Table 8.4 summarizes the available measurements of total mortality, grazing and non-grazing mortality rates obtained by the method proposed by Servais *et al.* (1985) for autochthonous bacterial communities in a large range of aquatic environments. In view of the large differences between the aquatic systems, which range from very oligotrophic to highly eutrophic, the range of overall mortality rates is surprisingly narrow (from 0.004 to 0.060/h). Recently, Billen *et al.* (1991) reviewed growth and mortality measurements collected from a large range of aquatic environments, which were characterized by difference of five orders of magnitude in their overall bacterial production rates.

These results showed a range of only two orders of magnitude for specific growth and mortality rates, which are both in approximately the same range (0.001–0.1/h). Billen *et al.* (1988) have shown that the relative value of growth and mortality rates consistently explains the observed variations of bacterial biomass during phytoplankton blooms. Similarly they show that increasing bacterial populations are characterized, as expected, by higher growth rates than mortality rates, while the reverse is true when bacterial biomass declines.

Enteric bacteria are often unable to grow in aquatic natural environments. Their rate of mortality has been severely overestimated in the past because their loss of culturability interfered with the assessment of disappearance, as discussed on page 104. Therefore, despite many reports on the disappearance of enteric bacteria in natural waters monitored by plate count, very little unbiased data on mortality rates in aquatic environments are available. Table 8.5 presents some data obtained by the ^{3}[H]thymidine method, which can detect the mortality of enteric bacteria without interference of

Table 8.5. *Mortality rates of enteric bacteria in Mediterranean coastal sea water (Barcelona)*

Date	Strain	Mortality rates (10^{-3}/h) measured by:	
		Decrease of radioactivity after ^{3}[H]thymidine labelling[1]	Loss of culturability
16 June 1987	*Escherichia coli*	19	nd
15 July 1987	*Escherichia coli*	38	nd
17 September 1987	*Escherichia coli*	14	38
2 November 1987	*Escherichia coli*	23	36
15 December 1987	*Escherichia coli*	14	nd
19 December 1988	*Escherichia coli*	08	nd
24 January 1989	*Escherichia coli*	19	36
21 July 1988	*Klebsiella pneumoniae*	14	nd

[1] Method, Martinez *et al.* 1989; nd, not done.

their loss of culturability. Comparisons between these mortality rates and those obtained with viable plate counts show that the latter are largely over-estimated. Martinez *et al.* (1989) found a two-fold overestimation (Table 8.5); working with various strains of faecal bacteria in the North Sea, Garcia-Lara *et al.* (1991) and Servais and Menon (1991) found a much higher overestimation: up to fifteen-fold. These authors also showed that the loss of culturability was severely influenced by the selec-tivity of the agar medium used. Most work on the survival of enteric bacteria has been carried out with filtered or autoclaved waters. If the same comparison between the loss of viability by plate counts and direct specific methods for counting morphologically intact enteric bacteria is made in sterile waters a much larger discrepancy is found. Intact bacteria often appear to survive for very long periods under sterile conditions, while their loss of culturability is very rapid (Roszak *et al.* 1984b; Colwell *et al.* 1985; Grimes and Colwell 1986). Mortality rates of enteric bacteria, calculated from the ^{3}[H]thymidine method are quite similar to the rates found for autochthonous bacteria, suggesting that the same processes are responsible (Table 8.5).

The importance of grazing in the overall mortal-ity rate varies from 20 to 90 per cent (see Table 8.4). Data presented by Servais *et al.* (1989) suggest that grazing dominates the mortality flux during productive periods, while non-grazing mortality is more important, both in relative and in absolute terms, during unproductive seasons.

Survival strategies

In this section we will briefly examine some of the physiological processes occurring when bacteria are prevented from growing quickly, either by lack of nutrients (starvation) or because of unfavourable physical conditions.

Much of the work on these survival processes was done with laboratory cultures, where bacteria are taken away from a number of natural mortality processes, such as grazing and other types of predation. We will, however, try to discuss the survival mechanisms from the perspective of the bacteria in their natural environment.

Pursuing growth

Obviously, growth is the best way to cope with mortality processes. Pursuing growth as long as possible, and in spite of decreased nutrient concen-trations, is therefore a common strategy for sur-vival. Shifts to high affinity substrate uptake sys-tems under starvation conditions was observed for several bacteria from aquatic environments (Gessey and Morita 1979; Davis and Robb 1985). When external substrates are no longer available, internal reserves can be mobilized. Lipids, poly-β-hydroxybutyrate and carbohydrates disappear from the cell in the early stages of starvation (Oliver and Stringer 1984; Hood *et al.* 1986; Malmcrona-Friberg *et al.* 1986). These reserves can be used for some time for ensuring the energetic requirements, not only for maintenance but also for pursuing

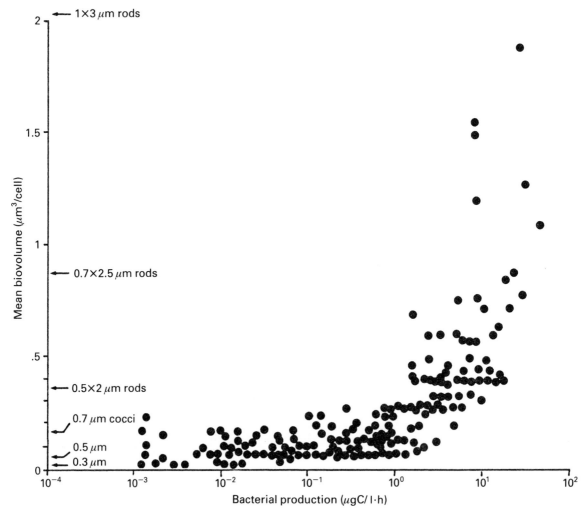

Fig. 8.3. Mean cell biovolume of authochthonous bacterial communities of a large range of aquatic environments of different 'richness'. Overall bacterial production, as measured by [³H]thymidine incorporation, is taken as an index level of eutrophy or oligotrophy of the system (Billen *et al.* 1991).

growth. As stressed by Novitsky and Morita (1978), the term 'growth' is ambiguous in this respect, as it might be understood as either referring to an increase in biomass or to an increase in number. Observations on a marine psychrophilic *Virio ANT-300* showed that neither increase is necessarily parallel. When these bacteria were suspended in artificial sea water at low cell density they showed an initial 200-fold increase in cell numbers before declining at very slow rate. This large increase in number was shown to occur mainly through reductive division, i.e. at the expense of internal reserves and capsule material, and

was accompanied by a change in size and shape from a 1×4 μm rod to a 0.4 μm coccus (Novitsky and Morita 1976, 1977, 1978).

Formation of ultramicrocells is a general survival strategy of bacteria in oligotrophic natural environments (Morita 1985). The processes, first described for the Antarctic *Vibrio ANT-300*, have also been observed with other marine bacterial strains (Kjelleberg *et al.* 1982; Amy and Morita 1983; Kurath and Morita 1983). It was also shown to occur with two strains of pathogenic bacteria released into sea water (Baker *et al.* 1983). On the other hand, ultramicrocells are reported to constitute the most important part of the bacterial community in many oligotrophic environments, including surface sea water (except during phytoplankton blooms) (Zimmerman and Meyer-Reil 1974; Billen *et al.* 1991), the deep sea (Tabor *et al.* 1981) and the soil (Bae *et al.* 1972; Bae and Casida 1973; Casida 1977; Olsen and Bakken 1983). The dominance of small bacterial forms in oligotrophic aquatic systems is well illustrated by the data in Fig. 8.3 (Billen *et al.* 1991).

The potential advantages for cells to be small are perhaps two-fold. By giving the cells more surface area per unit volume their ability to scavenge substrates is enhanced. Their small size also helps to avoid ingestion by predators. Indeed, it is known that flagellates select the larger size classes of their prey (Andersson *et al.* 1986). Similarly, ciliates grazed mostly on the 1–2 μm size ranges of bacteria (Fenchel 1980). Suspension-feeding metazoans are generally inefficient at filtering the smallest planktonic bacteria (Jorgensen 1984). Ultramicrocells can therefore constitute a form of refuge with some degree of protection against grazing. On the other hand, the rapid multiplication resulting from the process of reductive division considerably increases the survival probability of the genetic message.

Resting forms

Some bacteria can produce persistent cell forms with a specific morphology and chemical composition; these are more resistant to physical or chemical injury. These forms can survive temperature, radiation, starvation and toxicant stresses that would be lethal for the vegetative form. Endospore formation by *Clostridia* or *Bacillus* represents an extreme case of very resistant survival forms, the production of which involves profound morphological and biochemical transformations (Stanier *et al.* 1987). Other persistent forms produced by microbes upon starvation or stress include exospores, cysts and dormant or unculturable cells. Sudo and Dworkin (1973) defined resting cells as: 'those cells in which division does not occur, endogenous respiration is absent or much decreased, and in which formation of the resting stage is part of the natural life cycle of the organism. The resting cell is usually, but not necessarily, morphologically distinct from the growing or vegetative form.' Recent work on resting cells has emphasized the basic convergences more than the morpho-

logical differences between these various resistant forms (Sudo and Dworkin 1973).

As an example, the production of specific proteins, induced by starvation and conferring resistance to heat or chemical stresses, has been described both in *E. coli* and a marine vibrio (Jenkins *et al.* 1988; Nystrom *et al.* 1988; Schultz *et al.* 1988). Other convergences between the various resting forms produced by bacteria include reduction of the water content and miniaturization of the cell.

Conclusion

Bacteria in natural environments form a very active component of the microflora. Typical growth rates calculated for the overall bacterial community are 0.1–1/day in waters (Pace 1988; Billen *et al.* 1991) and only a little slower (0.1–0.5/day) in soils (Williams 1985). Grazing by flagellates or other eukaryotes, virus or bacterial attack and spontaneous lysis account for an overall mortality rate of bacteria that, on average, matches their growth rates. These processes are not *a priori* selective for authochthonous, actively growing bacterial populations. Therefore it can be estimated that an allochthonous strain of bacteria introduced in an aquatic system, where it cannot grow, would be reduced to 10^{-6} of its initial abundance after 2–20 weeks (after 4–20 weeks in soils).

This conclusion contradicts the well-known diversity of the bacterial flora of natural systems. One gram of garden soil, or one millilitre of natural water can easily be shown to contain nearly all bacterial types described in the Bergey manual when an adequate enrichment procedure is applied, even if this type of organism has not been obviously active for years in the original environment.

This paradox is apparent: the processes responsible for the overall mortality of bacteria contribute to a rapid turnover of bacterial biomass. However, various mechanisms ensure a much longer survival of the genetic information carried by non-growing bacteria. To a large extent, the fates of the biomass and the genetic message are not parallel. Of significance in this respect is the fact that some mortality processes bear the opportunity for spreading genetic information. Transduction after phage lysis or transformation after spontaneous lysis of bacteria are two examples discussed in detail elsewhere in this book.

The introduction to this chapter expressed concern over the survival of GEMs in the natural environment. This chapter shows that the situation is the worst that could be expected. A rapid biomass turnover might lead to rapid disappearance of the specific metabolic capability of deliberately introduced bacterial populations if these are not able to grow rapidly enough in the new environment. On the other hand, a dangerous genome unintentially introduced into the natural environment could persist, and even spread, despite the lack of effective growth of the microbe carrying it.

References

Albright, L.J., Sherr, E.B., Sherr, B.F., and Fallon, R.D. (1987). Grazing of ciliated protozoa on free and particle-attached bacteria. *Marine Ecology Progress Series*, **38**, 137–49.

Amy, P.S., and Hiatt, M.D. (1989). Survival and detection of bacteria in an aquatic environment. *Applied and Environmental Microbiology*, **55**, 788–93.

Amy, P.S., and Morita, R.Y. (1983). Starvation-survival patterns of sixteen freshly isolated open-ocean bacteria. *Applied and Environmental Microbiology*, **45**, 1109–15.

Andersen, P., and Fenchel, T. (1985). Bacteriology by microheterotrophic flagellates in seawater samples. *Limnology and Oceanography*, **30**, 198–202.

Anderson, I.C., Rhodes, M.W., and Kator, H.I. (1983). Seasonal variation in survival of *Escherichia coli* exposed *in situ* membrane diffusion chambers containing filtered and non-filtered estuarine water. *Applied and Environmental Microbiology*, **45**, 1877–83.

Andersson, A., Larsson, U. and Y., Hagström, A. (1986). Size selective predation by a microflagellate on pelagic bacteria. *Marine Ecology Progress Series*, **33**, 51–7.

Azam, F., Renchel, T., Field, J.G., Meyer-Reil, L.A., and Thingstad, F. (1983). The ecological rôle of water-column microbes in the sea. *Marine Ecology Progress Series*, **10**, 257–63.

Bae, H.C., and Casida, L.E. Jr. (1973). Responses of indigenous micro-organisms to soil incubations as viewed by transmission electron microscopy of cell thin sections. *Journal of Bacteriology*, **113**, 1463–73.

Bae, H.C., Cota-Robles, E.H., and Casida, L.E.Jr. (1972). Microflora of soil viewed by transmission electron microscopy. *Applied Microbiology*, **23**, 637–48.

Baker, R.M., Singleton, F.L., and Hood, M.A. (1983). Effects of nutrient deprivation of *Vibrio cholera*. *Applied and Environmental Microbiology*, **46**, 930–40.

Bakken, L.R., and Olsen, R.A. (1987). The relationship between cell size and viability in soil bacteria. *Microbial Ecology*, **13**, 103–14.

Barcina, I., Arana, I., Iriberri, J., and Egea, L. (1986a). Influence of light and natural microbiota of the Butron river on *E. coli* survival. *Antonie van Leeuwenhoek*, **52**, 555–66.

Barcina, I., Arana, I., Iriberri, J., and Egea, L. (1986b). Factors affecting the survival of *E. coli* in a river. *Hydrobiologia*, **141**, 249–53.

Barnet, Y.M. (1980). The effect of rhizobiophages on populations of *Rhozobium trifolii* in the root zone of clover plants. *Canadian Journal of Microbiology*, **26**, 572–6.

Bergh, O., Borsheim, K.Y., Bratbak, G., and Heldal, M. (1989). High abundance of viruses found in aquatic environments. *Nature*, **340**, 467–8.

Billen, G., Servais, P., and Fontigny, A. (1988). Growth and mortality in bacterial population dynamics of aquatic environments. *Archiv für Hydrobiologie Ergebnisse der Limnologie*, **31**, 173–83.

Billen, G., Servais, P., and Becquevort, S. (1991). Dynamics of bacterioplankton in oligotrophic and eutrophic aquatic environments: *Bottom-up* versus *top-down* control. *Hydrobiologia*, **207**, 37–42.

Borsheim, K.Y. (1984). Clearance rates of bacteria-sized particles by freshwater ciliates, measured with monodisperse fluorescent latex beads. *Oecologia*, **63**, 286–8.

Brayton, P.R., Tamplin, M.L., Huq, A., and Colwell, R.R. (1987). Enumeration of *Vibrio cholerae* 01 in Bangladesh waters by fluorescent-antibody direct viable count. *Applied and Environmental Microbiology*, **53**, 2862–5.

Buck, J.D. (1979). The plate count in aquatic microbiology. In *Native aquatic bacteria: enumeration activity and ecology*, ASTM STP 695, (Ed. Costerton, J.W. and Colwell, R.R.), pp. 19–28. American Society for Testing and Materials, Philadelphia, USA.

Campbell, L., and Carpenter, E.J., (1986). Diel patterns of cell division in marine *Synechococcus* spp. (cyanobacteria): use of frequency of dividing cells technique to measure growth rate. *Marine Ecology Progress Series*, **32**, 139–48.

Carlucci, A.F., and Pramer, D. (1960). An evaluation of factors affecting the survival of *Escherichia coli* in seawater. IV. Bacteriophages. *Applied Microbiology*, **8**, 254–6.

Carlucci, A.F., Scarpino, P.V., and Pramer, D. (1961). Evaluation of factors affecting survival of *Escherichia coli* in seawater. V. Studies with heat- and filter-sterilised seawater. *Applied Microbiology*, **9**, 400–4.

Caron, D.A., Goldman, J.C., Andersen, O.K., and Dennett, M.A. (1985). Nutrient cycling in a microflagellate food chain. II. Population dynamics and carbon cycling. *Marine Ecology Progress Series*, **24**, 243–54.

Casida, L.E. (1977). Small cells in pure cultures of

Agromyces ramosus and in natural soil. *Canadian Journal of Microbiology*, **23**, 214–16.

Casida, L.E. (1980a). Death of *Micrococcus luteus* in soil. *Applied and Environmental Microbiology*, **39**, 1031–4.

Casida, L.E. (1980b). Bacterial predators of *Micrococcus luteus* in soil. *Applied and Environmental Microbiology*, **39**, 1035–41.

Chambers, C.W., and Clarke, N.A. (1966). Control of bacteria in nondomestic water supplies. *Advances in Applied Microbiology*, **8**, 105–43.

Chaudhry, G.R., Toranzos, G.A., and Bhatti, A.R. (1989). Novel method for monitoring genetically engineered micro-organisms in the environment. *Applied and Environmental Microbiology*, **55**, 1301–4.

Colwell, R.R., Brayton, P.R., Grimes, D.J., Roszak, D.B., Huq, S.A., and Palmer, L.M. (1985). Viable but non-culturable *Vibrio cholerae* and related pathogens in the environment: implications for release of genetically engineered micro-organisms. *Biotechnology*, **3**, 817–20.

Cynar, F.J., and Sieburth, J.McN. (1986). Unambiguous detection and improved quantification of phagotrophic in aplochlorotic nanoflagellates using fluorescence microspheres and concomitant phase contrast and epifluorescence microscopy. *Marine Ecology Progress Series*, **32**, 61–70.

Daley, R.J., and Hobbie, J.E. (1975). Direct count of aquatic bacteria by a modified epifluorescent technique. *Limnology and Oceanography*, **20**, 875–82.

Danso, S.K.A., Keya, S.O., and Alexander, M. (1975). Protozoa and the decline of *Rhizobium* populations added to soil. *Canadian Journal of Microbiology*, **21**, 884–95.

Davis, C.L., and Robb, F.T. (1985). Maintenance of different mannitol uptake systems during starvation in oxidative and fermentative marine bacteria. *Applied and Environmental Microbiology*, **50**, 743–8.

Davis, P.G., and Sieburth, J.McN. (1984). Estuarine and oceanic microflagellate predation of actively growing bacteria: estimation by frequency of dividing-divided bacteria. *Marine Ecology Progress Series*, **19**, 237–46.

De Flaun, M.F., Paul, J.H., and Davis, D. (1986). Simplified method for dissolved DNA determination in aquatic environments. *Applied and Environmental Microbiology*, **52**, 654–9.

De Flaun, M.F., Paul, J.H., and Jeffrey, W.H. (1987). Distribution and molecular weight of dissolved DNA in subtropical estaurine and oceanic environments. *Marine Ecology Progress Series*, **38**, 65–73.

Ducklow, H.W., and Hill, S.M. (1985). The growth of heterotrophic bacteria in the surface waters of warm core rings. *Limnology and Oceanography*, **30**, 239–59.

Enzinger, R.M., and Cooper, R.C. (1976). Rôle of bacteria and protozoa in the removal of *Escherichia coli* from estuarine waters. *Applied and Environmental Microbiology*, **31**, 758–63.

Esteve, I., Guerrero, R., Montesimo, E., and Abella, C. (1983). Electron microscopy study of the interaction of epibiontic bacteria with *Chromatium minus* in natural habitats. *Microbial Ecology*, **9**, 57–64.

Evans, J., Barnet, Y.M., and Vincent, J.M. (1979). Effect of a bacteriophage on the colonisation and nodulation of clover roots by a strain of *Rhizobium trifolii*. *Canadian Journal of Microbiology*, **25**, 968–73.

Fenchel, T. (1980). Suspension feeding in ciliated protozoa: feeding rates and their ecological significance. *Microbial Ecology*, **6**, 13–25.

Fenchel, T. (1982). Ecology of heterotrophic microflagellates. VI. Quantitative occurrence and importance as bacterial consumers. *Marine Ecology Progress Series*, **9**, 35–42.

Fenchel, T.M., and Jorgensen, B.B. (1977). Detritus foods chain of aquatic ecosystems: the rôle of bacteria. In *Advances in microbial ecology*, Vol. 1, (Ed. M. Alexander), pp. 1–58. Plenum Press, New York.

Festl, H., Ludwig, W., and Schleifer, K.H. (1986). DNA hybridization probe for *Pseudomonas fluorescens* group. *Applied and Environmental Microbiology*, **52**, 1190–4.

Fitts, R., Diamond, M., Hamilton, C., and Neri, M. (1983). DNA–DNA hybridization assay for detection of *Salmonella* spp. in foods. *Applied and Environmental Microbiology*, **46**, 1146–51.

Fliermans, C.B., Schneider Cain, P., and Schmidt, E.L. (1975). Direct measurement of bacterial stratification in Minnesota Lakes. *Archiv für Hydrobiologie*, **76**, 248–55.

Fry, J.C. (1982). The analysis of microbial interactions and communities *in situ*. In *Microbial interactions and communities*, 1, (Ed. A.T. Bull and J.H. Slater), pp. 103–52. Academic Press, London.

Fuhrman, J.A., and Azam, F. (1982). Thymidine incorporation as a measure of heterotrophic bacterioplankton production in marine surface waters: evaluation and field results. *Marine Biology*, **66**, 109–20.

Fuhrman, J.A., and McManus, G.B. (1984). Do bacteria-sized marine eukaryotes consume significant bacterial production? *Science*, **224**, 1257–60.

Fukui, M., and Takii, S. (1989). Reduction of tetrazolium salts for sulfate-reducing bacteria. *FEMS Microbiology Ecology*, **62**, 13–20.

Garcia-Lara, J., Menon, P., Servais, P., and Billen, G. (1991). Mortality of fecal bacteria in seawater. *Applied and Environmental Microbiology*, **57**, 885–8.

Gast, V. (1985). Bacteria as a food source for microzooplankton in the Schei Fjord and Baltic Sea with special reference to ciliates. *Marine Ecology Progress Series*, **22**, 107–20.

Geesey, G.G., and Morita, R.Y. (1979). Capture of arginine at low concentrations by a marine psychrophilic bacterium. *Applied and Environmental Microbiology*, **38**, 1092–7.

Gocke, H. (1977). Untersuchungen über die heterotrophe Aktivität in der Zentralen Ostsee. *Marine Biology*, **40**, 87–94.

Goulder, R., Blanchard, A.S., Sanderson, P.L., and Wright, B. (1980). Relationships between heterotrophic bacteria and pollution in an industrialized estuary. *Water Research*, **14**, 591–601.

Gray, T.R.G. (1977). Survival of vegetative microbes in soil. In *The survival of vegetative microbes*, (Ed. T.R.G. Gray and J.R. Postgate), pp. 327–64. Society for General Microbiology, Cambridge University Press.

Gray, T.R.G., and Postgate, J.R. (1977). Editor's preface. In *The survival of vegetative microbes*, (Ed. T.R.G. Gray and J.R. Postgate). Society for General Microbiology, Cambridge University Press.

Grimes, D.J., and Colwell, R.R. (1986). Viability and virulence of *Escherichia coli* suspended by membrane chamber in semitropical ocean water. *FEMS Microbiology Letters*, **34**, 161–5.

Guerrero, R., Pedròs-Aliò, C., Esteve, I., Mas, J., Chase, D., and Margulis, L. (1986). Predatory prokaryotes: predation and primary consumption evolved in bacteria. *Proceedings of the National Academy of Science (USA)*, **83**, 2138–42.

Heldal, M., and Bratbak, G. (1991). Production and decay of viruses in aquatic environments. *Marine Ecology – Progress series*, **72**, 205–12.

Hentzschel, G. (1980) Wechselwirkungen bakteriolytischer und saprophytischer Bakterien aus der Nordsee. *Mitteilungen aus Institute für Allgemeine Botanik, Hamburg*, **17**, 113–24.

Hill, W.E., Madden, J.M., McCardell, B.A., Shah, D.B., Jagow, J.A., Payne, W.L., and Boutin, B.K. (1983a). Foodborne enterotoxigenic *Escherichia coli*: detection and enumeration by DNA colony hybridization. *Applied and Environmental Microbiology*, **45**, 1324–30.

Hill, W.E., Payne, W.L., and Aulisio, C.C.G. (1983b). Detection and enumeration of virulent *Yersina enterocolitica* in food by DNA colony hybridization. *Applied and Environmental Microbiology*, **46**, 636–41.

Hoff, K.A. (1988). Rapid and simple method for double staining of bacteria with 4,6 diamidino-2-phenylindole and fluorescein isothiocyanate-labeled antibodies. *Applied and Environmental Microbiology*, **54**, 2949–52.

Holben, W.E., Jansson, J.K., Chelm B.K., and Tiedje, J.M. (1988). DNA probe method for the detection of specific micro-organisms in the soil bacterial community. *Applied and Environmental Microbiology*, **54**, 703–11.

Hollibaugh, J.T., Fuhrman, J.A., and Azam, F. (1980). Radioactive labelling of natural assemblages of bacterioplankton for use in trophic studies. *Limnology and Oceanography*, **25**, 172–81.

Hood, M.A., Guckert, J.B., White, D.C., and Deck, F. (1986). The effect of nutrient (carbon) deprivation on the levels of lipids carbohydrates, DNA, RNA, and protein in *Vibrio cholera*. *Applied and Environmental Microbiology*, **52**, 1419–21.

Hoppe, H.G. (1976). Determination and properties of actively metabolizing heterotrophic bacteria in the sea, investigated by means of microautoradiography. *Marine Biology*, **36**, 291–302.

Iturriaga, R. (1979). Bacterial activity related to sedimenting particulate matter. *Marine Biology*, **53**, 157–67.

Jenkins, D.E., Schultz, J.E., and Matin, A. (1988). Starvation induced cross protection against heat or H_2O_2 challange in *Escherichia coli*. *Journal of Bacteriology*, **170**, 3910–14.

Jones, J.G., and Simons, B.M. (1975). An investigation of errors in direct counts of aquatic bacteria by epifluorescence microscopy, with reference to a new method for dyeing membrane filters. *Journal of Applied Bacteriology*, **39**, 317–29.

Jorgensen, C.B. (1984). Effect of grazing: metazoan suspension feeders. In *Heterotrophic activity in the sea*, (Ed. J.E., Hobbie and P.J.leB. Williams), pp. 445–64. Plenum Press, New York.

Kaspar, C.V., and Tartera, C. (1989). Methods for detecting microbial pathogens in food and water. In *Methods in microbiology*, (Ed. J.R. Norris). Academic Press (in press).

King, K.R., Hollibaugh, J.T., and Azam, F. (1980). Predator–prey interactions between the larvacean *Oikopleura dioica* and bacterioplankton in enclosed water columns. *Marine Biology*, **56**, 49–57.

Kjelleberg, S., Humphrey, B.A., and Marshall, K.C. (1982). The effect of interfaces on small starved marine bacteria. *Applied and Environmental Microbiology*, **43**, 1166–72.

Kogure, K., Simidu, U., and Taga, N. (1979). A tentative direct microscope method for counting living marine bacteria. *Canadian Journal of Microbiology*, **27**, 415–20.

Kogure, K., Simidu, U., and Taga, N. (1980). Distribution of viable marine bacteria in neritic seawater around Japan. *Canadian Journal of Microbiology*, **26**, 318–23.

Kogure, K., Simidu, U., Taga, N., and Colwell, R.R. (1987). Correlation of direct viable counts with heterotrophic activity for marine bacteria. *Applied and Environmental Microbiology*, **53**, 2332–7.

Kurath, G., and Morita, R.Y. (1983). Starvation-survival physiological studies of a marine *Pseudomonas* sp. *Applied and Environmental Microbiology*, **45**, 1206–11.

Landry, M.R., Haas, L.W., and Fagerness, V.L. (1984). Dynamics of microbial plankton communities: experiments in Kaneohe Bay, Hawaii. *Marine Ecology Progress Series*, **16**, 127–33.

Leduc, M., Kasra, R., and van Heijenoort, J. (1982). Induction and control of the autolytic system of *Escherichia coli*. *Journal of Bacteriology*, **152**, 26–34.

Lucas, M.I., Painting, S.J., and Muir, D.G. (1986). Estimates of carbon flow through bacterioplankton in the S. Benguela upwelling region based on ³H-thymidine incorporation and predator-free incubations. In *Proceedings of the Second International Colloquium of Marine Bacteriology*, Brest, 1984, pp. 375–83 (actes de Colloques 3). Centre National de la Recherche Scientifique/IFREMER.

McCambridge, J., and McMeekin, T.A. (1979). Protozoan predation of *Escherichia coli* in estuarine waters. *Water Research*, **13**, 659–61.

McManus, G.B., and Fuhrman, J.A. (1986). Bacterivory in seawater studied with the use of inert fluorescent particles. *Limnology and Oceanography*, **31**, 420–6.

McManus, G.B., and Fuhrman, J.A. (1988). Control of marine bacterioplankton populations: measurement and significance of grazing. *Hydrobiologia*, **159**, 51–62.

Maki, J.S., and Remson, C.C. (1981). Comparison of two direct-count methods for determining metabolizing bacteria in fresh water. *Applied and Environmental Microbiology*, **41**, 1132–8.

Malmcrona-Friberg, K., Tunlid, A., Marden, P., Kjelleberg, S., and Odham, G. (1986). Chemical changes in cell envelope and poly-β-hydroxybutyrate during short term starvation of a marine bacterial isolate. *Archiv Microbiologie*, **144**, 340–5.

Martinez, J., Garcia-Lara, J., and Vives-Rego, J. (1989). Estimation of *E. coli* mortality in seawater by the decrease of ³H-label and electron transport system activity. *Microbial Ecology*, **17**, 219–25.

Meyer-Reil, L.A. (1978). Autoradiography and epifluorescence micro-combined for the determination of number and spectrum of actively metabolizing bacteria in natural waters. *Applied and Environmental Microbiology*, **36**, 506–12.

Minear, R.A. (1972). Characterization of naturally occurring dissolved organophosphorous compounds. *Environmental Science and Technology*, **6**, 431–7.

Moebus, K. (1980). A method for the detection of bacteriophages from ocean water. *Heligolander Meeresuntersuchungen*, **34**, 1–14.

Moebus, K., and Nattkemper, H. (1981). Bacteriophage sensitivity patterns among bacteria isolated from marine waters. *Heligolander Meeresuntersuchungen*, **34**, 375–85.

Montagna, P.A. (1984). *In situ* measurement of meiobenthic grazing rates on sediment bacteria and edaphic diatoms. *Marine Ecology Progress Series*, **18**, 119–30.

Morita, R.Y. (1985). Starvation and miniaturisation of heterotrophs, with special emphasis on maintenance of the starved viable state. In *Bacteria in the natural environments: the effect of nutrient conditions*, (Ed. M. Fletcher and G. Floodgate), pp. 111–30. Academic Press, New York.

Newell, S.Y., Sherr, B.F., Sherr, E.B., and Fallon, R.D. (1983). Bacterial response to the presence of eukaryotic inhibitors in water from a coastal marine environment. *Marine Environmental Research*, **10**, 147–57.

Novitsky, J.A., and Morita, R.Y. (1976). Morphological characterization of small cells resulting from nutrient starvation of a psychrophilic marine *Vibio*. *Applied and Environmental Microbiology*, **32**, 616–22.

Novitsky, J.A., and Morita, R.Y. (1977). Survival of a psychrophilic vibrio under long-term nutrient starvation. *Applied and Environmental Microbiology*, **33**, 635–41.

Novitsky, J.A., and Morita, R.Y. (1978). Possible strategy for the survival of marine bacteria under starvation conditions. *Marine Biology*, **48**, 289–95.

Nusbaum, I., and Garver, R.M. (1955). Survival of coliform organisms in Pacific Ocean coastal waters. *Sewage and Industrial Wastes*, **27**, 1383–90.

Nystrom, T., Albertson, N., and Kjellberg, S. (1988). Synthesis of membrane and periplasmic proteins during starvation of a marine *Vibrio* sp. *Journal of General Microbiology*, **134**, 1645–51.

Obst, U., Hubner, I., Wecker, M., and Bitter-Suerman, D. (1989). Immunological method using monoclonal antibodies to detect enterobacteriaceae in drinking water. *Aqua*, **38**, 136–42.

Oger, C., Hernandez, J.F., Delattre, J.M., Delabroise, A.H., and Krupsky, S. (1987). Etude par épifluorescence de l'évolution de la microflore totale dans une eau minérale enbouteillée. *Water Research*, **21**, 469–74.

Ogram, A.V., and Sayler, G.S. (1988). The use of gene probes in the rapid analysis of natural microbial communities. *Journal Industrial Microbiology*, **3**, 281–92.

Oliver, J.D., and Stringer, W.F. (1984). Lipid composition of a psychrophilic marine *Vibrio* sp. during starvation-induced morphogenesis. *Applied and Environmental Microbiology*, **47**, 461–6.

Olsen, R.A., and Bakken, L.R. (1983). Electron microscopical investigations of bacterial cells released from soil. *Third international symposium on microbial ecology*, Abstract, p. 81.

Pace, M.L. (1988). Bacterial mortality and the fate of bacterial production. *Hydrobiologia*, **159**, 41–9.

Palmer, L.M., Baya, A.M., Grimes, D.J., and Colwell, R.R. (1984). Molecular genetic and phenotypic alteration of *Escherichia coli* in natural water microcosms containing toxic chemicals. *FEMS Microbiological Letters*, **21**, 169–73.

Peters, H., Jurs, M., Jann, B., Jann, K., Timmis, K.N., and Bitter-Suerman, D. (1985). Monoclonal antibodies to enterobacterial common antigen and to *Escherichia coli* lipopolysaccharide outer core: demonstration of an antigenic determinants shared by enterobacterial common antigen and *E. coli* K5 capsular polysaccharide. *Infection and Immunity*, **50**, 459–66.

Peterson, B.J., Hobbie, J.E., and Haney, J.F. (1978). *Daphnia* grazing on natural bacteria. *Limnology and Oceanography*, **23**, 1039–44.

Pillai, T.N.V., and Ganguly, A.K. (1972). Nucleic acid in the dissolved constituents of seawater. *Journal of the Marine Biological Association of India*, **14**, 384–90.

Pomeroy, L.R. (1974). The ocean's food web, a changing paradigm. *Bioscience*, **24**, 499–504.

Porter, K.G., and Feig, Y.S. (1980). Use of DAPI for identifying and counting aquatic microflore. *Limnology and Oceanography*, **25**, 943–8.

Porter, K.G., Feig, Y.S., and Vetter, E.F. (1983). Morphology, flow regimes and filtering rates of *Daphnia*, *Ceriodaphnia*, and *Bosmina* fed natural bacteria. *Oecologia*, **58**, 156–63.

Porter, K.G., Sherr, E.B., Sherr, B.F., Pace, M., and Sanders, R.W. (1985). Protozoa in planktonic food webs. *Journal of Protozoology*, **32**, 409–15.

Postgate, J.R. (1977). Death in macrobes and microbes. In *The survival of vegetative microbes*, (Ed. T.R.G. Gray and J.R. Postgate). Society for General Microbiology, Cambridge University Press, Cambridge.

Postgate, J.R., and Hunter, J.R. (1962). The survival of starved bacteria. *Journal of General Microbiology*, **29**, 233–67 (*Erratum*, **34**, 473).

Postgate, J.R., Crumpton, J.E., and Hunter, J.R. (1961). The measurement of bacterium viability by slide culture. *Journal of General Microbiology*, **24**, 15–24.

Pretorius, W.A. (1962). Some observations on the rôle of coliphages in the number of *Escherichia coli* in oxidation ponds. *Journal of Hygiene*, **60**, 279–81.

Prieur, D., Troussellier, M., Romana, A., Champoux, S., Mevel, G., and Baleux, B. (1987). Evolution of bacterial communities in the Gironde Estuary (France) according to a salinity gradient. *Estuarine, Coastal and Shelf Science*, **24**, 95–108.

Proctor, L.M., and Fuhrman, J.A. (1990). Viral mortality of marine bacteria and cyanobacteria. *Nature*, **343**, 60–2.

Riemann, B., Nielsen, P., Jeppesen, M., Marcussen B., and Fuhrman, J.A. (1984). Diel changes in bacterial biomass and growth rates in coastal environments, determined by means of thymidine incorporation into DNA, frequency of dividing cells (FDC), and microautoradiography. *Marine Ecology Progress Series*, **17**, 227–35.

Rogers, H.J. (1979). The function of bacterial autolysis. In *Microbial polysaccharides and polysaccharases*, (Ed. R.C.W. Berkley, G.W., Gooday, and D.C. Ellwood), pp. 237–68. Society for General Microbiology, Academic Press, London.

Rogers, H.J., and Forsberg, C.W. (1971). Rôle of autolysis in the killing of bacteria by some bactericidal antibiotics. *Journal of Bacteriology*, **108**, 1235–43.

Roszak, D.B., and Colwell, R.R. (1987). Survival strategies of bacteria in the natural environment. *Microbiological Reviews*, **51**, 365–79.

Roszak, D.B., Grimes, D.J., and Colwell, R.R. (1984a). Viable but non-culturable aquatic bacteria to act as a recipient of plasmid DNA. *Applied and Environmental Microbiology*, **54**, 115–17.

Roszak, D.B., Grimes, D.J., and Colwell, R.R. (1984b). Viable but non-recoverable stage of *Salmonella enteritidis* in aquatic systems. *Canadian Journal of Microbiology*, **30**, 334–8.

Ryssov-Nielsen, H. (1975). Measurement of the inhibition of respiration in activated sludge by a modified deterination of TCC dehydrogenase activity. *Water Research*, **9**, 1179–85.

Sanders, P., and Porter, K.G. (1986). Use of metabolic inhibitors to estimate protozooplankton grazing and bacterial production in a monomitic eutrophic lake with an anaerobic hypolimnion. *Applied and Environmental Microbiology*, **52**, 101–7.

Sayler, G.S., Shields, M., Tedford, E.T., Breen, A., Hooper, S.W., Sirotkin, K.M., and Davis, J.W. (1985). Application of DNA–DNA colony hybridization to the detection of catabolic genotypes in environmental samples. *Applied and Environmental Microbiology*, **49**, 1295–303.

Schlegel, H.G. (1986). *General Microbiology*, (6th edn), Cambridge University Press, Cambridge.

Schmidt, E.L. (1974). Quantitative autoecological study of micro-organisms in soil by immunofluorescence. *Soil Science*, **118**, 141–9.

Schultz, J.E., Later, G.I., and Matin, A. (1988). Differential regulation by cyclic AMP of starvation protein synthesis, in *Escherichia coli*. *Journal of Bacteriology*, **170**, 3903–9.

Servais, P., and Menon, P., (1991). Fate of autochthonous and fecal bacteria in marine ecosystem. *Kieler Meeresforschungen*, **8**, 290–6.

Servais, P., Billen, G., and Vives-Rego, J. (1985). Rate of bacterial mortality in aquatic environments. *Applied and Environmental Microbiology*, **49**, 1448–54.

Servais, P., Billen, G., Martinez, J., and Vives-Rego, J. (1989). Estimating bacterial mortality by the disappearance of ^{3}H-labeled intracellular DNA. *FEMS Microbiology Ecology*, **62**, 119–26.

Sherr, B.F., Sherr, E.B., and Berman, T. (1983). Growth, grazing and ammonia excretion rates of a heterotrophic microflagellate fed with four species of bacteria. *Applied and Environmental Microbiology*, **45**, 1196–201.

Sherr, B.F., Sherr, E.B., Andrews, T.L., Fallon, R.D., and Newell, S.Y. (1986). Trophic interactions between protozoa and bacterioplankton in estuarine water analyzed with selective metabolic inhibitors. *Marine Ecology Progress Series*, **22**, 169–79.

Sherr B.F., Sherr, E.B., and Fallon, R.D. (1987). Use of monodispersed fluorescently labeled bacteria to estimate *in situ* protozoan bacterivory. *Applied and*

Environmental Microbiology, **53**, 958–65.

Sherr, B.F., Sherr, E.B., and Rassoulzadegan, F. (1988). Rates of digestion of bacteria by marine phagotrophic protozoa: temperature dependence. *Applied and Environmental Microbiology*, **54**, 1091–5.

Shilo, M. (1984). *Bdellovibrio* as a predator. In *Current perspectives in microbial ecology*, (Ed. M.J. Klug and C.A. Reddy), pp. 334–9). American Society for Microbiology, Washington DC.

Spencer, R. (1957). A possible example of geographical variation in bacteriophage sensitivity. *Journal of General Microbiology*, **17**, Proc, 11.

Spencer, R. (1960). Indigenous marine bacteriophages. *Journal of Bacteriology*, **79**, 614.

Stanier, R.Y., Adelberg, E.A., and Ingraham, J.L. (1987). *General Microbiology*, (5th edn.), McMillan, London.

Stewart, G.J., and Carlson, C.A. (1985). The biology of natural transformation. *Annual Review of Microbiology*, **40**, 211–35.

Sudo, S.Z., and Dworkin, M. (1973). Comparative biology of prokaryotic resting cells. *Advances in Microbial Physiology*, **9**, 153–219.

Tabor, P.S., and Neihof, R.A. (1982a). Improved method for determination of respiring individual micro-organisms in natural waters. *Applied and Environmental Microbiology*, **43**, 1249–55.

Tabor, P.S., and Neihof, R.A. (1982b). Improved microautoradiographic method to determine individual micro-organisms active in substrate uptake in natural waters. *Applied and Environmental Microbiology*, **44**, 945–53.

Tabor, P.S., and Neihof, R.A. (1984). Direct determination of activities for micro-organisms of Chesapeake Bay populations. *Applied and Environmental Microbiology*, **48**, 1012–19.

Tabor, P.S., Ohwada, K., and Colwell, R.R. (1981). Filterable marine bacteria found in the deep sea: distribution, taxonomy and response to starvation. *Microbial Ecology*, **7**, 67–83.

Turley, C.M., Newell, R.C., and Robins, D.B. (1986). Survival strategies of two small marine ciliates and their rôle in regulating bacterial community structure under experimental conditions. *Marine Ecology Progress Series*, **33**, 59–70.

Ward, B.B. (1984). Combined autoradiography and immunofluorescence for estimation of single cell activity by ammonium-oxidizing bacteria. *Limnology and Oceanography*, **29**, 402–10.

Wiggins, B.A., and Alexander, M. (1985). Minimum bacterial density for bacteriophage replication: implications for significance of bacteriophages in natural ecosystems. *Applied and Environmental Microbiology*, **49**, 19–23.

Wikner, J., Andersson, A., Normark, S., and Hagstrom, A. (1986). Use of genetically marked minicells as a probe in measurement of predation on bacteria in aquatic environments. *Applied and Environmental Microbiology*, **52**, 4–8.

Williams, S.T. (1985). Oligotrophy in soil: fact or fiction? In *Bacteria in their natural environments*, (Ed. M. Fletcher and G.D. Floodgate). Academic Press, New York.

Wright, R.T., and Coffin, R.B. (1984). Measuring microzooplankton grazing on plantonic marine bacteria, by its impact on bacterial production. *Microbial Ecology*, **10**, 137–49.

Xu, H.S., Roberts, N., Singleton, F.L., Attwell, R.W., Grimes, D.J., and Colwell, R.R. (1982). Survival and viability of non-culturable *Escherichia coli* and *Vibrio cholerae* in the estuarine and marine environment. *Microbial Ecology*, **8**, 313–23.

Xu, H.S., Roberts, N.C., Adams, L.B., West, P.A., Siedeling, R.J., Huq, A., Huq, M.I., Rahman, R., and Colwell, R.R. (1984). An indirect fluorescent antibody staining procedure for detection of *Vibrio cholerae* serovar 01 cells in aquatic environmental samples. *Applied and Environmental Microbiology*, **52**, 297–301.

Zimmerman, R., and Meyer-Reil, L.A. (1974). A new method for fluorescence staining of bacterial populations on membrane filters. *Kieler Meeresforschungen*, **30(1)**, 24–7.

Zimmerman, R., Iturriaga, R., and Becker-Birck, J. (1978). Simultaneous determination of the total number of aquatic bacteria and the number thereof involved in respiration. *Applied and Environmental Microbiology*, **36**, 926–35.

Zobell, C.E. (1936). Bactericidal action of seawater. *Proceedings of the Society for Experimental Biology and Medicine*, **34**, 113–16.

9

Environmental release of *Bacillus thuringiensis*

M.P. Meadows

Introduction

MICROBIAL INSECTICIDES, of which *Bacillus thuringiensis* comprises 80–90 per cent of the current market value of $100 million per annum, occupy a small part of the total $20 billion per annum spent on insecticides globally (Daoust 1990). However, use of microbial pest control agents, in particular *B. thuringiensis*, looks likely to increase dramatically in future. Growing pressure to reduce the use of some synthetic chemical insecticides, because of their environmental impact and the increasing problem of pest resistance to synthetic chemicals (Georghiou and Mellon 1983), combined with advances in the research and technology of *B. thuringiensis*, indicate that its market share and hence, release in the environment, are due to increase. Application of recombinant DNA technology to *B. thuringiensis* gives the potential for new strains with increased potency and wider host spectra. The possible environmental impact of the use of a control agent on such a scale must be assessed, particularly as technology to produce recombinant strains is well advanced and transgenic plants containing *B. thuringiensis* genes are already being field-tested (Delanney *et al.* 1989). Fortunately, there is a well-documented history of the use of *B. thuringiensis* in the environment as it has been released as an insecticide in increasing quantities for over three decades on crops, ornamentals, forest trees and stored grains. There is a good background of knowledge from the decades of pure and applied research that have accompanied the exploitation of *B. thuringiensis*.

Much effort has gone into isolating new strains of *B. thuringiensis* from the environment and also into the study of the exchange of genes encoding insecticidal toxin production between strains of *B. thuringiensis*.

This chapter will briefly describe the biology of *Bacillus thuringiensis* and cover in more detail its natural occurrence and ecology. An understanding of its ecology will help in the modelling and prediction of the fate and possible effects of the release of genetically modified strains of *B. thuringiensis*. Release of *B. thuringiensis* into the environment will be reviewed to show why its well-documented history of use and particular properties make it a suitable candidate as a model for study of the release of other genetically manipulated micro-organisms. Finally, the possible environmental impact of using recombinant strains of *B. thuringiensis* and insect-resistant plants expressing *B. thuringiensis* genes will be discussed.

Advantages of microbial control

Use of microbial control agents has specific advantages over synthetic chemical insecticides. Microbial agents are usually highly specific, affecting only the target pests, allowing beneficial insects and natural predators of the pests to survive. There is little or no environmental impact associated with use of microbial agents and they are safe for the workers handling them. *B. thuringiensis* is the most widely used of these agents. Its major advantages are that it can be grown in bulk by conventional industrial liquid fermentation, making it relatively

Table 9.1. *Classification[a] of* Bacillus thuringiensis *by pathotype and crystal protein genes*

Pathotype	A	B	C	D	E
Host range	Lepidoptera	Diptera	Coleoptera	Lepidoptera and Diptera	Not toxic at present
Gene type	CryIA(a)	CryIVA	CryIIIA	CryIIA	
	CryIA(b)	CryIVB	CryIIIB		
	CryIA(c)	CryIVC			
	CryIB	CryIVD			
	CryIC	CytA[b]			
	CryID				
Representative varieties	kurstaki HD-73	israelensis	tenebrionis	kurstaki HD-1	dakota
	alesti	kyushuensis	san diego	galleriae HD-29	canadensis
	dendrolinus	morrisoni PG14	EG2158	aizawai ICI	indiana

[a] Modified from Ellar *et al.* 1986 and Höfte and Whiteley 1989.
[b] Cytolytic and haemolytic.

inexpensive to produce, although it still costs more to produce than most synthetic chemical insecticides; it is stable in storage, is pathogenic to many of the world's major crop pests (Krieg and Langenbruch 1981; Burges and Daoust 1986); and, once applied, it does not persist for long in the environment. In addition, reports of resistance developing in a target pest in the field are rare (McGaughey 1985; Tabashnik *et al.* 1990)

Bacillus thuringiensis

B. thuringiensis is a Gram-positive, aerobic, rod-shaped bacterium indistinguishable from *Bacillus cereus* by most taxonomic criteria (Baumann *et al.* 1984), except for its outstanding feature – production during sporulation of a proteinaceous parasporal crystal. On completion of growth, cells lyse, releasing spores and crystals. Depending on the strain, the crystal, which can comprise 30 per cent of the cell dry weight, can have highly specific toxic properties. Many strains kill the larvae of certain lepidopterous species, others kill Diptera (reviewed in Federici *et al.* 1990), coleopteran species (Krieg *et al.* 1983; Herrnstadt *et al.* 1986) or nematodes (Bottjer *et al.* 1985). There are also many strains with no known toxic activity. The biology and use of *B. thuringiensis* have been reviewed extensively (Burges and Hussey 1971; Bulla *et al.* 1980; Burges 1981a; Aronson *et al.* 1986; Burges 1986a,b).

Many *B. thuringiensis* strains, with different host spectra, have been identified. These strains have traditionally been classified into serotypes or subspecies based on their flagellar antigens (Bonnefoi

and de Barjac 1963), which show only limited correlation with the insecticidal activity in different hosts. As more new strains are isolated and the known host range of *B. thuringiensis* increases, it has become more informative to use a classification based on the insecticidal activity, which is after all, the distinctive and useful feature of *B. thuringiensis* (at the present time). Table 9.1 presents a combination of the classification systems based on pathogenicity (i.e. activity of the crystals in the presence of other factors, such as the spore) as suggested by Krieg *et al.* (1983) and Ellar *et al.* (1986), and the system based on the structural relatedness and activity spectra of the crystal proteins themselves (Höfte and Whiteley 1989). At the time of writing, 15 distinct crystal protein (*cry*) genes have been identified. Thirteen of these genes are divided into four major classes and a number of subclasses on the basis of structural similarities and anti-insect activity. There is also a protein produced by *B. thuringiensis* var. *israelensis* that exhibits non-specific cytolytic activity against a variety of invertebrate and vertebrate cells. This protein gene is structurally unrelated to the *cry* genes and so is placed in a separate group, *cytA*.

Mode of action

B. thuringiensis is a stomach poison and thus must be ingested to be effective. In most cases, infection resulting from the spore plays a secondary role in killing larvae. The crystal is an inert protoxin, which dissolves in the larval midgut releasing one or more large polypeptides (mol. wt approximately 27 000–140 000). These polypeptides are processed by gut enzymes to release insecticidal crystal proteins, also known as δ-endotoxins, largely in the range mol. wt 25 000–65 000 (Bulla *et al*. 1980; Fast 1981; Faust and Bulla 1982; Ellar *et al*. 1985). The activated toxins act at the outer membrane of the midgut epithelial cells of susceptible insects. The toxins interact with phospholipids in var. *israelensis* in mosquitoes (Thomas and Ellar 1983) and with glycoproteins in other varieties of *B. thuringiensis* in lepidopteran larvae (Ellar *et al*. 1985). These interactions are believed to generate non-specific hydrophilic pores in the cell membrane, which destroy the regulation of ion exchange and solutes, such as glucose, thus disturbing the osmotic balance, with the result that cells swell and lyse (Knowles and Ellar 1987). Simultaneously, the muscles of the gut and mouthparts are paralysed and feeding stops. Death follows in 10 minutes–3 days, according to dose ingested and species of host. Specific high affinity binding sites for several *B. thuringiensis* toxins exist on the midgut epithelium of susceptible insects (Hofmann *et al*. 1988) and this might explain the highly specific nature of these proteins.

Natural occurrence

B. thuringiensis is often found associated with insects, but epizootics (widely diffuse and rapidly spreading infections) are rare in the outdoor environment (Burges 1973). It can be found enzootic (i.e. ever-present but not at outbreak proportions) in insects in fields, forests and orchards, but epizootics usually occur only indoors where insects are kept in confined environments (Burges 1973). Indeed, one of the present major commercial strains, HD-1, was isolated from a mass-reared colony of the pink bollworm, *Pectinophora gossypiella* (Dulmage 1970).

Early studies found *B. thuringiensis* to be present only at low levels in some soils. For example, in the USA, only 0.5 per cent of the isolates from the *B. cereus* and *B. thuringiensis* morphological group were classfied as *B. thuringiensis* (DeLucca *et al*. 1981). A study of Japanese soils found 2.7 per cent of the same grouping to be *B. thuringiensis* (Ohba and Aizawa 1978). In the Philippines, only 12 out of 54 soil samples were found to harbour *B. thuringiensis* (Padua *et al*. 1982). However, these studies may have underestimated the distribution of *B. thuringiensis* in soil because of the difficulty of

distinguishing it in soil samples containing a high background level of other soil micro-organisms.

The development of the acetate selection technique (Travers *et al*. 1987) allowed a more accurate determination of the distribution of *B. thuringiensis*. This simple technique increases the efficiency of isolation of crystaliferous spore-forming bacteria in the presence of other spore formers. In a study of soils from the USA and 29 other countries covering five continents, *B. thuringiensis* was isolated from 785 of 1115 soil samples, a rate of isolation that indicated that it is a widespread soil micro-organism (Martin and Travers 1989). The percenage of samples from any one area containing *B. thuringiensis* varied depending on the continent from which the sample came. *B. thuringiensis* was found in 94 per cent of soil samples from Asia and Central and South Africa, 84 per cent of the samples from Europe and the lowest recoveries were from New Zealand and USA soils (56 and 60 per cent, respectively). *B. thuringiensis* was most frequent in samples from savannah, desert and agricultural land, but was also isolated from urban areas, forest, Arctic tundra and steppe. It was relatively less common in beaches, caves, some desert locations and rainforest. Surprisingly, for an entomopathogen, there appeared to be no correlation between presence of insects and number of *B. thuringiensis* isolates.

B. thuringiensis has also been isolated from the phylloplane of deciduous and coniferous trees in North America (Smith and Couche 1991). During a 3-year survey, 50–70 per cent of trees sampled harboured *B. thuringiensis*. Frequency and distribution of *B. thuringiensis* on leaves indicated that it could possibly be considered an epiphyte of the phylloplane in temperate regions. *B. thuringiensis* is also a natural occupant of grain dusts in stored product warehouses (Norris 1969; Burges and Hurst 1977). DeLucca *et al*. (1982) sampled both settled and respirable dust samples from elevators in grain stores. Of the settled dust samples, 55 per cent contained *B. thuringiensis*, which was also recovered from 17 per cent of the respirable dust samples.

In summary, *B. thuringiensis* can be readily isolated from insects, soil, stored product dusts, sericulture environments and leaves from deciduous and coniferous trees. It is quite rare to find samples of natural field materials that do not contain crystaliferous spore-forming bacteria (Meadows *et al*. 1990). This leads to the question 'Why is an entomopathogen so widely distributed in the environment?'

Ecology

Before estimating the risk of releasing a microorganism into the environment one needs to know the ecological role of that microbe, in short, the way that it interacts with its surroundings and the other biota. Perhaps surprisingly for a bacterium that has been used commercially and released in large quantities, relatively little is known about the ecology of *B. thuringiensis*. It is a natural pathogen of many insect species and, depending on the insect species and age, is sometimes able to multiply in their cadavers. It can, therefore, often be isolated in association with insects. As an enzootic it is maintained and spread through some insect populations, although it is not clear why it is also ubiquitous in soil. Spores might be released from an insect cadaver and disseminated by wind and water and simply be deposited in the soil. The bacterium might also be capable of growth and sporulation within the soil without infecting an insect. It may be present in soil animals that have yet to be studied or it may enter the soil when trees shed their leaves. In nature there may be a combination of two or more of these factors.

Studies have shown that *B. thuringiensis* is not capable of germination and growth in most soils, but a proportion of the spores can persist and remain viable for some time (Pruett *et al.* 1980; Petras and Cassida 1985; Martin and Reichelderfer 1989). Pruett *et al.* (1980) inoculated soil in the laboratory with spores and crystals of *B. thuringiensis* var. *galleriae*. The number of spores (estimated as colony-forming units, CFU) fell slowly to 24 per cent of the initial number over 135 days. Pathogenicity to the wax moth *Galleria melonella* fell rapidly to 12 percent in 15 days and to < 1 per cent in 135 days, indicating that crystals were degraded more rapidly than spores. The soil was not pasteurized during the isolation process, so presumably any vegetative cells present as the result of growth would have been detected. West *et al.* (1985) compared the effect of pH, nutrient availability and moisture content on the survival of *B. thuringiensis* and *B. cereus* spore inocula in soil in the laboratory. The most important factor was nutrient availability – *B. thuringiensis* did not grow

under most conditions in unamended soil. Numbers per gram remained stable in soil except when it had been saturated with moisture, where a slow 10-fold increase was detected, and in dry soil where a slow decrease of 60 per cent over 64 days occurred. In contrast, *B. cereus* increased approximately 10- to 100-fold in all but the driest soil. When the nutrient availability of the soil was increased, either by autoclaving or the addition of nutrients, growth of both bacilli was stimulated. In another study, viable log phase vegetative cells of *B. thuringiensis* added to untreated soil rapidly disappeared, a 91 per cent loss occurring in 24 hours. No spores were detected following the loss. The initial rapid disappearance was following by an exponential rate of loss between 1 and 10 days, after which cells were no longer detectable by immunofluorescent techniques (West *et al.* 1984). In contrast, when spores were added, their number remained unaltered for 91 days and, during this period, no germination took place. Scanning electron micrographs showed the possible degradation of *B. thuringiensis* vegetative cells by an actinomycete-like micro-organism. Casida (1988) has identified bacteria that degrade both *B. thuringiensis* spores and crystals, so it is possible that the reduction in *B. thuringiensis* numbers in soil may at least in part be due to predation by other micro-organisms.

The prolonged survival of *B. thuringiensis* spores in soil appears to be as a result of their inability to germinate in that soil. If the spores were to germinate, death from nutrient limitation would probably occur rapidly. This ability not to germinate may confer an advantage on *B. thuringiensis*, enabling it to persist in the relatively poor nutrient environment of soil, where growth, if possible at all, is usually slow. However, in the nutrient-rich environment of an insect cadaver it germinates and is able to grow and multiply rapidly, and produce large numbers of spores. Some evidence suggests that the ability to produce crystals that are toxic to some insects may be linked to the conditions required for spore germination: loss of plasmids encoding for the protein crystal can simultaneously alter the requirements for spore germination (Jarret 1985). Acrystalliferous or small crystal mutants have been found to respond to lower concentrations of germination-stimulating compounds and to germinate must faster than the crystal-producing parent strains (Jarret 1985).

Release into the environment

Use of *Bacillus thuringiensis* as a microbial insecticide

B. thuringiensis has been released into the environment in increasing quantities since 1933, but it was not successfully commercialized until the 1950s, when the then new technology of deep tank aerobic liquid fermentation was applied to the production of spore and crystal preparations. Commercial use involved large-scale releases of a few strains; other strains have been released in smaller quantities for experimental purposes and field trials.

Major releases of *B. thuringiensis* have taken place in North America, where it is used to control over 40 pest species in fields, forests, orchards, vineyards, parkland and gardens (Burges and Daoust 1986). The requirement for a safe, effective insecticide for forest regions containing residential areas provided an ideal opportunity for the application of a microbial insecticide, and Lepidopteran-active varieties of *B. thuringiensis* have been used in increasingly large quantities on forests for the last 30 years. The first aerial spray applications were in 1961 when spruce budworm, *Choristoneura fumiferana*, was treated in New Brunswick (Mott *et al.* 1961) and western black-headed budworm, *Acleris gloverana*, was treated in British Columbia (Kinghorn *et al.* 1961). Use of *B. thuringiensis* in North America has increased greatly over recent years. Between 1979 and 1983, 1–4 per cent of the sprayed forest was treated with *B. thuringiensis* in eastern Canada. By 1988, the figure had increased to 63 per cent. Again in eastern Canada, between 1985 and 1988, a total of 1 856 548 hectares were treated to control spruce budworm. In Ontario, *B. thuringiensis* was used on a total of 836 171 hectares to control jack pine budworm between 1985 and 1987, and on a total of 35 756 hectares to control hemlock looper in Newfoundland between 1985 and 1988 (Cunningham 1990). An approximation of the number of viable spores released into the environment during such applications can be made as follows. *B. thuringiensis* is applied at a rate of, for example, 30 billion international units (IU) per hectare to control spruce budworm (Morris 1982). There may be approximately $3–6 \times 10^4$ viable spores per IU (taking a 1974 formulation of Dipel^{TM}, Abbott Laboratories, as an example). This represents a release of 10^{15} viable spores per hectare.

B. thuringiensis var. *israelensis* has also been released in large quantities for disease vector control. *B. thuringiensis* var. *israelensis* was first discovered in Israel at a mosquito habitat (Goldberg and Margalit 1977). Its host range includes mosquitoes (Goldberg and Margalit 1977; Tyrell *et al.* 1979) blackflies (Undeen and Nagal 1978). Its first major release was in the river blindness eradication programme against blackflies in the Upper Volta region of Africa (Guillet 1984). The insects had begun to show serious resistance to synthetic pesticides used for their control. In the first year, 40 tons of *B. thuringiensis* var. *israelensis* formulation was released, providing successful control. Because of its relative expense, compared to the synthetic pesticides, it was used only in the dry season when water areas requiring treatment were at their lowest. In the wet season the organophosphate, chlorophoxim, was used but resistance to this chemical also became a problem (Guillet 1984). This resistance proved transitory and was abated by the use of *B. thuringiensis* var. *israelensis* the following summer, so that subsequently the bacteria and the chemical could be used in alternate seasons. In another application, *B. thuringiensis* var. *israelensis* has been released on a large scale in a mosquito-infested area in a southern region of Switzerland (Luthy 1989). Smaller scale releases have been used for the control of mosquitoes and blackflies, for example along the Rhine in Switzerland, Thorney Island in the UK, in California and on a large scale in the south eastern USA and Australia.

Fate in the environment

B. thuringiensis rarely spreads by epizootics either naturally or when applied as an insecticide. This has some disadvantages in terms of control of pest insects (see p. 126) but the potential for environmental impact is lower. There has been much study of the persistence of *B. thuringiensis* in the field following application, but little study of its movement. For example, Martin and Reichelderfer (1989) studied field persistence using antibiotic-resistant marked strains. Spores and crystals were applied at a rate of 10^{12} spores/acre on potato and corn. Recovery from leaves declined to 10 per cent in 1 week and 1 per cent in three weeks on potato, and to 10 per cent in 3 weeks on corn, with 5 per cent retained throughout winter. In soil, spores declined to 10 per cent in 3 weeks (in good agreement with laboratory studies of West *et al.*

(1985) and Petras and Casida (1985), see p. 123) after which the persistence curve remained on a plateau for 15 months. Some increase in spores was noted after frost, probably due to release of nutrients from the soil, which confirms that *B. thuringiensis* can grow in soil under certain conditions (West *et al.* 1984). No vertical movement deeper than 6 cm occurred and movement outside the plot was less than 10 m along drainage courses. No evidence of genetic exchange was noted, although potential for this to take place might increase if multiplication of *B. thuringiensis* were to occur in soil (Jarrett and Stephenson 1990, see p. 128).

Non-target organisms may play a role in dissemination of *B. thuringiensis*. For example, when fathead minnows were exposed in the laboratory to *B. thuringiensis* var. *israelensis* (2.2×10^5 CFU/ml), they accumulated spores within 1 hour of exposure, producing 4.0×10^6 CFU per fish (Snarski 1990). The major route of entry was by ingestion. There were no toxic effects on the fish. When the fish were transferred to clean water the bacterial count dropped rapidly by 1000-fold in 1 day, but spores were detected in faeces for over 2 weeks. Thus, fish might play a major role in transport of *B. thuringiensis* applied as an insecticide, although the transported bacteria are too few to have a significant effect on mosquito control. It is also possible that *B. thuringiensis* applied on land could be dispersed by non-target organisms such as scavenging or predatory birds and mammals ingesting it in infected insects.

Persistence

B. thuringiensis is sensitive to the ultra-violet (UV) component of sunlight, although the spores and crystals may be deactivated at different rates (Jarrett 1980). The half life of *B. thuringiensis* spores when applied to foliage may be only a few days (Pinnock *et al.* 1971) but when in soil, spores receive an element of protection from UV light and are able to survive for much longer. Studies (see p. 123) have indicated that *B. thuringiensis* does not grow in certain soils but can persist at low levels as viable spores, which are unlikely to germinate. Spores produced in soil through multiplication of added vegetative cells survive for only a short time (Petras and Casida 1985). Thus spores that are isolated from soil may not have originated from growth in the soil. *B. thuringiensis* can also lose efficacy in foliar applications due to wash-off by

rain (Frankenhuyzen and Nystrom 1989). In addition, the persistence of *B. thuringiensis* var. *israelensis* in water is low (Margalit and Bobroglio 1984).

Persistence of *B. thuringiensis* spores and crystals on leaves can be increased by altering the formulation of the insecticidal product (Couche and Ignoffo 1981). For example, encapsulation of spores and crystals of *B. thuringiensis* HD-1 in a starch matrix (Dunkle and Shasha 1988) significantly improved their half life (McGuire *et al.* 1989). No loss of activity was recorded over an 8-day period and activity, expressed as an LC_{50} on the European corn borer, *Ostrinia nubilalis*, decreased by less than 50 per cent in 16 days. This compares to a half life in the field of around 2 days for unprotected *B. thuringiensis* (Pinnock *et al.* 1971).

Recombinant DNA technology has been used to increase persistence of toxic activity within the current guidelines of the United States Environment Protection Agency (EPA). The gene coding for the production of a single crystal protein has been cloned into a strain of *Pseudomonas fluorescens* (Gelernter and Quick 1989). The cells produce the toxin during growth; they are then harvested and killed by a chemical treatment. The resulting non-viable, spore-free product has a two-fold increased persistence of activity to the diamondback moth, *Plutella xylostella*. In 1990 this became the first pesticide developed using recombinant DNA technology to receive regulatory approval for field testing on a farm (Gelernter 1990).

Safety

The safety record of *B. thuringiensis* is impeccable. By 1981, it had been registered in products by four different companies, giving four-fold replication of safety tests (Burges 1981b). There have been no reports of harm associated with its use for pest control. It has proved totally safe for use even on food crops harvested the same day for human consumption. No problems have been reported in the food industry and it has been absent from cases of food poisoning. There have been no reports of *B. thuringiensis* having any effect on non-target organisms in the field. The only two reservations concerning its use are production by some strains of a soluble toxin, β-exotoxin, which has a less specific range of activity and is toxic to some mammals. (Sebesta *et al.* 1981), and the solubilized δ-endotoxin of *B. thuringiensis* var. *israelensis* which is cytolytic to a wide range of cells and is a potent mammalian toxin when administered by injection. However, in both cases this has not resulted in a health hazard. In the first case, as a precaution, the presence of β-exotoxin is forbidden in commercial products using spores and crystals as the active components; strains can be selected that do not produce β-exotoxin. In the second case, the un-solubilized *B. thuringiensis* var. *israelensis* δ-endotoxin produces no cytolytic effects (Thomas and Ellar 1983; Armstrong *et al.* 1985) and the δ-endotoxin is only solubilized under alkaline conditions, such as those found in many insect midguts.

Environmental impact

To the author's knowledge there have been no reports of adverse environmental impact caused by the release of *B. thuringiensis*.

Improvement in efficacy

Limitations

Growth in use of *B. thuringiensis* has been limited by certain disadvantages (Burges 1986a). In crop protection, only lepidopterous and celeopterous pests are killed, requiring different control measures for other pests, such as aphids. The limited host range of *B. thuringiensis* (Krieg and Langenbruch 1981) restricts it from protecting crops with a complex of lepidopterous species. Some species with boring larvae may not ingest a lethal dose before reaching the safety of inner, untreated, plant tissue. Larvae feeding on an open leaf surface may include species, such as *Spodoptera* spp., with low susceptibility to the present, most common commercial strain, HD-1. In all cases the toxin must be ingested to take effect, so application must be thorough, particularly at feeding sites of the most susceptible stages, young larvae. These surfaces are often the underside of leaves – difficult to reach with sprays (Jarrett *et al.* 1978). Thus, many of its failures have been due to poor application, giving *B. thuringiensis* an undeserved reputation for unreliability. In addition, the cost of *B. thuringiensis*-based products is still greater than that of the often highly effective synthetic chemical insecticides. Lack of environmental persistence means that *B. thuringiensis* must often be reapplied during a growing season and such constraints have limited its market. However, several factors are now increasing its market potential, e.g. ecological and safety concerns over the environmental impact of some synthetic chemicals and increasing incidence of pest resistance to nearly all classes of synthetic chemicals (Georghiou and Mellon 1983). Application of traditional and newer technologies to *B. thuringiensis* have potential for overcoming its limitations.

Strain development
Natural isolates
The Dulmage International Cooperative Programme (Dulmage 1981) was the first attempt to catalogue systematically the pathogenicity of a large number of *B. thuringiensis* isolates. Most of the isolates in this study came from insects, stored product residues or sericulture and the study revealed isolates with greater potencies against important crop pests than the best isolates then available. There has been an intensive effort, particularly more recently, by commercial concerns, to isolate new strains of *B. thuringiensis* with increased potency and/or wider, or different host ranges. Such searches have reaped rich benefits, with the isolation of strains that have widened the host spectrum of the bacterium from lepidopterous insects to dipterous (Goldberg and Margalit 1977) and coleopterous insects (Krieg *et al.* 1983; Herrnstadt *et al.* 1986; Donovon *et al.* 1988). The discovery by Travers *et al.* (1987) that *B. thuringiensis* can be readily selected from a large background of other soil micro-organisms further galvanized the isolation of new strains of *B. thuringiensis*.

Recombinant DNA technology applied to Bacillus thuringiensis
Another major development with potential for improvement of *B. thuringiensis* was the discovery that δ-endotoxins are single gene products, encoded on extrachromosomal units of DNA (plasmids) (Gonzalez and Carlton 1980; Gonzalez *et al.* 1981; Jarrett 1983). A conjugation-like process between *B. thuringiensis* strains (Gonzalez *et al.* 1982) provides a mechanism for transfer of plasmids at a frequency high enough to allow use of insects, the most sensitive and accurate means of determining activity, as a primary screen for desirable recombinants (Jarrett and Burges 1986). The genetics and molecular biology of *B. thuringiensis* have been reviewed elsewhere (Gonzalez and Carlton 1984; Carlton and Gonzalez 1985; Whitelely and Schnepf 1986; Aronson and Beckman 1987; Höfte and

Whiteley 1989). Recently, electroporation, a high voltage discharge through a cell suspension, resulting in a transient permeability in the cell membrane, allowing entry of DNA into cells (Chassy and Flickinger 1987), has been applied to *B. thuringiensis* (Bone and Ellar 1989; Masson *et al.* 1989). Thus, the opportunity is now available to tailor strains with improved potency against target pests and/or with host ranges more suitable than the ranges of currently available strains for controlling pest complexes on particular crops (Jarrett and Burges 1986).

The major problem of applying the toxin exactly where it will be eaten by caterpillars is being addressed by incorporating the toxin gene into transgenic plants to produce insect-resistant crops. Genes can be inserted using an *Agrobacterium tumefaciens*-based transformation system (Barton *et al.* 1987; Fischoff *et al.* 1987; Vaeck *et al.* 1987) into, for example, tobacco, potato, cotton and tomato. Recent development of the use of high velocity microprojectiles (Klein *et al.* 1988; McCabe *et al.* 1988) may make it possible to transform other crops, especially monocotyledons (e.g. corn), which cannot be transformed by the *Agrobacterium* system. The exciting developments in application of recombinant DNA technology to the expression of *B. thuringiensis* genes in plants have been reviewed elsewhere (Meeusen and Warren 1989; Vaeck *et al.* 1989). There have been notable achievements with the expression of *B. thuringiensis* δ-endotoxin genes in plants to produce insect-resistant cultivars (Fischoff *et al.* 1987; Vaeck *et al.* 1987). Levels of expression have been improved recently by the incorporation of modified *B. thuringiensis* coding sequences to produce expression to 0.1 per cent of the total protein in potato (Perlack and Fischoff 1990). These demonstrate increased resistance to insect damage. Tests are planned for several locations in the USA to assess insect tolerance under field conditions (Perlack and Fischoff 1990).

Use of natural isolates compared to recombinants
Unmodified strains and recombinants resulting from plasmid curing and conjugation are on the lists of permissible strains for environmental release in the USA and some other countries. This is because plasmid curing and conjugation can be regarded as natural events in which genes themselves are unaltered, providing no greater environmental risk than any other natural recombination

event. The US Environmental Potection Agency has specifically included *B. thuringiensis* strains, developed by plasmid curing and by plasmid transfer with natural unmodified strains, for the purpose of release for risk assessment under new *B. thuringiensis* registration standards issued in 1988 (Carlton *et al.* 1990). Thus, there are great advantages in the development and use of such natural isolates. For example, Ecogen Inc., in the USA, were able to develop, field-test and register three products within a 4-year period from project initiation, with total registration costs of less than $200 000 per product, including all the required toxological testing (Carlton *et al.* 1990). The strains in these products had been developed using a combination of plasmid curing, to eliminate plasmids harbouring less active genes from one parental strain, and conjugal transfer to introduce a plasmid from another parental strain, carrying a highly active gene against the relevant target pests. Many of the modifications in host range and potency that are required for the increased application of *B. thuringiensis* can be carried out using such technology.

Recent advances in both introduction of recombinant plasmids via electroporation and development of novel vectors (Lereclus *et al.* 1989; Schurter *et al.* 1989) enable application of recombinant DNA methodologies to *B. thuringiensis* itself. Several *Eschericia coli/Bacillus* shuttle vectors have been generated that facilitate development of novel *B. thuringiensis* recombinant strains (Gawron-Burke *et al.* 1990).

Use of recombinant DNA technology would provide certain advantages over natural isolates. A limitation to the use of *B. thuringiensis* is the need for repeated foliar applications resulting from inactivation by UV light, losses due to rainfall or dilution of the application as the plant grows. Genes could be cloned into a phylloplane bacterium for protection of leaves against foliar feeders. Using a similar approach soil-inhabiting, root-colonizing bacterium, *Pseudomonas fluorescens*, has been engineered to express a *B. thuringiensis* insecticidal protein (Watrud *et al.* 1985), providing a biological delivery system for specific pesticides active against specific soil-borne pests. One of the major disadvantages of the development of recombinant organisms is that extensive testing must be carried out before these modified organisms could be released into the environment. Whereas possible risks of environmental impact of unmodified *B. thuringiensis* are minimal, partly because of its low level of persistence, a genetically modified *P. fluorescens* may persist and multiply in a favourable soil environment. Potential environment risk of the reproduction and spread of such as organism would need to be carefully assessed (Watrud *et al.* 1985).

Bacillus thuringiensis as a model for the release of genetically modified organisms

Recent advances in recombinant DNA technology offer prospects for creating new varieties of organisms that can have enormous benefits for health, agriculture and environmental management. However, there are potentially adverse consequences for natural and man-made ecosystems (e.g. agricultural and urban) that must be addressed (Beltz *et al.* 1983; Brown *et al.* 1984; Levin 1984; Milewski and Tobin 1984). Progress on the application of recombinant DNA technology in agriculture is being hindered by lack of knowledge about the fate and effects of bioengineered organisms in the environment: there are several study areas that have been defined to increase such knowledge (Halvorson *et al.* 1985). Thus, factors affecting the persistence of introduced organisms, their reproduction rate, their rate of gene transfer to indigenous organisms, and their movement away from the site of application are of prime importance. Their effects on the balance and functioning of the exposed ecosystem must be assessed before a release can be considered (Trevors *et al.* 1986). *B. thuringiensis* would be a good model organism for such studies for the following reasons: natural isolates have been released into the environment in large quantities for over 30 years without reports of adverse environmental impact; the toxin proteins are highly specific and are highly unlikely to affect non-target organisms; the safety record of *B. thuringiensis* in use is second to none; host range is narrow and there are well-documented methods for the detection, isolation and enumeration of *B. thuringiensis*. In addition, recommended antibiotic or heavy metal resistance markers (Lindow and Panopoulos (1983) can easily be introduced using conjugation or electroporation to allow detection of released strains.

Plasmid transfer in the environment

Plasmids coding for δ-endotoxin genes are transferred at a high frequency between strains of *B.* *thuringiensis* and *B. cereus* by a conjugation-like process when grown in mixed culture (Gonzalez *et al.* 1982). Plasmids from *B. thuringiensis* have been transferred into *B. subtilis* and the soil-inhabiting, human pathogen *B. anthracis* (Klier *et al.* 1983). Self-transmissible cryptic plasmids in four *B. thuringiensis* subspecies have been identified (Reddy *et al.* 1987). Therefore, under laboratory conditions, plasmids can be shuttled among strains of *B. cereus*, *B. anthracis* and *B. thuringiensis*. This could conceivably take place in the environment, allowing for horizontal spread of a new genotype. Jarrett and Stephenson (1990) have demonstrated plasmid transfer between strains of *B. thuringiensis* growing in infected lepidopterous larvae in the laboratory. This is the first report to indicate how plasmid transfer between strains of *B. thuringiensis* might occur in the environment. Transfer rates of plasmids coding for crystal production and antibiotic resistance were high, reaching levels similar to those obtained in laboratory broth cultures. In addition, in broth cultures, *B. thuringiensis* was able to transfer plasmids into spore-forming bacteria from soil samples (Jarrett and Stephenson 1990. Thus, it has been shown that plasmids can be transferred between strains of *B. thuringiensis* and soil-inhabiting bacteria; plasmid transfer in soil has not been demonstrated.

Potential for pest resistance

The increasing use of *B. thuringiensis* offers potential benefits to agriculture and the environment. However, one of the possible consequences of increases in release is the possibility of resistance developing in the target pest.

Natural isolates

Use of synthetic chemical insecticides has led to pest resistance and there are many reports of resistance causing failure of control in the field. For example, the Colorado potato beetle, *Leptinotarsa decemlineata*, is resistant to a wide range of chemicals, including the arsenicals, organochlorines, carbamates, organophosphates and pyrethroids (Forgash 1985). The diamondback moth, *Plutella xylostella*, the most common worldwide vegetable pest, has developed resistance to 46 insecticides throughout the world (Miyata *et al.* 1986). In contrast, there are only two reports to date of significant pest resistance to *B. thuringiensis* applied in the field (McGaughey 1985; Tabashnik *et al.* 1990). Populations of the Indianmeal moth, *Plodia interpunctella*, collected from stored grain bins treated with *B. thuringiensis* were found to have low levels of resistance to *B. thuringiensis*. Much higher resistance was selected within a few generations under laboratory conditions (McGaughey 1985). Tabashnik *et al.* (1990) have reported field development of resistance in *Plutella xylostella* to foliar applications of commercial formulations of spore/crystal complexes of *B. thuringiensis* var. *kurstaki*. Laboratory bioassay of larvae showed that the LC_{50} and LC_{95} values for a field populations of *P. xylostella* treated repeatedly with *B. thuringiensis* were 25–33 times greater than the respective LC_{50} and LC_{95} values for two susceptible laboratory colonies. Mortality in the field was 34–35 per cent in two resistant populations, compared with 90–100 per cent in two susceptible laboratory colonies.

It is noteworthy that, despite the large-scale application of *B. thuringiensis* in forestry, there have been no reports of resistance in the target forest pests. Possible reasons for lack of high resistance have been reviewed by Burges (1971). Boman (1981) developed the theoretical argument that, as insects have encountered *B. thuringiensis* in nature there must be some level of resistance already in the population and that the low levels of expression are due to some selective disadvantage to those resistant individuals.

Wild strains of *B. thuringiensis* often contain more than one δ-endotoxin gene (Jarrett 1985). There are indications of specific receptors in the insect gut for each toxin protein (see p. 129). If each of the toxins in a wild strain acts on a different receptor then the chances of resistance developing to all the toxins should be slight (see p. 130). This may also contribute to the rarity of pest resistance to wild *B. thuringiensis*. Sneh and Schuster (1983) studied the effect of feeding *Spodoptera littoralis* sublethal concentrations of a highly-active strain of *B. thuringiensis*. No resistance was shown by larvae issued from survivors but, surprisingly, these larvae had more susceptibility to the δ-endotoxin than corresponding generations of untreated larvae. In another study, Vazquez-Garcia (1983) was unable to induce resistance to *B. thuringiensis* var. *israelensis* in *Culex quinquefasciatus* in the laboratory.

Ferro (1989) provided a well-argued case against the likelihood of development of resistance to *B. thuringiensis*-based products applied as insecticides. He used the Colorado potato beetle as a model insect, due to its ability rapidly to develop resistance to synthetic chemicals. *B. thuringiensis* var. *san diego* effectively controls the beetle under field conditions (Ferro and Gelernter 1989; Zehnder and Gelernter 1989). The likelihood of an insect developing resistance depends largely on the degree of selective pressure applied. When *B. thuringiensis* is applied as an insecticide, a proportion of the population always survives (particularly late instars, as they are the least susceptible). Many survive without selection for resistance individuals as foliar coverage is not complete, leaving refuges in the field where some larvae will escape exposure. *B. thuringiensis* does not persist in the field: exposure time is short as the preparation remains active for less than 48 hours. In addition, natural predators of the pest survive, allowing minimal application of the bioinsecticide. Thus, the exposure of the pest population to the selective agent is such that susceptible individuals are likely to survive and breed.

Possible mechanisms of resistance

Study of the mode of action of *B. thuringiensis* toxins may lead to understanding of possible mechanisms of resistance and hence to the better management of pest control. Interaction of toxins with high affinity binding sites on the brush border of midgut epithelial cells predominately determines host specificity of *B. thuringiensis* toxins (Hofmann *et al.* 1988). Toxicity is correlated to the binding of the various toxins to different receptor sites, so each toxin has distinct binding sites. Studies on the resistance mechanism at the toxin binding level have been carried out in only one insect species to date. Such studies (Cardoen *et al.* 1990; Van Rie *et al.* 1990) have demonstrated that the Indianmeal moth, *Plodia interpunctella*, can develop resistance to δ-endotoxin by changing binding characteristics of the midgut receptors. Bioassays with purified crystal proteins demonstrated that, in laboratory-selected *P. interpunctella*, resistance to *cryIA(b)* toxin correlates with a 50-fold reduction in affinity of the membrane receptors for that toxin. However, there was an increased susceptibility to *cryIC* toxin in the same strain, which correlated to a small but significant increase of *cryIC* binding sites. Thus, resistance to one toxin does not necessarily imply resistance to another toxin that is effective in the same target pest.

Genetically modified organisms expressing *Bacillus thuringiensis* toxin

Genetically modified bacteria

There is probably a greater chance of selecting for resistance to one factor than to two or more independent factors present simultaneously. Thus, if genetically modified strains of *B. thuringiensis* are used in pest control, it may be better to use strains with more than one cloned δ-endotoxin gene. The resistance mechanism to *B. thuringiensis* appears to be a change in the receptor and not a non-specific effect, e.g. proteolytic cleavage of the active toxin by gut enzymes (Johnson *et al*. 1990). If only one toxin is present there might be more chance of selecting resistant individuals. Stone *et al*. (1989), using a genetically engineered *Pseudomonas fluorescens* strain containing the 134 000 mol. wt δ-endotoxin of *B. thuringiensis* HD-1 were able, under laboratory conditions, to select for resistance to this endotoxin by *Heliothis virescens*. Resistance was 13–20-fold over that in non-selected lines in seven generations. The selected insects were also resistant to a commercial product of HD-1 (Dipel[TM]). After 23 generations, resistance was 75-fold to *P. fluorescens* and 50-fold to Dipel[TM]. Evidence to date suggests there is no cross-resistance to different *B. thuringiensis* toxins (McGaughey and Johnson 1987; Van Rie *et al*. 1990) in the Indianmeal moth (see p. 129).

As the technology of *B. thuringiensis* progresses and the products become more effective and more widely used, so the greater the selective pressure and the potential for resistance. Thus, application of the preparations must be carefully managed so that the inherent advantages of *B. thuringiensis* can be maintained. Resistance development would be reduced if there was less than a nominal 100 per cent kill of the pest but control must be above the economic threshold value. In addition, resistance development could be reduced by using *B. thuringiensis* strains containing a number of δ-endotoxins, or the sequential use of genetically modified organisms containing δ-endotoxins that bind to different insect receptors.

Genetically modified plants

Another example of potentially increasing selective pressure on the insect is found with transgenic plants that are constitutively expressing *B. thuringiensis* toxin protein. The insect population would be exposed to *B. thuringiensis* for a long time at all stages of growth, so selecting for resistant individuals. Any resulting resistant insect populations would not be controlled by the transgenic plant or by applications of *B. thuringiensis* as an insecticide containing the same toxins. Additionally, if cross-resistance were to develop to other strains of *B. thuringiensis* the advantages of a biological insecticide would be lost.

The selective pressure on the pest species could be reduced by constitutive expression of the insect control genes to protect only target tissues – tissues where the more susceptible larval stages feed. For example, a number of lepidopterous pests that attack cotton are susceptible to some *B. thuringiensis* δ-endotoxins. Two of the most severe pests in the mid-south USA are *Heliothis virescens* and *H. zea*, which feed on the cotton bolls. If the *B. thuringiensis* genes placed in cotton are linked to constitutive promoters, and are therefore expressed in most of the plant's tissues, any *Heliothis* larvae feeding on the crop in any of the three to four generations per season would be subject to intensive selection pressure. Cotton can compensate for a high level of damage and the most important structures to protect are the young bolls (Burges and Daoust 1986). Efforts could be concentrated on cotton lines that express the toxin gene only in the boll tissues. Only one generation feeds on the bolls and the selection pressure would be discontinuous (Gould 1988). *B. thuringiensis* also has antifeedant properties and the larvae feeding on the bolls may receive a sublethal dose and then possibly move to other tissues. Another approach might be to express the toxin genes non-constitutively. Progress has been made in the latter direction with non-*B. thuringiensis*-derived genes. For example, Sanchez-Serrano *et al*. (1987) have produced expression of a proteinase-inhibiting gene from potato in transgenic tobacco only on response to mechanical damage to the leaves.

The incorporation in transgenic plants of more than one toxin gene may decrease the chances of resistance development. However, genetic models indicate that widespread planting of a cultivar with two resistance factors may in some cases offer good insect control initially, but because of the intense selective pressure exerted by combined factors, the insects may still adapt within a few generations (Cox and Hatchett 1986; Gould 1986). If 80 per cent of the plants contain two resistance factors, and 20 per cent contain none, then sexually reproducing pests are likely to adapt much more slowly (Gould 1986). Alternatively, a proportion of plants of another species could be interplanted in a crop.

Concluding remarks

B. thuringiensis forms a major part of the family of microbial control agents that offer alternatives to the use of synthetic chemical insecticides. However, it has been used with varying degrees of success, although its consistent feature has been its safety and lack of adverse environmental impact. To increase its market share, *B. thuringiensis* requires a wider host range and longer persistence on foliage. To this end much research has taken place and technological advances in both the use of natural isolates and recombinant strains mean that the future is bright for *B. thuringiensis*. As the amount of research increases, so will its applications and reliability.

With the increased use of *B. thuringiensis*, the potential for development of pest resistance also increases. The selective pressure of transgenic plants constitutively expressing *B. thuringiensis* insecticidal protein might further increase the chances of resistance developing. Thus, is the immediate future, the microbial product may be the better commercial route to gain environmental acceptance, at least until pest resistance is understood and management strategies have been developed.

Use of recombinant DNA technology bears great promise for agriculture, but only if tempered with thorough research and preparation on possible environmental risks. *B. thuringiensis* can play a major role in the application of biotechnology to agriculture in two ways. First, as a pest control agent applied as wild or recombinant microorganisms or as genes in transgenic plants. Secondly, as a model micro-organism for study of the environmental effect and risk assessment for the release of genetically engineered micro-organisms (GEMs). Potential use of genetically modified strains of *B. thuringiensis* suggests that more research is performed in many areas before accurate risk assessments can be made. There is a great opportunity for a multidisciplinary approach to further studies of its ecology, role in the environment, spread and fate in the environment, and the expression of its genes in plants. The study of *B. thuringiensis* now encompasses a large family of bacteriologists, microbial ecologists and physiologists, insect pathologist, entomologists, molecular biologists and plant biotechnologists. The knowledge and experience gained by such wide ranging investigations should greatly benefit the successful application of biotechnology in agriculture.

Acknowledgements

I would like to thank H. Denis Burges and Paul Jarrett for their helpful comments and suggestions concerning the manuscript.

References

Armstrong, J.L., Rohrmann, G.F., and Beaudreau, G.S. (1985). Delta-endotoxin of *Bacillus thuringiensis* subspecies *israelensis*. *Journal of Bacteriology*, **161**, 39–46.

Aronson, A.I., and Beckman, W. (1987). Transfer of chromosomal genes and plasmids in *Bacillus thuringiensis*. *Applied and Environmental Microbiology*, **53**, 1525–30.

Aronson, A.I., Beckman, W., and Dunn, P. (1986). *Bacillus thuringiensis* and related insect pathogens. *Microbiology Reviews*, **50**, 1–24.

Barton, K.A., Whiteley, H.R., and Ning-Sun Yang (1987). *Bacillus thuringiensis* δ-endotoxin expressed in transgenic *Nicotiana tabacum* provides resistance to lepidopteran insects. *Plant Physiology*, **85**, 1103–9.

Baumann, L., Okamoto, K., Unterman, B.M., Lynch, M.J., and Baumann, P. (1984). Phenotypic characterisation of *Bacillus thuringiensis* and *Bacillus cereus*. *Journal of Insect Pathology*, **44**, 329–41.

Beltz, F., Levin, M., and Rogul, M. (1983). Safety aspects of genetically-engineered microbial pesticides. *Recombinant DNA Technology Bulletin*, **6**, 135–41.

Boman, H.G. (1981). Insect responses to microbial infections. In *Microbial control of pests and plant diseases 1970–1980*, (Ed. H.D. Burges), pp. 192–222. Academic Press, London.

Bone, E.J., and Ellar, D.J. (1989). Transformation of *Bacillus thuringiensis* by electroporation. *FEMS Microbiology Letters*, **58**, 171–8.

Bonnefoi, A., and de Barjac, H. (1963). Classification des souches du groupe *Bacillus thuringiensis* par la determination de l'antigene flagellaire. *Entomophaga*, **8**, 223–9.

Bottjer, K.P., Bone, L.W., and Gill, S.G. (1985).

Nematoda: Susceptibility of the egg to *Bacillus thuringiensis* toxins. *Experimental Parasitology*, **60**, 239–44.

Brown, T.H., Colwell, R.K., Lenski, R.E., Levin, B.R., Lloyd, M., Regal, P.J., and Simberloff, D. (1984). Report on workshop on possible ecological and evolutionary impacts of bioengineered organisms released in the environment. *Ecological Society of America Bulletin*, **65**, 436–8.

Bulla, L.A. Jr, Bechtel, D.B., Kramer, K.J., Shethna, Y.I., Aronson, A.I., and Fitz-James, P.C. (1980). Ultrastructure, physiology and biochemistry of *Bacillus thuringiensis*. In *Critical reviews in microbiology*, Vol. 8, (Ed. H.D. Isenberg), pp. 147–204. CRC Press, Florida.

Burges, H.D. (1971). Possibilities of pest resistance to microbial control agents. In *Microbial control of insects and mites*, (Ed. H.D. Burges and N.W. Hussey), pp. 445–7. Academic Press, London.

Burges, H.D. (1973). Enzootic diseases of insects. In *Regulation of insect populations by microorganisms*, (Ed. L.A. Bulla, Jr), pp. 31–49). Annals of the New York Academy of Sciences, **217**, New York.

Burges, H.D. (1981a). (Ed.) *Microbial control of pests and plant diseases 1970–1980*, pp. 1–949. Academic Press, London.

Burges, H.D. (1981b). Safety, safety testing and quality control of microbial pesticides. In *Microbial control of pests and plant diseases 1970–1980*, (Ed. H.D. Burges), pp. 737–67. Academic Press, London.

Burges, H.D. (1986a). Production and use of pathogens to control insect pests. *Journal of Applied Bacteriology Symposium Supplement*, 127–37.

Burges, H.D. (1986b). Impact of *Bacillus thuringiensis* on pest control with emphasis on genetic engineering. *MIRCEN Journal*, **2**, 101–20.

Burges, H.D., and Daoust, R.A. (1986). Current status of the use of bacteria as biocontrol agents. In *Fundamental and applied aspects of invertebrate pathology*, (Ed. R.A. Samson, J.M. Vlak, and D. Peters), pp. 514–17. Agricultural University, Wageningen.

Burges, H.D., and Hurst, J.A. (1977). Ecology of *Bacillus thuringiensis* in storage moths. *Journal of Invertebrate Pathology*, **30**, 131–9.

Burges, H.D., and Hussey, N.W. (1971). Past achievements and future prospects. In *Microbial control of insects and mites*, (Ed. H.D. Burges and N.W. Hussey), pp. 687–709. Academic Press, London.

Cardoen, J., Van Rie, J., Van Mallaert, H., and Peferoen, M. (1990). Mechanisms of insect resistance to *Bacillus thuringiensis* crystal proteins. In *Proceedings of the Vth international colloquium on invertebrate pathology and microbial control*, Adelaide, Australia, August 20–24th 1990, p. 288. Society for Invertebrate Pathology.

Carlton, B.C., and Gonzalez, J.M., Jr. (1985). Plasmids and delta-endotoxin production in different subspecies of *Bacillus thuringiensis*. In *Molecular biology of microbial differentiation*, (Ed. J.A. Hoch and P. Setlow), pp. 246–52. American Society for Microbiology Publications, Washington, DC.

Carlton, B.C., Gawron-Burke, C., and Johnson, T.B. (1990). Exploiting the genetic diversity of *Bacillus thuringiensis* for the creation of new pesticides. *Proceedings of the Vth international colloquium on invertebrate pathology and microbial control*, Adelaide, Australia, 20–24th August 1990, pp. 118–22. Society for Invertebrate Pathology.

Casida, L.E. (1988). Response in soil of *Cupriavidus necator* and other copper-resistance bacterial predators of bacteria to the addition of water-soluble nutrients, various bacterial species, or *Bacillus thuringiensis* spores and crystals. *Applied and Environmental Microbiology*, **54**, 2161–6.

Chassy, B.M., and Flickinger, J.L. (1987). Transformation of *Lactobacillus casei* by electroporation. *FEMS Microbiology Letters*, **44**, 173–7.

Couche, T.L., and Ignoffo, C.M. (1981). Formulation of insect pathogens. In *Microbial control of pests and plant diseases 1970–1980*, (Ed. H.D. Burges), pp. 621–34. Academic Press, London.

Cox, T.S., and Hatchett, J.H. (1986). Genetic model for wheat/Hessian fly (Diptera: Cecidomyiidae) interaction: strategies for deployment of resistance genes in wheat cultivars. *Environmental Entomology*, **15**, 24–31.

Cunningham, J.C. (1990). Use of microbial for control of defoliating pests of conifers. *Proceedings of the Vth international colloquium in invertebrate pathology and microbial control*, Adelaide, Australia, 20–24th August 1990, pp. 164–8. Society for Invertebrate Pathology.

Daoust, R.A. (1990). Commercialization of bacterial insecticides. *Proceedings of the Vth international colloquium on invertebrate pathology and microbial control*, Adelaide, Australia, 20–24th August 1990, pp. 7–11. Society for Invertebrate Pathology.

Delanney, X., LaVallee, B.J., Proksch, R.K., Fuchs, R.L., Sims, S.R., Greenplate, J.T., Marrone, P.G., Dodson, R.B., Augustine, J.J., Layton, J.F., and Fischoff, D.A. (1989). Field performance of transgenic tomato plants expressing the *Bacillus thuringiensis* var. kurstaki insect control protein. *Bio/Technology*, **7**, 1265–9.

DeLucca, A.J., Simonson, J.G., and Larson, A.D. (1981). *Bacillus thuringiensis* distribution in soils of the United States. *Canadian Journal of Microbiology*, **27**, 865–70.

DeLucca, A.J., Palmgren, M.S., and Ciegler, A. (1982). *Bacillus thuringiensis* in grain elevator dusts. *Canadian Journal of Microbiology*, **28**, 452–6.

Donovan, W.P., Gonzalez, J.M., Pearce Gilbert, M., and Dankscik, K. (1988). Isolation and characterization of EG2158, a new strain of *Bacillus thuringiensis* toxic to coleopteran larvae, and nucleotide sequence of the gene. *Molecular and General Genetics*, **214**, 365–72.

Dulmage, H.T. (1970). Insecticidal activity of HD-1, a new isolate of *Bacillus thuringiensis* variety *alesti*. *Journal of Invertebrate Pathology*, **15**, 232–9.

Dulmage, H.T. (1981). Insecticidal activity of isolates of *Bacillus thuringiensis* and their potential for pest control. In *Microbial control of pests and plant diseases 1970–1980*, (Ed. H.D. Burges), pp. 193–222. Academic Press, London.

Dunkle, R.L., and Shasha, B.S. (1988). Starch-encapsulated *Bacillus thuringiensis*: a potential new method for increasing environmental stability of entomopathogens. *Environmental Entomology*, **17**, 120–6.

Ellar, D.J., Thomas, W.E., Knowles, B.H. Ward, S., Todd, J., Drobniewski, F., Lewis, J., Sawyer, T., Last, D., and Nichols C. (1985). Biochemistry, genetics and mode of action of *Bacillus thuringiensis* δ-endotoxins. In *Molecular biology of microbial differentiation*, (Ed. J. Hoch and P. Setlow), pp. 230–40. American Society for Microbiology Publications, Washington DC.

Ellar, D.J., Knowles, B.H., Drobniewski, F.A., and Haider, M.Z. (1986). The insecticidal specificity and toxicity of *Bacillus thuringiensis* δ-endotoxins may be determined respectively by an initial binding to membrane-specific receptors followed by a common mechanism of cytolysis. In *Fundamental and applied aspects of invertebrate pathology*, (Ed. R.A. Samson, J.M. Vlak, and D. Peters), pp. 7–10. Agricultural University, Wageningen.

Fast, P.G. (1981). The crystal toxin of *Bacillus thuringiensis*. In *Microbial control of pests and plant diseases 1970–1980*, (Ed. H.D. Burges), pp. 223–48. Academic Press, London.

Faust, R.M., and Bulla, L.A., Jr (1982). Bacteria and their toxins as insecticides. In *Microbial and viral pesticides*, (Ed. E. Kurstak), pp. 75–208. Dekker, New York.

Federici, B.A., Luthy, P., and Ibarra, J.E. (1990). The parasporal body of *Bacillus thuringiensis* subspecies *israelensis*: structure, protein composition and toxicity. In *Bacterial control of mosquitoes and blackflies*, (Ed. H. de Barjac and S. Sutherland). Rutgers University Press, New Brunswick, New Jersey.

Ferro, D.N. (1980). Potential for insect resistance to *Bacillus thuringiensis*. *Proceedings of the Agbiotech conference*, Arlington, Virginia, March 28–30, 1989, pp. 363. Conference Management Corporation, Connecticut, USA.

Ferro, D.N., and Gelernter, W.D. (1989). Toxicity of a new strain of *Bacillus thuringiensis* to Colorado potato beetle (Coleoptera: Chrysomelidae). *Journal of Economic Entomology*, **82**, 750–5.

Fischoff, D.A., Bowdish, K.S., Perlack, F.J., Marrone, P.G., McCormick, S.M., Niedermeyer, J.G., Dean, D.A., Kusano-Kretzmer, K., Mayer, E.J., Rochester, D.E., Rogers, S.G., and Fraley, R.T. (1987). Insect tolerant transgenic tomato plants. *Bio/Technology*, **5**, 807–13.

Forgash, A.J. (1985). Insecticide resistance in the Colorado potato beetle. *Proceedings of the symposium on the Colorado potato beetle, XVIIth international congress of entomology*, (Ed. D.N. Ferro and R.H. Voss), pp. 33–53. Research Bulletin 704, Massachussetts Agricultural Experimental Station, Amhurst, Massachusetts, USA.

Frankenhuyzen, K. van, and Nystrom, C. (1989). Residual toxicity of a high potency formulation of *Bacillus thuringiensis* to spruce budworm (Lepidoptera: Tortricidae). *Journal of Economic Entomology*, **82**, 868–72.

Gawron-Burke, C., Chambers, J., Jelen, A., Donovan, W., Rupar, M., Jany, C., Slaney, A., Baum, J., English, L., and Johnson, T. (1990). Molecular biology and genetics of *Bacillus thuringiensis*. *Proceedings Vth international colloquium on invertebrate pathology and microbial control*. Adelaide, Australia, 20–24th August 1990, pp. 456–60. Society for Invertebrate Pathology.

Gelernter, W.D. (1990). MPV Bioinsecticide: A bioengineered, bioencapsulated product for control of lepidopteran larvae. *Proceedings Vth international colloquium on invertebrate pathology and microbial control*. Adelaide, Australia, 20–24th August 1990, p. 14. Society for Invertebrate Pathology.

Gelernter, W.D., and Quick, T.C. (1989). The MCap™ delivery system: a novel approach for enhancing efficacy and foliar persistence of biological toxins. *Proceedings and abstracts, Society for Invertebrate Pathology XXIIth annual meeting*. University of Maryland, USA, 20–24th August 1989, p. 24.

Georghiou, G.P., and Mellon, R.B. (1983). Pesticide resistance in time and space. In *Pest resistance to pesticides*, (Ed. G.P. Georghiou and T. Saito), pp. 1–46. Plenum Press, New York.

Goldberg, L.H., and Margalit, J. (1977). A bacterial spore demonstrating rapid larvicidal activity against *Anopheles sergentii, Uranotaenia unguiculata, Culex unuvittatus, Aedes aegypti* and *Culex pipiens. Mosquito News*, **37**, 355–8.

Gonzalez, J.M., Jr, and Carlton, B.C. (1980). Patterns of plasmid DNA in crystaliferous and acrystaliferous strains of *Bacillus thuringiensis. Plasmid*, **3**, 92–8.

Gonzalez, J.M., Jr, and Carlton, B.C. (1984). A large transmissible plasmid is required for crystal toxin production in *Bacillus thuringiensis* variety *israelensis*. *Plasmid*, **11**, 28–38.

Gonzalez, J.M., Jr, Dulmage, H.T., and Carlton, B.C. (1981). Correlation between specific plasmids and endotoxin production in *Bacillus thuringiensis*.

Plasmid, 5, 351–6.

Gonzalez, J.M., Jr, Brown, B.J., and Carlton, B.C. (1982). Transfer of *Bacillus thuringiensis* plasmids coding for δ-endotoxin among strains of *Bacillus thuringiensis* and *Bacillus cereus*. *Proceedings of the National Academy of Sciences, USA*, 79, 6951–5.

Gould, F. (1986). Simulation models for predicting durability of insect resistant germplasm: Hessian fly (Diptera: Cecidomyiidae)- resistant winter wheat. *Environmental Entomology*, 15, 11–23.

Gould, F. (1988). Evolutionary biology and genetically engineered crops. *BioScience*, 38, 26–33.

Guillet, P. (1984). Development and field evaluation of *Bacillus thuringiensis* H-14 against blackflies (Review of research activities within the special programme). WHO TDR/BCV-SWG-7/84.

Halvorson, H.O., Pramer, D., and Rogul, M. (eds.) (1985). *Engineered organisms in the environment: scientific issues*, pp. 1–239. American Society for Microbiology Publications, Washington DC.

Herrnstadt, C., Soares, G.G., Wilcox, E.R., and Edwards, D.L. (1986). A new strain of *Bacillus thuringiensis* with activity against coleopteran insects. *Bio/Technology*, 4, 305–8.

Hofmann, C.H., Vandergruggen, H., Höfte, H., Van Rie, J., Jansens, S., and Van Mallaert, H. (1988). Specificity of *Bacillus thuringiensis* δ-endotoxins is correlated with the presence of high-affinity binding sites in the brush border membrane of target insect midguts. *Proceedings of the National Academy of Science (USA)*, 85, 7844–8.

Höfte, H., and Whiteley, H.R. (1989). Insecticidal crystal proteins of *Bacillus thuringiensis*. *Microbiology Reviews*, 53, 242–55.

Jarrett, P. (1980). *Persistence of* Bacillus thuringiensis on tomatoes. Annual Report, Glasshouse Crops Research Institute, pp. 117.

Jarrett, P. (1983). Comparison of plasmids from 12 isolates of *Bacillus thuringiensis* H-serotype 7. *FEMS Microbiology Letters*, 16, 55–60.

Jarrett, P. (1985). Potency factors in the delta-endotoxin of *Bacillus thuringiensis* var. *aizawi* and the significance of plasmids in their control. *Journal of Applied Bacteriology*, 58, 437–48.

Jarrett, P., and Burges, H.D. (1986). *Bacillus thuringiensis*: tailoring the strain to fit the pest complex on the crop. In *Biotechnology in crop improvement and protection*, (Ed. P. Day), pp. 259–64. British Council for Crop Protection Monograph Number 34.

Jarrett, P., and Stephenson, M. (1990). Plasmid transfer between strains of *Bacillus thuringiensis* infecting *Galleria mellonella* and *Spodoptera littoralis*. *Applied and Environmental Microbiology*, 56, 1608–14.

Jarrett, P., Burges, H.D., and Matthews, G.A. (1978). Penetration of controlled droplet spray of *Bacillus thuringiensis* into chrysanthemum beds compared with high volume spray and thermal fog. *Proceedings of the Symposium on Controlled Droplet Application*, pp. 75–81. British Crop Protection Council, Croyden, Surrey.

Johnson, D.E., Brookhart, G.L., Kramer, K.F., Barnett, B.D., and McGaughey, W.H. (1990). Resistance to *Bacillus thuringiensis* by the Indian meal moth, *Plodia interpunctella*: comparison of midgut proteinases from susceptible and midgut larvae. *Journal of Invertebrate Pathology*, 55, 235–44.

Kinghorn, J.M., Fisher, R.A., Angus, T.A., and Heimpel, A.M. (1961). Aerial spray trails against the blackheaded budworm in British Columbia. *Department of Forestry Bimonthly Report*, 17, 3–4.

Klein, T.M., Gradziel, T., Fromm, M.E., and Sanford, J.C. (1988). Factors influencing gene delivery into *Zea mays* cells by high-velocity microprojectiles. *Bio/Technology*, 6, 559–63.

Klier, A., Bourgouin, C., and Rapoport, G. (1983). Mating between *Bacillus cereus* and *Bacillus thuringiensis* and transfer of cloned crystal genes. *Molecular and General Genetics*, 191, 257–62.

Knepper, R.G., and Walker, E.D. (1989). Effect of *Bacillus thuringiensis israelensis* (H-14) in the isopod *Asellus forbesi* and spring *Aedes* mosquitoes in Michigan. *Journal of the American Mosquito Control Association*, 5, 596–8.

Knowles, B.J., and Ellar, D.J. (1987). Colloid-osmotic lysis is a general feature of the mechanism of action of *Bacillus thuringiensis* δ-endotoxins with different insect specificities. *Biochimica et Biophysica Acta*, 924, 509–18.

Krieg, A., and Langenbruch, G.A. (1981). Susceptibility of arthropod species to *Bacillus thuringiensis*. In *Microbial control of pests and plant diseases 1970–1980*, (Ed. H.D. Burges), pp. 837–98. Academic Press, London.

Krieg A., Huger, A., Langenbruch, G., and Schnetter, W. (1983). *Bacillus thuringiensis* variety *tenebrionis*: a new pathotype effective against larvae of Coleoptera. *Journal of Applied Entomology*, 96, 500–8.

Lereclus, D., Arantes, O., Chaufaux, J., and Lecadet, M.-M. (1989). Transformation and expression of a cloned δ-endotoxin gene in *Bacillus thuringiensis*. *FEMS Microbiology Letters*, 60, 211–18.

Levin, B.R. (1984). Changing views of the hazards of recombinant DNA manipulation and the regulation of these procedures. *Recombinant DNA Technology Bulletin*, 7, 107–14.

Lindow, S.W., and Panopoulous, N.J., (1983). *Request for permission to test* Pseudomonas syringae *pv.* syringae *and* Erwinia herbicola *carrying specific deletions in ice nucleation genes under field conditions as biocontrol agents of frost injury to plants*. Revised protocol for Recombinant DNA Committee, National Institute of Health, Washinton DC.

Luthy, P. (1989). Large-scale use of *Bacillus thuringiensis* H-14 in a mosquito-infected area in a southern region of Switzerland. *Proceedings and abstracts, Society for*

Invertebrate Pathology, XXIIth annual meeting. University of Maryland, USA, 20–24th August 1989, p. 82.

Margalit, J., and Bobroglio, H. (1984). The effect of organic materials and solids in water on persistence of *Bacillus thuringiensis* var. *israelensis* serotype H-14. *Zeitschrift für Angewandte Entomologie*, **97**, 516–20.

Martin, W.F., and Reichelderfer, C.F. (1989). *Bacillus thuringiensis*: persistence and movement in field crops. *Proceedings and abstracts, Society for Invertebrate Pathology, XXIIth annual meeting.* University of Maryland, USA, 20–24th August 1989, pp. 25.

Martin, P.A.W., and Travers, R.S. (1989). Worldwide abundance and distribution of *Bacillus thuringiensis* isolates. *Applied and Environmental Microbiology*, **55**, 2437–42.

Masson, L., Prefontaine, G., and Brousseau, R. (1989). Transformation of *Bacillus thuringiensis* vegetative cells by electroporation. *FEMS Microbiology Letters*, **60**, 273–8.

McCabe, D.E., Swain, W.F., Martinell, B.J., and Christou, P. (1988). Stable transformation of soybean (*Glycine max*) by particle acceleration. *Bio/Technology*, **6**, 923–6.

McGaughey, W.H. (1985). Insect resistance to the biological insecticide. *Bacillus thuringiensis. Science*, **229**, 193–5.

McGaughey, W.H., and Johnson, D.E. (1987). Toxicity of different serotypes and toxins of *Bacillus thuringiensis* to resistant and susceptible Indianmeal moths (Lepidoptera: Pyralidae). *Journal of Economic Entomology*, **80**, 1122–6.

McGuire, M.R., Shasha, B.S., and Bartelt, R.J. (1989). Development and evaluation of starch formulations of *Bacillus thuringiensis. Proceedings and abstracts, Society for Invertebrate Pathology, XXIIth annual meeting*, University of Maryland, USA, August 20–24 1989, p. 23.

Meadows, M.P., Ellis, D.J., Pethybridge, N.J., Bernhard, K., Burges, H.D., and Jarrett, P. (1990). Activity of new isolates of *Bacillus thuringiensis* against five insect pests. *Proceedings Vth international colloquium on invertebrate pathology and microbial control*. Adelaide, Australia, August 20–24th 1990, pp. 479. Society for Invertebrate Pathology.

Meeusen, R.L., and Warren, G. (1989). Insect control with genetically engineered crops. *Annual Review of Entomology*, **34**, 373–81.

Milewski, E., and Tobin, S.A. (1984). Development of guidelines for field testing of plants modified by recombinant DNA techniques. *Recombinant DNA Technology Bulletin*, **7**, 114–24.

Miyata, T., Saito, T., and Noppun, V. (1986). Studies on the mechanism of Diamondback moth resistance to insecticides. In *Diamondback moth management: proceedings of the first international working group*, (Ed. N.F. Taleker), pp. 347–57. AVRDC, Taiwan.

Morris, O.N. (1982). Bacteria as pesticides: forest applications. In *Microbial and viral pesticides*, (Ed. E. Kurstak), pp. 239. Marcel Dekker, New York.

Mott, D.G., Angus, T.A., Heimpel, A.M., and Fisher, R.A. (1961). Aerial application of Thuricide against spruce budworm in New Brunswick. *Department of Forestry Bimonthly Progess Report*, **17**, 2.

Norris, J.R. (1969). The ecology of serotype 4B of *Bacillus thuringiensis. Journal of Applied Bacteriology*, **32**, 261–7.

Ohba, M., and Aizawa, K. (1978). Seriological identification of *Bacillus thuringiensis* and related bacteria in Japan. *Journal of Invertebrate Pathology*, **32**, 303–9.

Padua, L.E., Gabriel, B.P., Aizawa, K., and Ohba, M. (1982). *Bacillus thuringiensis* isolated from the Philippines. *Philippine Entomologist*, **5**, 199–208.

Perlack, F.J., and Fischoff, D.A. (1990). Expression of *Bacillus thuringiensis* insect control proteins in genetically modified plants. *Proceedings Vth international colloquium on invertebrate pathology and microbial control*, Adelaide, Australia, 20–24th August 1990, pp. 461–5. Society for Invertebrate Pathology.

Petras, S.T., and Casida, L.E. (1985). Survival of *Bacillus thuringiensis* spores in soil. *Applied and Environmental Microbiology*, **50**, 1496–501.

Pinnock, D.E., Brand, R.J., and Milstead, J.E. (1971). The field persistence of *Bacillus thuringiensis* spores. *Journal of Invertebrate Pathology*, **18**, 405–11.

Pruett, C.J.H., Burges, H.D., and Wyborn, C.H. (1980). Effect of exposure to soil on potency and spore viability of *Bacillus thuringiensis. Journal of Invertebrate Pathology*, **35**, 168–74.

Reddy, A., Battisti, L., and Thorne, C.B. (1987). Identification of self-transmissible plasmids in four *Bacillus thuringiensis* subspecies. *Journal of Bacteriology*, **169**, 5263–70.

Saiki, R.K., Scharf, S., Fallona, F., Mullis, K.B., Horn, G.T., Erlich, H.A., and Arnheim, N. (1985). Enzymatic amplification of β-globin genomic sequences and restriction site analysis for diagnosis of sickle cell anaemia. *Science*, **230**, 1350–4.

Saleh, S.M., Harris, R.F., and Allen, O.N. (1969). Method for determining *Bacillus thuringiensis* var. *thuringiensis* Berliner in soil. *Canadian Journal of Microbiology*, **15**, 1101–4.

Sanchez-Serrano, J.J., Keil, M., O'Conner, A., Schell, J., and Willmitzer, L. (1987). Wound-induced expression of a potato proteinase inhibitor II gene in transgenic tomato plants. *EMBO Journal*, **6**, 303–6.

Schurter, W., Geiser, M., and Mathe, D. (1989). Efficient transformation of *Bacillus thuringiensis* and *Bacillus cereus* via electroporation: transformation of acrystalliferous strains with a cloned δ-endotoxin gene. *Molecular and General Genetics*, **218**, 177–81.

Sebesta, K., Farkas, J., Horska, K., and Vankova, J. (1981). Thuringiensis, the beta-exotoxin of *Bacillus thuringiensis*. In *Microbial control of pests and plant diseases 1970–1980*, (Ed. H.D. Burgess), pp. 249–82. Academic Press, London.

Smith, R.A., and Couche, G.A. (1991). The phylloplane as a source of *Bacillus thuringiensis* variants. *Applied and Environmental Microbiology*, **57**, 311–15.

Snarski, V.M. (1990). Interactions between *Bacillus thuringiensis* subspecies *israelensis* on Fathead Minnows, *Pinnephales promelas* Rafinesue, under laboratory conditions. *Applied and Environmental Microbiology*, **56**, 2618–22.

Sneh, B., and Schuster, S. (1983). Effect of exposure to sublethal concentration of *Bacillus thuringiensis* Berliner spp. *entomocidus* on the susceptibility to the endotoxin of subsequent generations of the Egyptian cotton leafworm *Spodoptera littoralis* Boisd. (Lep., Noctuidae). *Zeitschfift für Angelwandte Entomologie*, **96**, 425–8.

Stone, T.B., Sims, S.R., and Marrone, P.G. (1989). Selection of tobacco budworm for resistance to a genetically engineered *Pseudomonas fluorescens* containing the δ-endotoxin of *Bacillus thuringiensis* subspecies *kurstaki*. *Journal of Invertebrate Pathology*, **53**, 228–34.

Tabashnik, B.E., Cushing, N.L., Finson, N., and Johnson, M.W. (1990). Field development of resistance to *Bacillus thuringiensis* in Diamondback moth (Lepidoptera: Plutellidae). *Journal of Economic Entomology*, **83**, 1671–6.

Thomas, W.E., and Ellar, D.J. (1983). *Bacillus thuringiensis* variety *israelensis* crystal δ-endotoxin: effects on insect and mammalian cells *in vitro* and *in vivo*. *Journal of Cell Science*, **60**, 181–97.

Travers, R.S., Martin, P.A.W., and Reichelderfer, C.F. (1987). Selective process for the efficient isolation of soil *Bacillus* species. *Applied and Environmental Microbiology*, **53**, 1263–6.

Trevors, J.T., Barkey, T., and Bourquin, A.W. (1986). Gene transfer among bacteria in soil and aquatic environments: a review. *Canadian Journal of Microbiology*, **33**, 191–8.

Tyrell, D.J., Davidson, L.I., Bulla, L.A. Jr, and Ramoska, W.A. (1979). Toxicity of parasporal crystals of *Bacillus thuringiensis* subsp. *israelensis* to mosquitoes. *Applied and Environmental Microbiology*, **38**, 656–8.

Undeen, A.H., and Nagel, W.L. (1978). The effect of *Bacillus thuringiensis* ONR-60A strain (Goldberg) on *Simulium* larvae in the laboratory. *Mosquito News*, **38**, 524–7.

Vaeck, M., Raynaerts, A., Höfte, H., Jansens, S., De Beukeleer, M., Zabeau, M., Van Montagu, M., and Leemans, J. (1987). Transgenic plants protected from insect attack. *Nature*, **328**, 33–7.

Vaeck, M., Raynaerts, A., and Höfte, H. (1989). Protein engineering in plants: expression of insecticidal protein genes. *Cell Culture and Somatic Cell Genetics of Plants*, **6**, 425–39.

Van Rie, J., McGaughey, W.H., Johnson, D.E., Barnett, B.D., and Van Mallaert, H. (1990). Mechanism of insect resistance to the microbial insecticide *Bacillus thuringiensis*. *Science*, **247**, 72–4.

Vazquez-Garcia, M. (1983). Investigations of the potentiality of resistance to *Bacillus thuringiensis* serotype H-14 in *Culex quinquefasciatus* through accelerated selection pressure in the laboratory. PhD thesis, University of California.

Watrud, L.S., Perlack, F.J., Tran, M., Kusano, K., Mayer, E.J., Miller-Wideman, M.A., Obukowicz, M.G., Nelson, D.R., Kreitinger, J.P., and Kaufman, R.J. (1985). Cloning of the *Bacillus thuringiensis* subsp. *kurstaki* delta-endotoxin gene into *Pseudomonas fluorescens*: molecular biology and ecology of an engineered microbial pesticide. In *Engineered organisms in the environment: scientific issues*, (Ed. H.O. Halvorson, D. Pramer, and M. Rogul), pp. 40–6. American Society for Microbiology Publications, Washington DC.

West, A.W., Crook, N.E., and Burges, H.D. (1984). Detection of *Bacillus thuringiensis* in soil by immunofluorescence. *Journal of Invertebrate Pathology*, **43**, 150–5.

West, A.W., Burges, H.D., Dixon, T.J., and Wyborn, C.H. (1985). Survival of *Bacillus thuringiensis* and *Bacillus cereus* spore inocula in soil: effects of pH, moisture, nutrient availability and indigenous microorganisms. *Soil Biology and Biochemistry*, **17**, 657–65.

Whiteley, H.R., and Schnepf, H.E. (1986). The molecular biology of parasporal crystal formation in *Bacillus thuringiensis*. *Annual Review of Microbiology*, **40**, 549–76.

Zehnder, G.W., and Gelernter, W.D. (1989). Activity of the M-ONE formulation of a new strain of *Bacillus thuringiensis* active against the Colorado potato beetle (Coleoptera: Chrymosilidae): relationship between susceptibility and insect life stage. *Journal of Economic Entomology*, **82**, 756–61.

10

The use of genetically engineered virus insecticides to control insect pests

D.H.L. Bishop, J.S. Cory and R.D. Possee

Introduction

NATURALLY OCCURRING BACULOVIRUSES have been used since the last century to control insect pests. Unlike most chemical insecticides, they affect only a few species of insect. They only infect arthropods, having no effect on vertebrates or plants, nor do they pollute the environment or cause adverse reactions in soil or water (farmland, rivers, lakes, etc.). Because they are slow to exert an effect, the use of baculovirus insecticides has largely been superseded by chemical insecticides. However, in some situations viruses and other biological control agents are still preferred to chemicals. Examples include (i) areas that are environmentally sensitive (e.g. forests and water catchment areas); (ii) where the cost of chemical insecticides is prohibitive; or (iii) where the pest species has developed resistance to the available chemical insecticides.

A joint WHO–FAO meeting on insect viruses endorsed the use of baculoviruses as pest control agents (WHO–FAO 1973). The report noted that, in addition to their specific host ranges, baculoviruses exhibit good storage properties, are safe to handle, are relatively easy to produce and are widely distributed in nature, particularly among insects. More than a dozen baculoviruses have been employed commercially to control insect pests. Reports by Podgewait (1985) and Entwistle and Evans (1985) list some of the baculoviruses that have been used. The environmental safety of baculoviruses is a matter of record and a major factor when considering their further development

by genetic engineering. This safety record has been confirmed by experience gained over many decades of the use of naturally occurring baculovirus insecticides in agriculture and forestry.

The objective of genetically engineering baculovirus insecticides is to improve their speed of action. This is desirable because, during the normal infection process, a baculovirus undergoes several cycles of replication (Fig. 10.1). These cycles take time – several days, or weeks, depending on the virus, the host and the environmental conditions (e.g. temperature). By contrast, most chemical insecticides act quickly, killing the target insect (and often other, beneficial insects) in a matter of hours. Using genetic engineering procedures it should be possible to minimize the time taken for a viral insecticide to act by incorporating other genes (toxins, insect hormone genes, etc.) into the viral genome.

The principal target of a baculovirus, such as a nuclear polyhedrosis virus (NPV), is the larval (caterpillar) stage of the host. Viruses are ingested with food. The polyhedrin protein is removed by proteolytic digestion when the virus reaches the alkaline environment of the caterpillar midgut (Fig. 10.1). Infectious virus particles infect cells in the epithelium of the caterpillar's midgut. Virus DNA enters the cell nucleus and replication ensues. In Lepidoptera (e.g. for AcNPV) non-occluded virus particles are released from these cells to spread the infection, via the haemocoel, to cells in other tissues of the larva. Inclusion bodies (polyhedra) are produced late in the course of the infection. The virulence and extent of the virus

137

BACULOVIRUSES

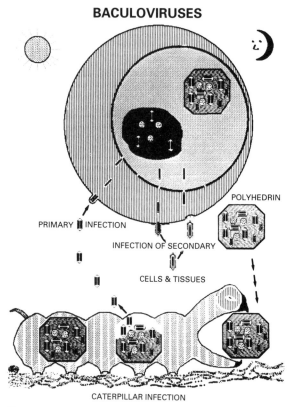

Fig. 10.1. Schematic course of an infection by an occluded baculovirus such as AcNPV. Polyhedrin inclusion bodies (PIBs) are ingested with food (foliage) and the viruses liberated to infect midgut cells. After replication, non-occluded progeny are released to infect other cells and organs. Cell and larval death results in the release of PIBS to the environment.

virus passes directly through their intestines without inactivation and is deposited with their faeces.

Control of *Panolis flammea*, the pine beauty moth, in forests in Scotland using a natural baculovirus insecticide

The use of natural baculovirus insecticides can be illustrated by recent research undertaken by staff of the Institute of Virology and Environmental Microbiology (Oxford) on the control of *Panolis flammea*, the pine beauty moth. In 1976 an outbreak of the pest occurred in Scottish forests of lodgepole pine (*Pinus contorta*), a north American species of pine that had been planted in Scotland on account of its ability to grow well in land containing peat. Since the 1976 outbreak the pest has spread to several regions of Scotland. Although controllable with chemical insecticides, many of the affected forests are in environmentally sensitive areas. A search was therefore undertaken for a more acceptable means of control. A nuclear polyhedrosis virus of *P. flammea* (PfNPV), isolated in 1979 from a natural epizootic of disease in the insect, has been investigated to determine if it can be used as an efficient and cost-effective means to control *P. flammea* infestations.

Initial field trials of the virus insecticide in Scottish forests started in 1981. Dose–response relationships were studied using virus administered from the ground. In the following years this was extended to small-scale aerial trials using a variety of spray equipment. The initial trials indicated that high levels of infection could be obtained using dosages in excess of 10^{11} viral polyhedral inclusion bodies PIBs) per hectare of 2000–3000 trees (Entwistle and Evans 1987).

To develop a virus as an insecticide it has to be produced cheaply and conveniently. Ideally it should be grown in its natural host to avoid any reduction in virulence. Unfortunately, because of the fastidious food requirements of *P. flammea*, this proved to be impossible. A closely related alternative host was therefore used, *Mamestra brassicae*, the cabbage moth (Kelly and Entwistle 1988). PfNPV grown in *M. brassicae* larvae was found to be as effective at controlling *P. flammea* infestations in Scottish forests as a virus grown in the host of origin.

The success of a baculovirus insecticide often depends on detailed and accurate knowledge of the

infection induces death of the host species. The replication process can be prolific, with up to 10^9 progeny polyhedra present in the corpse of an infected insect. Infected insects commonly exhibit behavioural abnormalities, which may be of benefit to the subsequent spread of virus to other larvae. For example, infected larvae may ascend to the tips of plants prior to death, thereby facilitating the distribution of virus over the foliage below. Virus is released from decaying larval corpses and can be spread by physical forces (e.g. rain splash). Viruses may also be distributed passively by animals that eat the remains of the insect and subsequently defecate at other sites. Virus ingestion by birds, rodents and beneficial insects such as carabids does not lead to infection of those species – the baculo-

behavioural characteristics of the target pest. This knowledge is needed to ensure that the larva encounters and ingests virus deposited by the spray procedure. The information required to ensure this includes knowledge of the egg-laying habits of the adult moth, the sites where caterpillars feed, the unit area of food consumed and data on the susceptibility of the pest to infection (in relation to the caterpillar developmental stages).

P. flammea posed specific problems in relation to establishing viral infections in the field. In the north of Scotland the pest is univoltine (single annual generation) and eggs hatch in time to coincide with the growth and extension of pine tree shoots. Young larvae of *P. flammea* are fastidious feeders and only eat newly emerged foliage; they feed off the growing tips of the new shoots. This feeding behaviour poses two problems in relation to spray applications. First, the caterpillars only consume a small quantity of the total surface area of the shoot and consequently they only ingest small quantities of the applied virus. Second, as the pine shoots are in the stage of active growth, the virus sprayed onto a shoot becomes diluted. The timing and conditions of aerial application of *P. flammea* NPV had to address these problems by providing maximum coverage of the growing shoots when the larvae emerged from eggs.

Research undertaken by staff of the Institute, in collaboration with the UK Forestry Commission and the Cranfield Institute of Technology, led to the development of an ultra-low-volume (ULV) spray equipment that produced small spray droplets that were effective at providing the required coverage (and consequent pest control). Semioperational trials were undertaken in 1986 on some 620 hectares of forest using helicopter-mounted ULV equipment in the north of Scotland. Excellent control was obtained using dosages of 2×10^{11} PIBs per hectare, even in sites where *P. flammea* infestations reached several thousand eggs per tree (P.F. Entwistle, J.S. Cory and C.J. Doyle, unpublished data). Although purified virus was used in the initial trials, production of such virus is labour-intensive (and wasteful). Consequently, both impure and semipure preparations were assessed in the 1986 trial, in addition to a commercial formulation of a closely related *M. brassicae* NPV. All combinations performed well in the semioperational conditions and demonstrated that pine beauty moth infestations could be completely controlled and at costs comparable to those involved with the

use of chemical insecticides. As the natural virus insecticide persists for some years after an initial administration (and amplification by the pest), long-term control is a further advantage.

Following this research and development, preparations of PfNPV have been tested for safety using regimes agreed with the UK Ministry of Agriculture, Fisheries and Food (MAFF) and the data have been submitted to MAFF for licencing the virus as an insecticide.

Risk assessment studies with natural virus insecticides

Baculoviruses, like all viruses, are obligate parasites and have to be ingested by a permissive host and gain entry to a cell before they can replicate. If there is no available host then the virus remains inert. In this regard baculoviruses differ from most bacteria or other free-living micro-organisms. In the absence of a permissive host, the inability of viruses to replicate *per se* inhibits them from escaping and multiplying *ad libitum*. In view of these characteristics, and their restricted host ranges, baculoviruses are ideal insecticides from an environmental point of view. They pose no risk to vertebrates, plants or invertebrates other than to specific species of arthropod.

Risk and environmental impact assessment of the use of native or exotic baculoviruses as insecticides have largely centred on two aspects.

1. Host range testing (i.e. the likelihood of infection of unintended hosts such as 'beneficial' insects).
2. Persistence and means of natural dissemination.

Field analyses have shown that virus may persist at a site of initial application and spread beyond that site by passive transfer (involving birds, rodents or predatory arthropods that feed on the infected caterpillars or their cadavers). Such a distribution may account for some of the natural epizootics (spread of virus through the insect population) of baculovirus infection that occur in nature. Laboratory-based host range testing entails selecting representative groups of arthropods of ecological or conservation interest (selected in consultation with others, e.g. the Nature Conservancy Council (NCC) and ideally from areas in which spraying will take place, or to which the virus, or the host species, might spread). Tests on numerous natural baculoviruses have shown that relatively few hosts other than the host of origin are susceptible to infection and die. In some cases only a single species can be infected with virus, in others host ranges may extend to representatives of several families of a particular Order of arthropods (e.g. members of various families of Lepidoptera). Of the non-target hosts in which a virus may replicate, the infection may not be lethal. Whether such hosts carry and maintain latent infections that are instrumental in establishing epizootics in more susceptible species is under investigation. So far no cross-infection has been reported in beneficial insects (e.g. bees) of a baculovirus insecticide that has been used for pest control.

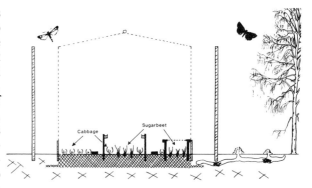

Fig. 10.2. Schematic diagram of the netted facility used for the field testing of genetically engineered AcNPV. As illustrated, the facility was designed to restrict the entry and egress of arthropods and other animals. The only animals likely to enter the facility were moles, via burrowing beneath the sunken fence. The cross-hatched area indicates the cultivated soil.

The use of genetically engineered baculoviruses as improved insecticides

Studies have been conducted in Oxford to assess the risks associated with the use of genetically engineered baculovirus insecticides. These studies employed an NPV of *Autographa californica*, which is amenable to genetic manipulation and is known to be capable of infecting pests of agricultural and horticultural importance.

Although viruses are not models for assessing the risks involved with genetically engineered, free-living organisms (such as bacteria), due to their host preferences, baculoviruses are excellent subjects with which to conduct a risk assessment programme because their infection requirements (presence of a permissive host species) allows restrictions to be imposed on the study that limit risk. To inhibit an engineered virus from replicating *in natura*, the field trials that were undertaken at Oxford between 1986 and 1989 were performed at a time of year (autumn/winter) when the natural permissive insect host was not present in the environment (infected caterpillars were supplied from the laboratory). In addition, the field facility used for the studies was designed to restrict egress of infected hosts and to limit entry to the site of other insects, rodents, moles or larger animals that might be involved in the dissemination of virus (Fig. 10.2). By such physical and temporal restraints, any risk to the environment due to the introduction of a genetically engineered virus (or its spread) was mimimized. Finally, after each study, the field site was disinfected and demonstrated to be free from virus before new insect species

emerged in the following spring.

The research programme at Oxford involving genetically engineered baculoviruses is primarily concerned with the assessment of risks associated with the introduction of an engineered virus into a natural habitat. It would be irresponsible to construct and release into the environment a genetically engineered baculovirus insecticide that expresses a new (foreign) gene without a thorough understanding of the possible consequences. A major issue in terms of risk is whether the host range of an engineered virus differs from that of the parent virus. To investigate this question extensive host range analyses have been undertaken in the laboratory for each of the engineered viruses we have prepared prior to seeking permission for field trials. A second consideration of risk concerns the possible spread of virus from the field site. Spread could be mediated by environmental conditions such as rain, by the movement of the host species, or by other animals (birds, rodents, predatory insects, etc.). By studying natural epizootics of viral infection, as well as those induced when viruses are applied to caterpillar infestations in the wild, Entwistle *et al.* (1983) demonstrated that such spread does occur. To limit spread of the engineered viruses, a netted field facility was used (Fig. 10.2), which restricted entry or egress of insects and other animals from the site.

In the investigations reported by Entwistle and associates (1983) it was never entirely possible to determine if the virus recovered from the environment originated from the applied virus, or whether it represented a naturally occurring isolate of the same virus. To address this issue a genetically marked virus was developed for the first (1986–7) study. The marker was an innocuous piece of non-coding DNA inserted into an intergenic, non-regulatory position in the genome. This allowed the recovered virus to be differentiated from other viruses or from other DNA samples.

Another consideration of potential risk is the question of the exchange of genetic information. Although DNA exchange is frequent and well-documented among bacteria (involving plasmids, conjugation, etc.), such exchange is probably a low frequency event for baculoviruses, except between closely related baculoviruses in ecologically intimate circumstances. The transfer of a foreign gene from a virus into a dead cell, or to non-germline cells would be an event without consequence. DNA transfer involving the acquisition by a baculovirus of genetic information from cells of a host species could occur (albeit infrequently) and produce an altered phenotype. If this occurs naturally then the question with an engineered virus is whether the frequency of its occurrence is increased, or whether particular DNA sequences are transferred due to the presence of the engineered DNA. The occurrence and frequency of such events can be sought by studying virus passaged in the laboratory in permissive cells, or in the permissive hosts (caterpillars). To date, no such transfer has been documented in our studies with any of the engineered insecticides.

Gene exchange between genetically compatible baculoviruses may occur when two closely related baculoviruses coinfect the same host. Marker rescue studies indicate that such rates could be > 1 per cent for closely related viruses (R.D. Possee, unpublished data). The likelihood of gene exchange occurring between such viruses will depend on whether the target host is already infected with a related virus, or whether another virus coinfects the same target host at the same time. The transfer of foreign DNA between baculoviruses can be analysed by studying the progeny virus recovered from dually infected cells (or by appropriately designed marker rescue studies). Data obtained from such studies indicate that gene transfer from a distantly related baculovirus only occurs at very low frequencies ($< 10^{-7}$), if at all. So far in our studies we have not been able to identify a UK baculovirus that is closely related to AcNPV,

or that is genetically compatible with AcNPV.

Another consideration of risk associated with an engineered organism is the question of its genetic stability, i.e. whether the introduced gene induces genetic instability. This can be investigated by genome analyses of virus passaged in the laboratory. No genetic instability has yet been identified with any of the virus constructions we have made or field tested.

In summary, each genetically engineered baculovirus that was prepared was subjected to a series of laboratory analyses to address the questions listed above. The laboratory investigations involved: (i) the construction of the candidate virus insecticide; (ii) verification of the precise genetic change by DNA sequence analyses; (iii) laboratory studies of the phenotype and genetic stability (mutability) of the viruses; (iv) host range and genetic compatibility determinations; and (v) analyses of the physical stability of the virus in simulated systems (plant surfaces and soil). Other than the expected genetic alteration designed to be incorporated into the viral genome, no unexpected change in the genome was detected and there was no genetic instability. The viral phenotypes (e.g. host ranges, physical stabilities) have also been as expected.

Regulatory issues

In each study the laboratory data were submitted to all the relevant regulatory authorities for independent evaluation, especially in relation to the characteristics of the site proposed for the release, the insect fauna of the area, the possibility for spread beyond the immediate ecosystem and the proposed physical constraints imposed at the release site. Permission to undertake the releases described below were given in the form of licences to conduct the studies under certain prescribed conditions (site arrangements, procedures to be followed for the introduction of the engineered virus, disinfection of the site after use, etc.). As baculoviruses are insecticides, the licences were issued by Ministry of Agriculture, Fisheries and Food (MAFF) under the Food and Environment Protection Act 1985 and Regulation of the Control of Pesticides Regulations 1986. Primary evaluation of the proposals was made by the Health and Safety Executive (HSE) Advisory Committee on Genetic Manipulation (ACGM), in consultation with MAFF, the Department of the Environment (DOE) and the NCC. In addition, other interested parties were informed, including the Natural Environment Research Council's senior management, the owners of the field site (Oxford University), senior University authorities, the University Safety Officer, the Oxford HSE factory inspector, the Vale of the White Horse Environmental Health Officer, the Environmental Services Committee of the Vale of the White Horse District Council and agencies such as Friends of the Earth. Press coverage was also sought (national and local papers, including, for the 1988 releases, a 'Public Notice' placed in local papers). Radio and television interviews aided the process of informing the general public of the proposed studies.

Field studies using genetically marked AcNPV

The manipulated virus

The first release involved a genetically marked AcNPV that was otherwise identical in every way to the parent AcNPV (Bishop 1986). The virus was designed so that it could be distinguished from all other baculoviruses and from genome sequences of the target, or another host. The marker was a synthetic oligonucleotide, 80 basepairs in length, inserted into the viral genome. The oligonucleotide was placed downstream of the AcNPV polyhedrin coding region.

The synthetic oligonucleotide was introduced into the AcNPV genome so that it was located just after the end of the polyhedrin gene coding sequence, but within the immediate 3' non-translated sequences (Bishop et al. 1988). It served only as a marker, or flag. The position of the marker was verified by restriction enzyme and sequence analyses. The marker contained translation stop codons in all six reading frames, no ATG codon and no other genetic signal likely to affect the replication, or gene product expression, of the virus. Due to the precise location of the insertion (a dispensible, intergenic, non-regulatory position in the AcNPV genome), the oligonucleotide neither added to, nor detracted from, the constitution or synthesis of any of the natural AcNPV gene products. This was verified by analyses of the viral phenotype (infectivity, proteins synthesized) by comparison to that of the unmodified virus.

The target pest

For the field studies it was decided to use caterpillars of the small mottled willow moth (*Spodoptera exigua*) as the target pest. This is a noctuid moth with practically a worldwide distribution; it is a seasonal immigrant to the UK. It eats a range of herbaceous dicotyledons, including sugar beet, and has a wide distribution in communities of low growing plants, both wild and cultivated. Larvae of this moth pass through five instars. The site chosen and used for the release of the marked virus (and the other genetically engineered derivatives in 1987 and 1988) was in a field of light loam soil bounded by agricultural land at the Oxford University field station at Wytham, Oxfordshire. The area has been extensively studied with regard to fauna and flora for many decades. Moth species native to the region were collected at night using two ultra-violet light traps. The moths were trapped throughout the late spring, summer and autumn of 1986 (and 1987).

Field trials

After obtaining the requisite permissions (see above) the first field trials with the marked AcNPV were undertaken (Bishop et al. 1988). In September of 1986, early third instar S. exigua larvae were fed overnight in the laboratory with diet containing ten LD_{50} PIB equivalents of the marked AcNPV incorporated into an artificial diet plug. The infected larvae (approximately 200) were refed with diet containing similar quantities of virus and, after they had eaten all the food, were taken in sealed containers to the site and placed on the sugar beet plants in the central, plexiglass-enclosed area of the containment facility (Fig. 10.3, top). Uninfected larvae were released onto nearby plants in an adjacent, totally netted enclosure. After 1 week all of the infected caterpillars were dead (Fig. 10.3, bottom). Dead caterpillars (approximately 55) were evident on many of the plant surfaces. Other dead caterpillars were observed in or on the surface of the soil. Samples of soil and foliage (cabbages, sugar beet and chickweeds that grew within the facility) were returned to the laboratory at 1–2 week intervals for analysis for virus. These analyses were continued over a 6-month period (until February 1987).

Virus was assayed on the recovered plants by allowing groups of permissive *Trichoplusia ni* larvae

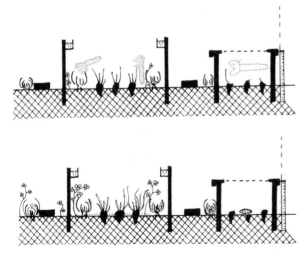

Fig. 10.3. Infection protocols in the netted facility used for the field trials of genetically engineered baculoviruses. In the top panel, infected (centre) and uninfected (right side) caterpillars were placed on sugar beet and allowed to feed *ad libitum*. By the end of the study viruses released from the infected caterpillars were recovered on foliage and in soil. The uninfected caterpillars caused substantial damage to the plants and pupated. At the end of the study plants and pupae were removed (and autoclaved) and the site disinfected.

(second instar) to eat all the foliage. The larvae were then reared on an artificial diet until pupation or death, when they were assayed for virus (polyhedra, DNA probing). As expected for plants that originally had only a limited number of infected *S. exigua* caterpillars on them (and consequently a limited amount of residual virus), not all the *T. ni* larvae became infected. However, virus was identified at every sampling period throughout the entire 6-month period. Virus was identified on plants that were present in the site at the initiation of the experiment (cabbages, sugar beet), as well as those (e.g. chickweed) that grew up in the facility after the caterpillars had died. The amounts of virus on the infected plants or in the soil were not quantified. Virus identified after replication in *T. ni* larvae retained the marker element in its original form. Chickweed recovered from outside the facility, or from plants collected outside the enclosed area did not yield virus.

The presence of virus in soil was assayed by planting cabbage seeds in soil samples brought back to the laboratory. The derived seedlings were then fed to *T. ni* larvae and the insects reared until pupation, or death, and similarly analysed for the

marked virus. Again, although only approximately 12 per cent of the caterpillars died of virus infections, marked virus was identified by this procedure at each sampling period. In summary, and as expected, the study demonstrated the persistence of the marked virus in the field site.

In the completely netted enclosure, where the uninfected caterpillars were placed, the larvae consumed substantial quantities of foliage for 3 weeks after the release. At the end of this period caterpillars were removed, all the plants were destroyed by autoclaving and the area was flooded with water to retrieve pupae. Virus was never isolated from these insects.

Postexperimental treatment

At the completion of the study in February 1987, the field site was cleared and foliage removed and sterilized to destroy any remaining virus. The site was treated with 5 per cent formalin on three separate occasions, twice in February and again in March 1987. Soil samples were taken before and after each treatment and used in the laboratory to propagate 200–400 cabbage seedlings. The seedlings were fed to *T. ni* larvae to act as a biological indicator for the presence of virus. The larvae were reared until pupation or death. All insects, infected or healthy, were processed to detect the presence of the marked virus as described above. The results showed that infectious virus was present in the soil before the first formalin treatment, in amounts sufficient to infect most of the *T. ni* larvae. No attempt was made to quantify the numbers of infectious virus. After the third formalin treatment no virus was detected in the soil samples; all the insects that fed on the cabbage remained healthy, pupated and developed normally. None contained any detectable quantity of the marked virus. These tests demonstrated that virus had survived in the soil throughout the 6-month period and that the site had been successfully disinfected by formalin treatment.

Studies with a marked, genetically crippled AcNPV

A second release of a genetically engineered baculovirus insecticide was undertaken in 1987. This time, the AcNPV polyhedrin gene and its transcription promoter were removed in their entirety from the virus and replaced by another (different) synthetic oligonucleotide (Bishop *et al.* 1988). The oligonucleotide sequence was positioned so that it would not add to, or detract from, the expression of any AcNPV gene product. This time no attempt was made to exclude ATG codons from the sequence. The size and constitution of the oligonucleotide were such that it would serve only as a unique marker for the recombinant virus. The sequence and position of the oligonucleotide in the viral genome was verified by sequence analyses. No alteration to the phenotype of the virus was detected by protein analyses, other than that predicted by the loss of the polyhedrin protein and the inability of the virus to form occlusion bodies. The host range and genetic and physical stability of the recombinant virus were determined as before. The data indicated that the virus was more restricted in host range than the parent virus, and that although genetically stable it was physically unstable (i.e. it degraded rapidly in soil or on foliage).

After review of the laboratory data, a licence was issued by MAFF to undertake a release under the same conditions and using the same protocols and facility that were employed for the occluded, marked recombinant AcNPV (Fig. 10.3). The objective was to determine if the polyhedrin-negative virus released from dead caterpillars persisted in the environment in soil and on plant surfaces. Caterpillars were infected in the laboratory, taken to the field site and released in the central enclosed area on to sugar beet. By 1 week all the caterpillars were dead. No virus capable of infecting second instar *T. ni* larvae was subsequently identified either on the plant surfaces or in soil samples. Even the corpses of dead caterpillars did not yield infectious virus. In summary, the data obtained with the marked, so-called 'self-destructive' virus, demonstrated that it did persist in the environment.

To determine whether the virus could be used as a biological control agent *per se*, a third release using the same virus was undertaken in 1988 to measure virus efficacy. Virus was placed on sugar beet leaves on which the adult *S. exigua* had laid eggs. The plants were then transferred from the laboratory to the field facility to monitor infection and survival rates. The dose–response data indicated that at high inoculum doses all the emerging larvae died from virus infection and that no infectious virus remained after 1–2 weeks.

Field trials using a 'fail safe' AcNPV containing a 'junk' gene

As the polyhedrin-negative virus did not persist in the environment it is a suitable vehicle for the expression of a foreign gene because it has a 'fail-safe' genotype (i.e. the virus will degrade in the environment after expression of a foreign gene). To investigate the expression of a foreign gene by such a recombinant baculovirus, a genetically crippled AcNPV was prepared with the bacterial enzyme β-galactosidase substituted in lieu of the coding region of the AcNPV polyhedrin gene. The enzyme was included as a 'junk' gene to allow its level of expression to be quantitated in infected caterpillars. Laboratory analyses established that the recombinant virus exhibited the same host range phenotype as the marked and genetically crippled virus described above. Also as described above, studies on the stability of the virus following successive passage in tissue culture, or in permissive hosts, demonstrated that the virus was genetically stable and did not lose the inserted gene upon passage. Physically, however, the virus exhibited the same susceptibility to degradation in soil or on leaf surfaces as the self-destructive virus described above. Following permission to perform field tests, infected caterpillars were placed on the leaves of sugar beet plants in the field facility at Wytham and the infections and recovery of virus monitored as a function of time. The caterpillars died from virus infection and the β-galactosidase enzyme was expressed.

Phenotype analyses of a 'fail safe' AcNPV containing *Bacillus thuringiensis* δ-toxin

A recombinant baculovirus has been prepared in which the AcNPV polyhedrin gene has been replaced with the δ-toxin of *Bacillus thuringiensis* (*B. thuringiensis* subspecies *kurstaki*, isolate HD-73). Details of the construction and properties of the virus have been reported elsewhere (Merryweather *et al.* 1990). The virus expresses the toxin in a form that is lethal to feeding insects that are susceptible to the toxin. Host range, genetic stability and physical stability analyses are in progress to assess the value of this recombinant for pest control.

Discussion

In considering the results of the studies reported here, it is important to emphasize that the objectives were to assess and minimize any risk that might be involved in the release into the environment of a genetically engineered virus insecticide. The study is part of a programme of scientific research, rather than the commercial development of a product. As such, the results and information derived from the studies apply only to baculovirus insecticides, and only to the particular virus that was used. Having said this to be scientifically accurate, there is good reason to believe that other baculoviruses, similarly marked or manipulated, will behave in a like manner.

Various factors were considered to minimize risk to the environment of the proposed releases. These included the choice of virus, the choice of the target pest, the design of the engineered viruses, the design of the field site and the procedures employed to conduct the experiment (including the final disinfection of the site). After preparing the engineered viruses, they were subjected to extensive laboratory analyses. From the data obtained we could not identify any measurable risk associated with the proposed experiments. This does not exclude the possibility of an unforeseen risk, although none was identified in the laboratory studies undertaken prior to the field trials. With the results obtained from the laboratory analyses, we therefore applied to the appropriate authorities for permission to proceed with the field trials. The data were independently evaluated and, after obtaining the requisite licences, the field trials were instigated and the predictions of the laboratory studies were verified.

The results from the field studies established that an innocuous genetic marker is a suitable tool for the identification of an engineered virus released into the environment. They also showed that it is possible to prepare a genetically crippled virus that is rapidly degraded in the environment after killing its caterpillar host. Such a 'fail safe' virus is a suitable candidate for further engineering, and for the preparation of viruses that act more quickly than the natural virus (through the incorporation of genes that determine insect-specific toxins, or hormones, or other genes).

Conclusions

The objectives of genetically engineering a baculovirus insecticide are to improve its speed of action while maintaining the host specificity and other attributes that make it a desirable alternative to the use of chemical pesticides. In our programme of research into the question of whether there are risks associated with the use in the environment of genetically engineered baculoviruses, we have undertaken five field studies. The first study involved a release (1986–7) of a genetically marked *Autographa californica* nuclear polyhedrosis virus (AcNPV). The results demonstrated that an innocuous piece of DNA, appropriately positioned in the AcNPV genome, was an effective means of tagging the virus without affecting its phenotype. It allowed the virus to be identified in bioassays of plant and soil samples. The second and third field trials (1987–8) demonstrated that a genetically crippled, polyhedrin-negative (so-called 'self-destructive') virus, also appropriately marked, could be made. This virus was unable to persist in the environment. Such engineered viruses constitute 'fail safe' substrates for field studies of engineered viruses. The field data obtained with this virus showed that it did not persist in the environment – in soil, on vegetation or within the corpses of caterpillars. The fourth and fifth releases (1988 and 1989) involved a polyhedrin-negative virus that contained a 'junk' gene (the gene coding for the *Escherischia coli* β-galactosidase enzyme) as a phenotypic marker. This study was undertaken to measure the expression level of a foreign gene in infected caterpillars in a field situation. New engineered viruses that express the *B. thuringiensis* δ-toxin and other potentially active products have been prepared and are under investigation in the laboratory. The overall goal of the programme is to develop new types of custom-designed virus insecticides that will be environmentally safe to deploy for pest control and for crop (forest) protection. The research involves the determination of any identifiable risks associated with the application of such viruses using both laboratory and field risk assessment and environmental impact analyses.

Acknowledgements

We thank Mr C.F. Rivers, Mr P.H. Sterling, Mrs M.E.K. Tinson, Mr T. Carty and Mr M.L. Hirst for their assistance in collecting and/or rearing the insects for the host range studies. The genetic manipulations were undertaken in part by Dr I. Cameron of this Institute and the host range tests by Ms Cynthia Allen. In 1986, moth collections and data analyses were undertaken by Mr S.M. Eley. In 1987, similar collections were undertaken by Ms C. Doyle and Mr A.C. Forkner of the Institute. Ultra-violet light measurements in the facility were conducted by Mr H.J. Killick. The support of these and other members of the Institute's staff, particularly Mr P.F. Entwistle and Mr J.S. Robertson, are gratefully acknowledged. Dr J.I. Cooper and his staff kindly provided the plants. We are indebted to Mr R. Broadbent and Mr R.M. MacKenzie who constructed the field facility.

References

Bishop, D.H.L. (1986). UK release of genetically marked virus. *Nature*, **323**, 496.

Bishop, D.H.L., Entwistle, P.F., Cameron, I.R., Allen, C.J., and Possee, R.D. (1988). Field trials of genetically engineered baculovirus insecticides. In *The release of genetically engineered microorganisms*, (Ed. M. Sussman, G.H. Collins, F.A. Skinner, and D.E. Stewart-Tull), pp. 43–79. Academic Press, London.

Entwistle, P.F., and Evans, H.F. (1985). In *Comprehensive insect physiology, biochemistry and pharmacology*, Vol. 12, (Ed. L.I. Gilbert and G.A. Kerkut), pp. 347–412. Pergamon Press, Oxford.

Entwistle, P.F., and Evans, H.F. (1987). Trials on the control of *P. flammea* with a nuclear polyhedrosis virus. In *Population biology and control of the Pine Beauty Moth*, (Ed. S.R. Leather, J.T. Stoakley, and H.F. Evans), Forestry Commission Bulletin 67, pp. 61–8. HMSO, London.

Entwistle, P.F., Adams, P.H.W., Evans, H.F., and Rivers, C.F. (1983). Epizootiology of a nuclear polyhedrosis virus (Baculoviridae) in European Spruce Sawfly (*Gilpinia hercyniae*): spread of disease from small epicentres in comparison with spread of baculovirus diseases on other hosts. *Journal of Applied Ecology*, **20**, 473–87.

Kelly, P.M., and Entwistle, P.F. (1988). *In vivo* mass production in the cabbage moth (*Mamestra brassicae*) of a heterologous (*Panolis*) and a homologous (*Mamestra*) nuclear polyhedrosis virus. *Journal of Virological Methods*, **19**, 249–56.

Merryweather, A.T., Weyer, U., Harris, M.P.G., Hirst, M., Booth, T., and Possee, R.D. (1990). Construction of genetically engineered baculovirus insecticides containing the *Bacillus thuringiensis* subsp. *kurstaki* HD-73 delta endotoxin. *Journal of General Virology*, **71**, 1535–44.

Podgewaite, J.D. (1985). Strategies for field use of baculoviruses. In *Viral insecticides for biological control*, (Ed. K. Maramorosch and K.E. Sherman), pp. 775–97. Academic Press, New York.

WHO–FAO (1973). Expert Group (GTRES): The use of viruses for the control of insect pests. *World Health Organization Technical Report Series*, **531**.

11

Spread and survival of genetically marked bacteria in soil

D.J. Drahos, G.F. Barry, B.C. Hemming, E.J. Brandt, E.L. Kline,
H.D. Skipper, D.A. Kluepfel, D.T. Gooden, and T.A. Hughes

Introduction

TO MAKE FULL USE of genetically modified
micro-organisms in commercial applications it is
critical to address two key areas: performance and
environmental effect. The performance of any microbe, modified or not, in the site from which it
was originally taken, or to which it is being newly
applied, is not always predictable (Cook and Baker
1983). Seemingly simple genetic changes can
seriously debilitate the candidate when it is faced
with the stiff competition of the microbial community that is keenly adapted to the occupied
niche. For example, the soil bacterium engineered
to more effectively control a crop disease must
satisfy rigid criteria of population density and
longevity. If modified, the new microbe must carry
out expression of the newly inserted genes at
crucial times of disease impact. No less important
is the potential influence an engineered microbe
might have upon the community it is re-entering.
Questions arise concerning microbial spread, carry-
over on subsequent crops and genetic exchange
with a variety of similar micro-organisms.

Any effort to adequately assess performance and
environmental effect requires a means to monitor
an appreciable portion of the introduced microbial
population over at least critical time periods. Traditionally, this has often been laborious and highly
variable. The recent development of several novel
genetic tools, such as the *lac*ZY and *lux* bacterial
marking systems (Turner 1985; Drahos *et al*. 1986,

1988; Schmetterer *et al*. 1986; Schauer 1988), have
provided a much greater degree of efficiency and
accuracy in monitoring strategies. This chapter will
focus on information gained through the use of the
*lac*ZY genetic system in an extensive 18-month
field study. Two successive field tests of a second
*lac*ZY-engineered biocontrol bacterium, and future
implications of the commercial development of
such engineered organisms, will also be reviewed.

Background to the *lac*ZY marker system

A relatively large proportion of soil bacteria closely
associated with plant roots are members of the
fluorescent pseudomonad group. A number of
these bacteria have been identified as playing an
active beneficial role in plant vigour and natural
disease defence (Burr *et al*. 1978; Kloepper *et al*.
1980a,b; Kloepper and Schroth 1981; Howie and
Echandi 1983; Xu and Gross 1986). The discovery
that these bacteria are incapable of using lactose as
a sole carbon and energy source (Hemming and
Drahos 1984) has led to the development of a rapid
and sensitive monitoring strategy with benefits
both in the practical, commercial applications of
such beneficial strains and in broadening our knowledge of their behaviour in the environment.

The monitoring handle is provided by the addition of two *Escherichia coli* K-12 *lac* operon
genes, *lac*Z and *lac*Y, referred to as *lac*Y. Expression of these genes provides for synthesis of

β-galactosidase and lactose permease, and permits efficient uptake and metabolism of the lactose substrate when *lac*ZY-marked pseudomonad bacteria are recovered on a defined minimal medium (Drahos *et al.* 1986; Fig. 11.1a). Additional selection for the target strain from the native rhizosphere (root-soil zone) population (typically $> 10^9$ colony-forming units (CFU)/g soil) may be based on intrinsic resistance to certain antibiotics in the strain of interest, or by scoring for the fluorescent phenotype. With this approach, as few as one marked pseudomonad bacterium per gram of soil can be efficiently detected. Equally important is the ability not only to monitor the marked bacterium, but also to follow additional new genes, which may be linked with the *lac*ZY genes during construction, for potential transfer to the indigenous population.

A high level of genetic stability has been achieved through the insertion of the *lac*ZY genes directly into the bacterial chromosome (Barry 1986, 1988). This method is based on the use of a disarmed Tn7 transposon system, in which the *lac*Z and *lac*Y genes are carried between the left and right termini of the original transposon (Fig. 11.1b). Transcomplementation of this construction by products of the Tn7 transposase genes results in a permanent, or one-way insertion of this *lac*ZY gene construct into the bacterial genome. As the complementing transposase genes are carried separately on unstable plasmid elements, they are rapidly lost, and leave behind the inserted segment, now incapable of moving itself to another location. Tn7 has a very high specificity for insertion, generally having only one conserved site per bacterial chromosome. This feature minimizes location

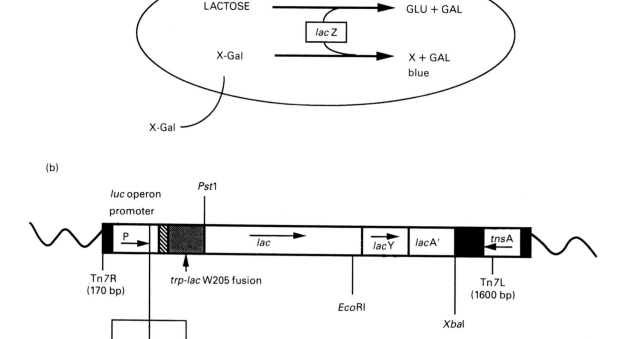

Fig. 11.1. (a) Function of the *lac*Z and *lac*Y genes. The *lac*Y gene product, lactose permease, is membrane-associated in both *E. coli* and *Pseudomonas fluorescens*, and is responsible for transport of lactose into the cell. Lactose permease is not required for transport of X-Gal (5-bromo-4-chloro-3-indolyl-β-D-galactopyranoside). The *lac*Z product is located in the cytoplasm and functions to cleave lactose and X-Gal into the catabolites indicated. (b) Structure of the Tn7–*lac*ZY element in Ps. 3732RNL11.

effects, particularly when comparing different marked isolates. Recent versions of the marking system, such as pMON7197 (Barry 1988), have provided a relatively rapid, single-step method, which has been used successfully to mark a wide variety of fluorescent pseudomonads and other lactose-incompetent bacteria.

Objectives of the *lac*ZY field studies

For all its perceived value, the real test of utility for the *lac*ZY tracking system could only come from actual field studies. Critical questions regarding the real world competitiveness of strains 'burdened' with the *lac*ZY gene system, the long-term maintenance and function of the marker, spread of the introduced microbes, survival and carry-over to subsequent crops in the same field and potential gene transfer to indigenous strains, all require substantial data from actual field studies. The tremendous variability in climatic conditions, populations of microbial communities, soil type and moisture content, and the dynamic interaction between a plant-associated microbe and its host during the growing season, greatly enhances the value of field-generated data over that obtained under more artificial conditions.

Ultimately, this technology will find full use only after it has been proved effective for native or engineered microbial agents designed to protect plants against soil-borne disease or other pests. The objective, therefore, in the first field study was to develop an extensive database, over a relatively long period, which addresses the key questions of survival, dissemination, gene exchange, effects of the marker gene on bacterial survival and effects of the engineered bacterium on plant development. Subsequent field tests, which extend this research and assess performance of certain *lac*ZY-engineered plant-beneficial microbes have recently been initiated.

The primary objective of the first small-scale field test using *lac*ZY-marked soil bacteria was to verify the ability of the *lac*ZY marker to provide an accurate, practical means for monitoring the survival and movement of a natural root-colonizing fluorescent pseudomonad over three cropping cycles (winter wheat, no-till soybeans and minimum-till winter wheat) using normal agronomic practices. To achieve this, key requirements of marker effectiveness and stability, test design and

layout, and monitoring strategy were addressed. Furthermore, this field test proved to be the first of its kind in the US, involving an engineered microbe containing genes from two different bacterial genera.

Design and construction of the engineered microbe Ps. 3732RNL11

The bacterium used in the first field study (Ps. 3732RN) was originally isolated from soil planted to corn in St. Charles, Missouri (Obukowicz *et al.* 1986). It was characterized by standard methods, and by gas chromatography (GC)/fatty acid analysis, as the fluorescent bacterium *Pseudomonas aureofaciens*. Extensive prerelease analysis demonstrated that Ps. 3732RN was both non-toxic and non-pathogenic to animals and plants, was unlikely to survive in river, lake or sewage water and had no significant impact on host plant growth or development (US Environmental Protection Agency Premanufacture Notification #P-87-1292 1987; Drahos *et al.* 1988). Having met what were considered to be basic prerequisites for an effective and acceptable monitoring study, the next step was to introduce the *lac*ZY tracking system.

Early versions of *lac*ZY-marked soil bacteria carried broad host range plasmid vectors containing the *E. coli lac*Z and *lac*Y genes. While generally effective in providing selection from natural soil samples (Drahos *et al.* 1986), this method failed to provide desired levels of gene stability and containment within the original host in some strains. As described above, rapid and efficient systems have been developed to insert the *lac*ZY genes directly and permanently into the host bacterial chromosome, based on a defective but complementable derivative of the transposon Tn7 (Barry 1986, 1988). By this method, the Tn7*lac*ZY gene element was transposed from an unstable plasmid to the chromosome of the target bacterium using the Tn7 transposition (*tns*) genes located in a separate section of a monocomponent Tn7/*lac*ZY marker plasmid pMON7197 (Barry 1988), to create the *lac*ZY-marked derivative Ps. 3732RNL11, used in the field study.

Laboratory analysis of Ps. 3732RNL11 was then conducted to conform the marker insertion, as well as the expression, stability and permanence of the *lac*Z and *lac*Y genes. A beneficial feature of the Tn7 system is its propensity to insert at a single, specific site in the bacterial chromosome (Lichtenstein and Brenner 1982). Southern hybridization of Ps. 3732RNL11 confirmed that this single site insertion did indeed occur, and growth kinetics in defined mimimal media demonstrated that no essential gene was inactivated during the insertion process. Adequate expression of the *lac*ZY element was shown by the identical growth rates of Ps. 3732RNL11 on either glucose or lactose as sole carbon sources. Similar growth rates of Ps. 3732RN and Ps. 3732RNL11 on minimal glucose medium demonstrated that expression of the *lac*ZY genes was not detrimental for the host under defined laboratory conditions.

Regulatory review and prerelease laboratory testing

In 1987, field inoculation of a recombinant microorganism in the US required review by several federal regulatory agencies, including the US Environmental Protection Agency (EPA) and the US Department of Agriculture (USDA). As the engineered organism Ps. 3732RNL11 is a benign strain, with the inserted *lac*ZY marker the only new genetic element, it was not considered a pesticidal agent. The application to field-test the engineered strain was therefore reviewed by the EPA Office of Toxic Substances (OTS) under the Toxic Substance Control Act (TSCA). As part of this procedure, Monsanto and Clemson submitted to the EPA–OTS as a Premanufacture Notification (PMN) in voluntary compliance with TSCA section 5 and the policy guidance of the Agency regarding biotechnology products subject to TSCA as set out in the *Coordinated framework for regulation of biotechnology* (Federal Register, June 26 1986). The submission was made on 18 June, 1987.

An integral part of this review process is an assessment of the potential hazard presented to human health and the environment as a consequence of field testing of engineered micro-organisms. The degree of hazard is considered a function of the toxicity/pathogenicity of the test organisms and the level of human and environmental exposure to them. Having fulfilled the essential characteristics described in the previous section, more extensive prefield test analysis of the marked strain was undertaken. This included reconfirmation of Ps. 3732RNL11 strain identity, non-pathogenicity, potential for survival under contained conditions,

stability of the introduced genes and containment within the host organism, as well as the efficacy of the monitoring system to track Ps. 3732RNL11 under greenhouse conditions in the presence of native field microbes (natural field soil). A more detailed account of this testing has been described (Drahos *et al.* 1988).

In addition to the EPA, the field test was also pre-reviewed by the USDA, to provide concurrence on the non-pathogen status of Ps. 3732RNL11, and by the Institutional Biosafety Committees at both Monsanto Company (St. Louis) and Clemson University (Clemson, South Carolina). The EPA review process included a 1-day public review by the EPA's Biotechnology Scientific Advisory Committee and public notification in the Federal Register. While seemingly formidable, close communication and cooperation between these regulatory agencies and the submitting parties kept the entire process relatively short. Approval from the EPA was received 19 October 1987, barely 4 months after the PMN was filed. In addition, this was the first field test of a recombinant microbe approved by the EPA Office of Toxic Substances. The only prior field test of a recombinant bacterium, the Ice⁻ *Pseudomonas syringae* strain (Lindow and Panopoulos 1988), which was created by deleting a gene segment, was approved under a different agency, the EPA Office of Pesticide Products.

Field test design and initiation

The actual field test of the *lac*ZY-engineered strain, Ps. 3732RNL11, was initiated on 2 November 1987, at Clemson's Edisto Research and Education Center, a 2300-acre facility near Blackville, South Carolina. A centrally located, 5-acre test area in the Edisto site was chosen and fenced to discourage the entry of animals. Within this area, a 1.5-acre test plot (Fig. 11.2) was laid out; this provided for a six-replicate test for each of three treatments; (i) non-inoculated wheat (A); (ii) wheat inoculated with the parental strain Ps. 3732RN (B); (iii) wheat inoculated with the *lac*ZY-marked strain Ps. 3732RNL11 (C). Individual treatment plots were 100 feet by 10 feet. The inoculated plot area was surrounded by a 30-foot border area planted with wheat 1 month prior to test initiation. In turn, this area was surrounded by a 15-foot plant-free zone, then by another 10-foot plant zone (grass).

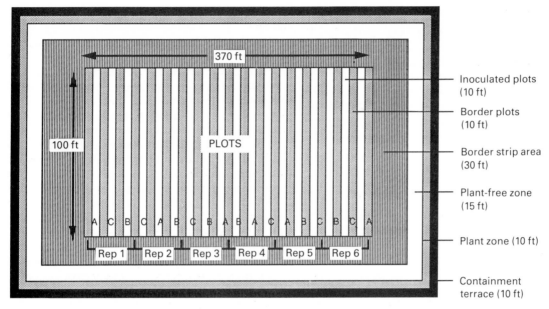

Fig. 11.2. Field test plot design for Ps. 3732RNL11. A, non-inoculated replicate plots; B, plots inoculated with Ps. 3732RN (parent strain); C, plots inoculated with Ps. 3732RNL11 (*lac*ZY-marked).

Finally, a containment terrace 10-feet wide and 1 foot high encompassed the entire test plot site to aid in retaining run-off water.

To monitor the survival and potential carry-over of the inoculated microbes, three crops, following normal agronomic rotations, were included in the test. The first crop of winter wheat was the only crop inoculated with microbes during the study. Application was as a liquid treatment directly on the seed in-furrow during the planting process. Wheat was planted in rows 7 inches apart and the eight central rows of the 16 rows in each treatment plot received the bacterial inoculum. This first crop of wheat was harvested approximately 31 weeks later, at which time seed yields were taken for all treatments. A second crop of soybean was then planted, without inoculation or tillage of the wheat field, in the early summer of 1988. The soybean was planted with 30-inch row spacings, such that two of the soybean rows fell within rows that had received the original bacterial treatment. After harvest of the soybean crop in early autumn, a third crop of winter wheat was planted, in the same location and spacing as used in the first wheat crop. As with the soybean crop, the third test crop (winter wheat) was planted without additional bacterial inoculation.

Concurrent with the first microbial application, an extensive monitoring programme, which eventually involved the collection of 11 156 individual plant and soil samples, was initiated. The sampling was designed to follow the progression of the inoculated bacteria both horizontally and vertically from the treated plant rows, and to monitor the bacteria for movement into border areas, run-off water and surface bodies of water near the test site. In addition, samples were taken specifically to monitor potential genetic transfer of the *lac*ZY genes from the inoculated Ps. 3732RNL11 strain to indigenous field bacteria. Samples were analysed on at least two specific defined media, which provided for both the monitoring of the *lac*ZY-marked strain and the comparison of the recovery of the marked strain (Ps. 3732RNL11) with that of the parent strain (Ps. 3732RN).

Summary of field test results

A significant number of the principal questions concerning performance, survival and genetic stability were addressed in the first winter wheat crop cycle. Among the most important was the issue of whether or not the *lac*ZY monitoring strategy would, in fact, function effectively in the field environment. This was answered in two steps:

1. Demonstrating that the 'fitness' of the *lac*ZY-engineered strain, presumably carrying non-advantageous genetic baggage, was not less than that of its 'unencumbered' parent.
2. Confirming that recovery based on the *lac*ZY monitoring 'handle' (growth on minimal lactose medium) is at least as effective as recovery based on traditional selection using antibiotic resistance alone.

Colonization levels of inoculated bacteria

The parent strain used in this study (Ps. 3732RN) is inherently resistant to the antibiotics rifampicin and nalidixic acid. These resistance characteristics do not seriously debilitate this bacterium or prevent it maintaining its environmental competitiveness, unlike many other biocontrol candidates (Compeau *et al.* 1988). Therefore, comparative analysis of soil samples taken from the roots of plants inoculated with the parent Ps. 3732RN with samples taken from plants inoculated with the *lac*ZY-marked strain Ps. 3732RNL11 provides an

indication of the relative 'fitness' of the engineered strain.

Results of this comparative analysis, performed over the 31-week course of the first winter wheat crop, is given in Figure 11.3. There is no statistically significant difference (below $P = 0.05$) between recovery of the parent and the *lac*ZY-engineered strain during the growing season, indicating that the *lac*ZY marker genes have little effect on environmental fitness. It is also clear from this analysis that the colonization pattern of the engineered and non-engineered strains were very similar. As had been predicted by earlier greenhouse studies (Drahos *et al.* 1988), a high initial population density was followed by a steady decline through the growing season. Survival of appreciable numbers of the inoculated bacteria through the dormant winter period was encouraging as this is considered to be an important feature of an effective biocontrol agent for this crop. A confidence in the statistics of sample measurement was established by assessing the levels of marked bacteria on 25 adjacent plants in the same row in plots inoculated with Ps. 3732RNL11. This provided a measurement for the uniformity of colonization across each plot.

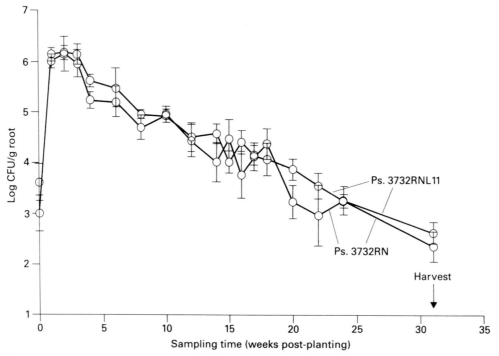

Fig. 11.3. Colonization level comparison. Error bars indicate ± one standard deviation. Harvest date was at week 31.

Four of the six replicate plots were sampled. In these, there was no significant difference between plant samples (below $P = 0.05$), indicating a generally uniform colonization level and reproducible sampling procedure.

Effectiveness of *lac*ZY marker in the field

The second step was to confirm the effectiveness of the *lac*ZY marker under field conditions, relative to more traditional recovery based on antibiotic resistance alone. This was accomplished by analysis of the same plant and soil samples on two different selective media: a 'rich' medium (Pseudomonas Agar F; Difco Co., Detroit, Michigan, USA) containing the selective antibiotics rifampicin and nalidixic acid (PAF/Rif/Nal), and a 'minimal' medium with lactose (1 per cent w/v) as the sole carbon source (M9/Lac/Rif). Early experiments demonstrated that a minimal lactose medium without rifampicin (M9/lac) could also be used with equal effectiveness by scoring the *lac*ZY-marked strains by their fluorescent pigment production under UV light (Drahos *et al.* 1986). A comparison of samples

taken from winter wheat plots inoculated with the *lac*ZY-engineered microbe (Ps. 3732RNL11), and analysed on these media is shown in Figure 11.4. There is no significant difference in the recovery of Ps. 3732RNL11 on either media, demonstrating the effective selective capacity of lactose utilization.

Bacterial dissemination

Assessments were made at regular intervals throughout the entire test period to determine the extent of dissemination of the inoculated bacteria. Measurements of horizontal movement were performed by recovering soil and root-wash samples from the non-inoculated border rows, which were spaced at 7-inch intervals from the inoculated wheat plants. It was planned that samples were to be taken from border rows progressively further from the inoculated rows as the marked strain (Ps. 3732RNL11) and the parent strain (Ps. 3732RN) moved, or were carried passively, during the growing season. By design, movement as far as 24 border-row distances (approximately 15 feet) could have been monitored before encounter-

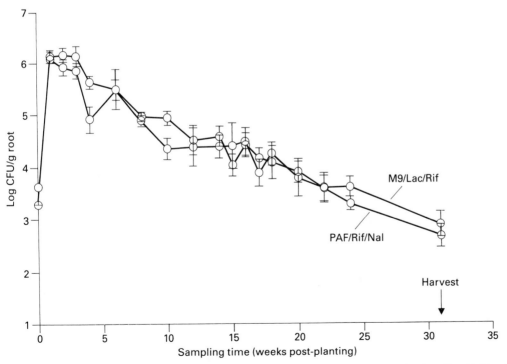

Fig. 11.4. Comparative recovery of Ps. 3732RNL11 on rich medium (PAF/Rif/Nal) and minimal medium (M9/Lac/Rif). Error bars indicate ± one standard deviation.

ing another inoculated row area. It was found, however, that less than 3 per cent of the 5700 border-row plant and soil samples taken during the entire test contained detectable numbers of the inoculated strains (Fig. 11.5). In two of these positive samples, a very small number of bacterial colonies (<5) were found in soil taken from the area between the 7-inch and the 14-inch border row of plots inoculated with Ps. 3732RNL11. These have been indicated as positive in Figure 11.5, although the samples were not isolated from the standard root washings. Each analysis was performed such that as few as one marked bacterium in 0.1 g root (or g soil) would be detected. In addition, samples were taken regularly from the plant-free zone area, surface run-off water and pond water located approximately 1000 feet from

the test site. The inoculated bacteria were never detected in these locations.

Vertical dissemination was measured 5 weeks after planting. For this, ditches 2 feet wide were dug adjacent to wheat rows inoculated with Ps. 3732RNL11 in each of four test plots. Horizontal core samples were then taken at 2, 6, 12, 18 and 24 inches from the test plot surface. Ps. 3732RNL11 was detected only in samples above 18 inches that also contained roots from the inoculated plants. This indicates that the *lacZY*-marked strain had not significantly migrated below the root zone.

Prior to harvest, samples of upper wheat plant parts (leaf, stem and grain) were taken, homogenized and tested for the possible presence (either internal or external) of Ps. 3732RNL11. As ex-

Sample	Week 0	1	2	3	4	6	8	10	12
Inoculated plants	+	+	+	+	+	+	+	+	+
7-inch border rows	−	−	*+	−	−	−	−	−	+
14-inch border rows	−	−	−	−	−	−	−	−	*+
Plant-free zone	−	−	−	−	−	−	−	−	−
Pond water	−	−	−	−	−	−	−	−	−

Sample	Week 14	15	16	17	18	20	22	24	31
Inoculated plants	+	+	+	+	+	+	+	+	+
7-inch border rows	−	−	*+	−	*+	−	*+	−	*+
14-inch border rows	−	−	*+	−	−	−	−	−	−
Plant-free zone	−	−	−	−	−	−	−	−	−
Pond water	−	−	−	−	−	−	−	−	−

Fig. 11.5. Field detection of Ps. 3732RNL11 at various locations in the test site during the first winter wheat growing season. +, Ps. 3732RNL11 detected; − not detected; * detected very infrequently.

pected from earlier growth chamber tests, inoculated bacteria were not found in this material. This demonstrated that this root-colonizing strain had not migrated systematically into the upper plant, nor had it become established on plant leaf or stem surfaces due to splashing or upward dissemination.

Yield of the inoculated wheat and subsequent non-inoculated crops

Approximately 31 weeks after planting, the first winter wheat crop was harvested. The wheat yield from the non-inoculated plots was essentially identical to the yield from plots that had received either Ps. 3732RN or Ps. 3732RNL11 (Fig. 11.6a). While there is some trend towards a slightly higher yield from the bacteria-inoculated plots, this was not statistically significant. Nor would a yield increase have been expected, as the strain selected for this tracking test is a benign root-associated microbe, and not observed to be plant-beneficial.

During the 18-month test period, two subsequent crops, no-till soybeans followed by minimal-till winter wheat, were planted on the same site. Yields from these crops was also generally uniform across the plots, without statistically significant differences (Fig. 11.6b and 11.6c).

Measurement of genetic exchange in the field

As described, the *lac*ZY genes were inserted into the bacterial chromosome along with the non-functional 'ends' of the Tn7 transposon (see Fig 11.1b). This provided both a selection for such microbes expressing the *lac*Z and *lac*Y gene products, and unique genetic material (Tn7 'ends') closely linked to the inserted *lac*ZY genes. The presence of this unique material could be detected by hybridization analysis, using radioactively labelled probes specific to a Tn7 end segment.

Using these features it was possible to test for the transfer of the inserted *lac*ZY-Tn7 material from the inoculated strain Ps. 3732RNL11 to native soil microbes. Any native soil microbe that received, and expressed, the recombinant *lac*ZY-Tn7 genes could be detected by its acquired capacity for lactose metabolism, and its concomitant hybridization with a Tn7-specific probe. Therefore, root washings and soil samples, taken from plants inoculated with Ps. 3732RNL11, were first plated on minimal lactose medium, which selected all bacteria capable of efficient lactose metabolism. Early

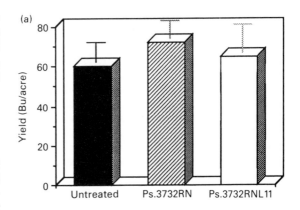

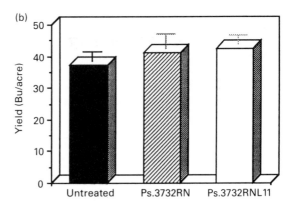

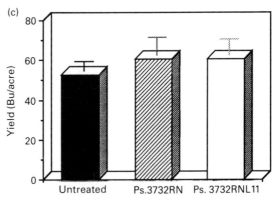

Fig. 11.6. Yield comparison of the three crops in the field study. (a) Yield of inoculated first winter wheat crop; (b) non-inoculated soybean crop; (c) non-inoculated second winter wheat crop. Error bars indicate ± one standard deviation.

tests of field samples from the test site measured the level of such indigenous strains between 10^4 and 10^6 CFU/g soil. A sampling date was chosen such that the expected level of the *lac*ZY-marked inoculated strain Ps. 3732RNL11 would comprise approximately 5–10 per cent of the indigenous lactose-competent population. This time point (which occurred approximately 10–15 weeks post-planting) was judged the most optimal, as the greatest population of the inoculated marked strain had been present in close proximity to very similar indigenous strains for a relatively long period of time. At later sampling dates, populations of Ps. 3732RNL11 would have further declined and would not have been sufficient to provide an internal positive control for the hybridization experiments. These root samples were plated on the semiselective M9/lac media, from which 10 058 individual colonies were picked and separately cultured. DNA from each culture was individually prepared and spotted on nitrocellulose hybridization filters. These were then hybridized to a [^{32}P]labelled DNA probe, which consisted only of the portion of the Tn7 right 'end' present in the

marked Ps. 3732RNL11 strain. No evidence was found for genetic exchange of the inserted *lac*ZY-Tn7 element with the native microbial population. All 557 lactose-competent isolates that also hybridized with the Tn7 probe proved to be Ps. 3732RNL11, following analysis of rifampicin and nalidixic acid resistance, chromosomal fingerprint patterns (Drahos *et al.* 1985), and GC/fatty acid profiles (Miller 1982). At this detection level, the frequency of genetic exchange of this marker system in the field is less than 9.8×10^{-8} events/g soil.

Carry-over of the inoculated strains to subsequent crops

Following harvest of the first winter wheat crop, the test plot was planted with no-till soybean, without any new inoculation of bacteria. The planting pattern was arranged such that two-soybean rows, which were spaced 30 inches apart, fell within the same rows that had contained the previously inoculated winter wheat plants. Sampling and analysis of these soybean plants during the

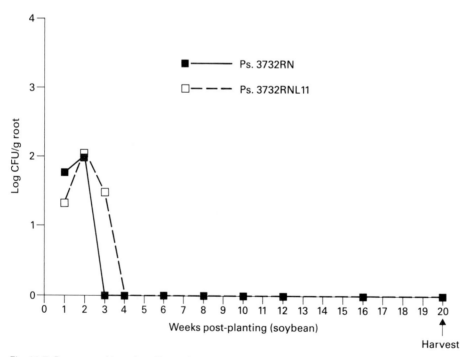

Fig. 11.7. Presence of inoculated bacteria on the soybean crop. Viable counts from PAF/Rif/Nal medium. Harvest date was at week 20.

growing season demonstrated that the carry-over of the wheat-inoculated bacteria to the subsequent crop in the normal rotation is very minimal, and of short duration. As shown in Figure 11.7, some of the inoculated bacteria were detected during the first 4 weeks after soybean planting. However, the test bacteria colonizing the soybean roots never reached a level greater than 0.1 per cent of the levels attained on the inoculated wheat plants. In addition, detection of these strains was non-uniform, variable across the plots and only oc-curred on soybean roots from rows that had con-tained the previously inoculated wheat. Soybean samples from border areas remained free of the marked strains.

After harvest of the soybean crop, a second winter wheat crop (third crop in the series) was planted following minimal tillage. As before, seeds were planted in the same place as the rows from the initial wheat crop, and plant and soil samples were taken at regular intervals throughout the growing season. While none of the inoculated bacteria were detected from soybean samples taken after 4 weeks,

a potential 'bloom' of these strains, presumably adapted to wheat, could not be ruled out. How-ever, neither Ps. 3732RN nor Ps. 3732RNL11 were detected in samples taken from the 'inocu-lated' row areas. One Ps. 3732RNL11 bacterium was detected in a border row sample taken 4 weeks after planting the second wheat crop, but in no other sample after this. These results indicate that there is no significant perseverance of these inocu-lated bacteria beyond the initial year of inoculation. A summary diagram depicting the viable counts of the genetically engineered organism, Ps. 3732RNL11, throughout the entire 83 week field study is shown in Figure 11.8.

Implications for future field releases

The first field releases of genetically engineered microbes, such as the *lac*ZY-marked *Pseudomonas* strain described here, provide a tremendous oppor-tunity to evaluate performance and environmental effects under actual field conditions. From an

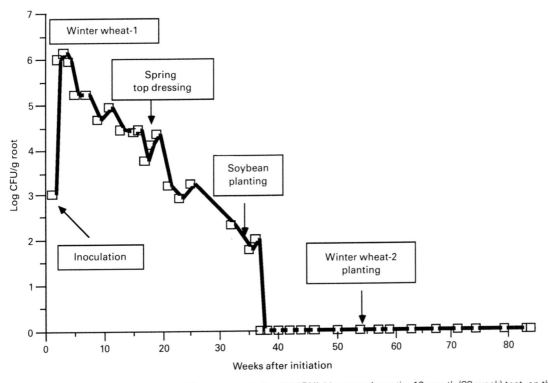

Fig. 11.8. Colonization levels of the *lac*ZY-marked strain Ps. 3732RNL11 across the entire 19-month (83-week) test, on the three crops, as indicated.

ecological perspective, the application of non-cumbersome and efficient tracking tools such as this allow difficult and pressing questions to be adequately addressed through actual field measurements. As a database becomes established, regulatory agencies are able to make a more informed and accurate assessment of the risk versus benefit in applying this technology on a broader scale. Certainly, an industrial organization, interested in developing an engineered microbe to a commercial level, will benefit greatly from an early and favourable regulatory approval and registration process, made possible through an informed, science-based review. With each succeeding study, the safety of the next field test, perhaps involving different strains of microbes, further genetic modification, or additional genes, becomes more predictable.

The information gained through this initial study was therefore applied in the next phase of testing involving Monsanto, Clemson University, and USDA–ARS (Agricultural Research Service) scientists located at Washington State University (WSU) in Pullman, Washington. This second set of field releases involved a microbial strain similar to Ps. 3732RNL11, but with a demonstrated ability to protect wheat roots against the soil-borne fungal pathogen Take-All (*Gaeumannomyces graminis* pv. *tritici*). This disease is significant in temperate climates where wheat and grass culture is intensive. The microbe, Ps. 2-79RNL3, was engineered to carry the same *lac*ZY marker system as before, which would now aid in evaluating the disease control performance of this candidate strain by providing a reliable tracking system to determine population levels of the microbe.

As this field test involved release of genetically engineered microbial pesticide, the EPA Office of Pesticide Products (EPA–OPP) was the principal review agency, working under the authority of the Federal Insecticide, Fungicide, Rodenticide Act (FIFRA). Drawing on the database from the initial study, the EPA–OPP was able to expedite its review and issue the necessary approval 75 days after receipt of the field test application. The 1-year test was initiated by the USDA–ARS scientists at WSU's Spillman farm near Pullman on 6 October 1988). Clemson scientists began an essentially identical study at their Edisto Research site on 2 November 1988.

As we gain experience in studies such as these, we acquire understanding and a more balanced perspective. As we develop our ability to effectively monitor and evaluate released micro-organisms, we gain a clearer picture of their ecological impact and utility. As regulatory authorities acquire a confidence in accurate risk assessment, and commercial organizations an assurance of investment return, the public gains the correct perception that engineered microbes may well be applied wisely, safely and with a potentially substantial benefit.

Conclusion

On 2 November 1987, Monsanto Company and Clemson University initiated a landmark field test of a genetically engineered microbial tracking system in a wheat field at the university's Edisto Research and Education Center near Blackville, South Carolina. The test proved to be the first release in the US of a live engineered bacterium carrying genes from two different strains, and required a thorough review by the US Environmental Protection Agency and the US Department of Agriculture over a 4-month period. The tracking system tested consists of the *lac*Z and *lac*Y genes of *E. coli* K-12 permanently inserted into the chromosome of a common *Pseudomonas* soil bacterium using a disarmed Tn7 transposition vector. The 18-month study extended over three crops and involved analysis of more than 11 000 individual samples of plant and soil material. Test results confirm the effectiveness and practicality of the *lac*ZY marker under field conditions and display a pattern of bacterial survival with high population on plant roots in the inoculated rows but remarkably limited dissemination of the marked bacteria even centimetres away. Genetic exchange of the inserted *lac*ZY marker with native field strains was not detected by a selection/hybridization protocol capable of detecting such transfer at a frequency of 9.8×10^{-8} events/g soil. There was minimal carry-over to subsequent non-inoculated crops and no significant impact on crop yield caused by the marked strains. While this test has proved challenging and costly, it has nevertheless provided an invaluable scientific database, and the promise of a broadly applicable monitoring strategy. It would perhaps by just as important for this test, or others like it, to build public confidence in the belief that genetically altered microbial pesticides may be effectively monitored and safely applied. Additional field tests have now been initiated using this marker system in microbes capable of providing a significant measure of protection against soil-borne fungal disease.

Acknowledgements

We would like to express our sincere thanks to Arthur Kelman for generously performing plant pathogenicity tests; Lidia Watrud, Mark Obuko-wicz, Fred Perlak, Kuniko Kusano-Mretzmer, Ernest Mayer, Ellen Lawrence, Margan Miller-Wideman, Minhtien Tran and Suzanne Bolten for providing the host bacterial strain Ps. 3732RN, as well as important initial data on strain toxicity and growth characteristics; Peter Burrows for statistical evaluations; and James Riley Hill, Harold Musen and Judy Solarz for their valued assistance in the study.

References

Barry, G.F. (1986). Permanent insertion of foreign genes into the chromosomes of soil bacteria. *Bio/Technology*, **4**, 446–9.

Barry, G.F. (1988). A broad-host range shuttle system for gene insertion into the chromosome of Gram-negative bacteria. *Gene*, **71**, 75–84.

Burr, T.J., Schroth, M.N., and Suslow, T. (1978). Increased potato yields by treatment of seed pieces with specific strains of *Pseudomonas fluorescens* and *P. putida*. *Phytopathology*, **68**, 1377–83.

Compeau, G., Al-Achi, B.J., Platsouka, E., and Levy, S.B. (1988). Survival of rifampin-resistant mutants of *Pseudomonas fluorescens* and *Pseudomonas putida* in soil systems. *Applied and Environmental Microbiology*, **54**, 2432–8.

Cook, R.J., and Baker, K.F. (1983). *The nature and practice of biological control of plant pathogens*. The American Phytopathological Society, St Paul, MN, USA.

Drahos, D., Brackin, J., and Barry, G. (1985). Bacterial strain identification by comparative analysis of chromosomal DNA restriction patterns. *Phytopathology*, **75**, 1381.

Drahos, D.J., Hemming, B.C., and McPherson, S. (1986). Tracking recombinant organisms in the environment: β-galactosidase as a selectable, non-antibiotic marker for fluorescent pseudomonads. *Bio/Technology*, **4**, 439–43.

Drahos, D.J., Barry, G.F., Hemming, B.C., Brandt, E.J., Skipper, H.D., Kline, E.L., Kluepfel, D.A., Hughes, T.A., and Gooden, D.T. (1988). Pre-release testing procedures: US field test of a *lacZY*-engineered soil bacterium. In *The release of genetically-engineered micro-organisms*, (Ed. M. Sussman, C.H. Collins, F.A. Skinner, and D.E. Stewart-Tull), pp. 181–91. Academic Press, London.

Hemming, B.C., and Drahos, D.J. (1984). β-galactosidase, a selectable non-antibiotic marker for fluorescent pseudomonads. *Journal of Cellular Biochemistry*, **Suppl. 8B**, 252.

Howie, W.J., and Echandi, E. (1983). Rhizobacteria: Influence of cultivar and soil type on plant growth and yield of potato. *Soil Biology Biochemistry*, **15**, 127–32.

Kloepper, J.W., and Schroth, M.N. (1981). Plant growth-promoting rhizobacteria and plant growth under gnotobiotic conditions. *Phytopathology*, **71**, 642–4.

Kloepper, J.W., Schroth, M.N., and Miller, T.D. (1980a). Effects of rhizosphere colonization by plant growth-promoting rhizobacteria on potato plant development and yield. *Phytopathology*, **70**, 1078–82.

Kloepper, J.W., Leong, J., Teintze, M., and Schroth, M.N. (1980b). Enhanced plant growth by siderophores produced by plant growth-promoting rhizobacteria. *Nature*, **286**, 885–6.

Lindow, S.E., and Panopoulos, N.J. (1988). Field tests of recombinant Ice- *Pseudomonas syringae* for biological frost control in potato. In *The release of genetically-engineered micro-organisms*, (Ed. M. Sussman, C.H. Collins, F.A. Skinner, and D.E. Stewart-Tull), pp. 121–38. Academic Press, London.

Lichtenstein, C., and Brenner, S. (1982). Unique insertion site of Tn7 in *E. coli* chromosome. *Nature*, **297**, 601–3.

Miller, L.T. (1982). Single derivatization method for routine analysis of bacterial whole-cell fatty acid methyl esters, including hydroxy acids. *Journal of Clinical Microbiology*, **16**, 584–6.

Obukowicz, M.G., Perlak, F.J., Kusano-Kretzmer, K., Mayer, E.J., Bolten, S.L., and Watrud, L.S. (1986). Tn5-mediated integration of the delta-endotoxin gene form *Bacillus thuringiensis* into the chromosome of root-colonizing pseudomonads. *Journal of Bacterialogy*, **168**, 982–9.

Schauer, A.T. (1988). Visualizing gene expression with luciferase fusions. *Trends in Biotechnology*, **6**, 23–7.

Schmetterer, G., Wolk, C.P., and Elhai, J. (1986). Expression of luciferases from *Vibrio harveyi* and *Vibrio fischeri* in filamentous cyanobacteria. *Journal of Bacteriology*, **167**, 411–14.

Turner, G.K. (1985). In *Bioluminescence and chemiluminescence: instruments and applications*, (Ed. K. Van Dyke), pp. 43–78. CRC Press Inc., Boca Raton, Florida, USA.

US Environmental Protection Agency Premanufacture Notification #P-87-1292 (1987). U.S. Environmental Protection Agency, Office of Toxic Substances, Washington DC.

Xu, G.W., and Gross, D.C. (1986). Selection of fluorescent pseudomonads antagonistic to *Erwinia carotovora* and suppression of potato seed piece decay. *Phytopathology*, **76**, 414–22.

12

Microbial ecology, genetics and risk assessment

M.J. Day and J.C. Fry

The reasons for assessment and studying microbial ecology

ALL TECHNOLOGIES bring their own individual mixture of promises and threats and the use of genetic engineering protocols to modify organisms is no exception. This chapter aims to illustrate the areas of scentific weakness that we perceive as being important for an assessment of risk. The information discussed is directed solely at genetically engineered micro-organisms (GEMs), although the principles may well apply to other organisms that have been subject to manipulation.

Should we be concerned about intergeneric gene transfer? Vast numbers of plants, animals and microbes die daily and, during their decay, present DNA directly to bacteria capable of taking it up. Thus, it is conceivable that DNA exchange occurs routinely in nature. There is a difference, however, between this natural presentation of non-specific DNA and DNA entering the environment within GEMs. DNA derived from GEMs has not been subjected to natural selection for its amplification. At no time during its synthesis did it compete with natural organisms or have natural selective pressures imposed upon it. The GEM DNA, in stable recombinant organisms, will be released into the environment in relatively large amounts. Thus, the evolution of GEMs, which are constructed in the laboratory, does not rely on chance, their fitness or on natural selection.

It is important to realize that the various abilities to release, mobilize, take up or receive gene sequences does not mean that there is an inherent risk from organisms that can do these things. These processes are all part of the normal evolutionary mechanisms utilized by bacteria. The risk, then, is associated solely with the particular engineered gene/DNA sequence, and not the transfer mechanism. Thus, assessment is made on the basis of an evaluation of the dangers (and benefits) resulting from the juxtaposition of genes from two previously unassociated genetic systems into a novel combination. The perceived need for analysis arises because this genetic association is not one that we believe occurs naturally. The need for an assessment of risk arises directly from our lack of understanding of the biological consequences from the release of such novel genotypes. In a way we are trying to 'run before we can walk' because we are attempting to exploit novel genetic combinations before we have a biological understanding of their effects in an uncontained ecological situation.

The problem of assessment is compounded by the fact that each GEM is inherently different in design detail and the genes cloned. As there are such divergences in their genetic origins and construction, it signifies that each, in the first instance, must be considered uniquely. Hence the case-by-case approach currently taken by most of the world's conscientious regulatory authorities. Common sense tells us that some GEMs will be benign and others not so. It is relatively easy to discriminate those at the extremes. It is the ones occupying the large middle ground that betray our lack of basic information. Once common themes are observed then a general basis for analysis can be adopted. It is worth stressing again that the transfer of genes *per se* does not equate with risk, an effect or even a problem. It is the lack of basic ecological and genetic information on survival, transfer, migration, persistence, growth and death of host cells and the inactivation of their DNA sequences, which presents an assessor with such difficult problems. It is this lack of ecological knowledge that ensures that risk assessment is discussed in terms of probabilities and not of certainties. Thus, the assessment of risk is an evaluation based on a balance between potential benefits versus perceived hazards.

Government authorities have a great deal of experience in giving approval for the environmental use of chemicals such as pesticides, herbicides and veterinary agents. There is also a substantial literature about how to assess the effects of such compounds on the environment (Domsch *et al.* 1983; Domsch 1984). The major difference between the release of GEMs and chemicals is that the latter are not self-reproducible. The information that produced the chemical is not a component of the chemical.

In most GEMs the biologically active component requires the cell to be metabolically active. This ensures the genome is maintained in an integral state and thus the genes are available for transfer into other organisms. It is not this potential for transfer into a range of unknown organisms that poses the problem in the assessment, it is the biological consequences of this event that cause concern.

The aim of this chapter is to consider briefly the aspects of these issues that need further work and analysis. From this understanding can come the quality assurance provisions that other industries are used to providing.

Gene transfer

Chapters 4, 5 and 6 cover these processes in great detail and explain the principles of each of the mechanisms involved. Table 12.1 summarizes the highest transfer frequencies reported in the laboratory and in the environment for these processes. Significantly, each of the processes is capable of transferring genes efficiently under certain conditions.

What relevance do laboratory-derived data have to any ecological situation? Until recently the general view was that it would be little. However,

Table 12.1. *Highest transfer frequencies obtained over about 24 h in laboratory experiments and those predicted[a] in situ*

Mechanism	Frequency	
	Laboratory	*In situ*
Conjugation		
Plasmid (self-transfer)	1	10^{-1b}
Plasmid mobilization/conduction	10^{-1}	10^{-3c}
Chromosomal gene	10^{-6}	10^{-8}
Plasmid retrotransfer	10^{-2}	10^{-4}
Chromosome retrotransfer	10^{-6}	10^{-8}
Transformation		
Plasmid	10^{-7}	10^{-9}
Chromosomal gene	10^{-5}	10^{-7}
Whole cells (plasmid)	10^{-4}	10^{-6}
Whole cells (chromosomal)	10^{-4}	10^{-6}
Transduction		
Plasmid	10^{-6}	10^{-6d}
Chromosomal (generalized)	10^{-5}	10^{-5e}
Chromosomal (specialized)	$\geq 10^{-1}$	10^{-3}
Plasmid (lysogen-mediated)	10^{-5}	10^{-5d}
Chromosomal (lysogen-mediated)	$\geq 10^{-1}$	10^{-5e}

[a] A reduction in frequency of at least 10^{-2} was noted in conjugation experiments done in the laboratory compared with crosses carried out unenclosed *in situ* (Fry and Day 1990). This factor was used in the calculations for this table.
[b] Unenclosed *in situ* data available for one plasmid (Bale *et al.* 1988).
[c] Enclosed *in situ* data Gealt *et al.* (1985).
[d] Enclosed *in situ* data Saye *et al.* (1987).
[e] Enclosed *in situ* data (after 4 days incubation it rose to about 10^{-2}) Morrison *et al.* (1978).

the analysis by Fry and Day (1990) clearly showed that plasmid transfer in the laboratory, in microcosms and *in situ* can differ by less than 100-fold. If such an extrapolation can be made for all other transfer mechanisms then frequencies of transfer *in situ* could be of the order predicted in the final column of Table 12.1.

Retrotransfer and mobilization

Conjugation is one of three systems that permit genes to be exchanged between bacteria. It is a parasexual process requiring cell-to-cell contact between donor and recipient cells (Willetts and Skurray 1980). The genetic information required for self-transfer is encoded on transfer proficient plasmids (Tra^+) and cells may be host to more than one plasmid (Bender and Cooksey 1986). These plasmids can also mobilize other plasmids (Tra^-) and chromosomal genes (see Chapter 4). Mobilization is a process by which smaller plasmids, under 30 kb (Thiry *et al.* 1984), which are not large enough to encode a conjugative system, may be transferred from one cell to another by the action of another Tra$^+$ plasmid.

IncP plasmids and some others can participate in the capture of genes from what would, in a normal mating, be considered the recipient cell (Mergeay 1990). This process is equivalent in efficiency to the normally orientated process. The net effect of retrotransfer or retrocapture, as this process is called, is a significant increase in the potential frequency with which genes in GEMs could be moved into indigenous bacteria. Until now our understanding of the processes required transfer of a conjugative plasmid into the GEM, which could then mediate further transfer of the novel sequence. Logically and practically we would expect these two steps to occur at a low frequency, but this assumption can no longer be made. The issue is further complicated because we do not know what proportion of natural plasmids are capable of mediating either (or both) of these processes. Neither do we know the host or transfer ranges of natural plasmids, or if these two characteristics always occur together. Thus, we need to find out what proportion of plasmids *in situ* are broad or narrow host range and which can mobilize themselves, others and chromosomal genes?

Transformation

Transformation is a process by which bacteria can acquire exogenous DNA (see Chapter 6). Some bacterial species selectively discriminate the DNA they will take up. In *Haemophilus* spp. the discriminatory mechanism recognizes the 'host' DNA by means of a short labelling sequence (Goodgal 1982). Many microbial species excrete/secrete DNA during their growth phase without significant associated cell lysis or death. Thus there is a possibility that gene transfer, between whole cells by DNAse-sensitive transformation systems, might be significant in natural habitats where there is cell-to-cell contact between actively growing and/or inactive cells. We have shown that pasteurized cells can act as donors and that they can do so at frequencies equivalent to a normal cell-to-cell mating (H.G. Williams, M.J. Day and J.C. Fry, unpublished data). We have also shown that the process can occur between viable growing cells; it is thus bidirectional. This process occurs at frequen-

cies of about 10^{-6} per recipient (Rochelle *et al.* 1988) for both plasmid DNA and chromosomal genes.

Transduction

Transduction is a process of phage-mediated gene transfer (Chapter 5). Many phages have a narrow host range. This is formally recognized by the ability of the bacteriophage to form plaques in a lawn of sensitive bacteria and not by their success in transferring genes. This ability to form plaques is usually confined to a few strains of one species. As this identification method does not recognize infections that do not produce plaques it is not a definitive test for transfer potential.

Some DNA phages can carry, either as a part or as full replacement of their genetic material, a fragment of host chromosome. A total replacement of the phage genome, by any part of the host cell genome, results in an almost equal likelihood of transfer for any gene. This is termed 'generalized transduction'. Other phages, e.g. the *Escherichia coli* bacteriophage lambda, are capable of transferring one or two genes that lie adjacent to the phage integration site (e.g. *gal* or *bio*). These so called 'specialized transducing phage' are formed as a result of an error in excision.

Bidirectionality of transfer mechanisms

There are two important points arising from this analysis. First, that there is a presumption that each of these processes is directional. Secondly, although each mechanism is easily definable in the laboratory they have not been considered to have an equal probability of occurring between members of a microbial community *in situ*. These presumptions ought now to be reconsidered. Finally, it must be remembered that these processes do not occur in isolation *in situ*, they act in concert.

Host factors

The transfer of genes from one cell into another does not mean that the incoming genes will survive the transfer and become established. There are several barriers to establishment (Wilkins 1990).

Restriction/modification

Restriction/modification (RM) is an enzymologically controlled process that permits a host cell to discriminate 'self' DNA from 'non-self' DNA. DNA that is enzymologically modified in one cell will survive the transfer to another similar cell. Cells carrying a different RM system will survey the incoming DNA and enzymologically cleave it at unmodified restriction target sites. Restriction can reduce the success of transfer by as much as 5–6 orders of magnitude (Roberts 1985).

Integration

If the new genetic sequence has homology with part of the new host genome then the normal recombinational processes of the cell will substitute a proportion of the material. Judging from laboratory experiments this will occur at a frequency of about 10 per cent (Hartman *et al.* 1960).

Establishment

If the new sequence is held on a replicon, and provided this is able to maintain itself within the new host cell, then the gene(s) will be stabilized. However, the maintenance of a plasmid through repeated cell cycles requires a coordination of plasmid replication with the host cell division. Without this coordination plasmid loss is high. There is a great variety of plasmid stability and replication mechanisms. These are important in deciding whether a plasmid will persist in the novel host.

Expression

For activity and maintenance of the plasmid within the host cell its gene sequences (replication, stability, phenotype, etc.) need to be expressed. It is well-known that gene sequences do not work equally well in different genetic backgrounds (Saunders and Saunders 1987). Thus, it is possible for replication to be adversely affected (very unstable) and for the novel genetically manipulated gene sequence not to be expressed. This will result in the loss of plasmid and/or the novel phenotype.

The potential for gene exchange within a microbial community is thus an aggregate of all these transfer processes.

Environmental factors

Temperature

Is there an equal distribution of plasmids with temperature optima for transfer at all biologically sensible temperatures or is there a bias in microbial populations for gene transfer systems towards certain temperatures? Could the temperature regime of the environment influence plasmid distribution, stability, the success of transfer, etc.?

Density factors

Factors influencing successful transfer by the various mechanisms are cell densities, numbers of phages, concentration of available DNA and nucleases, both free and bound to particulate matter and cells. Donor-to-recipient ratios are an important factor determining successful transfer (Rochelle *et al.* 1989).

Nutrient status

We know, for conjugation at least, that the nutrient strength of the mating medium positively influences transfer success. The high conjugal transfer frequencies observed in nutrient-rich habitats, like epilithon and rhizosphere, contrast with the very low frequencies observed in bulk soil, which is a nutritionally poor habitat. These data suggest that nutrient status affects transfer *in situ* as it does in laboratory experiments. However, we need more data from other aquatic and terrestrial habitats, and for other transfer mechanisms, to be sure of this conclusion.

Physiological state

Many bacteria in water and soil are not actively growing but are in a starved, moribund or otherwise stressed state (Morita 1982, 1985; Williams 1985). It has also been suggested that many bacteria can be viable but non-culturable (Colwell 1987). Recent work (Breittmayer and Gauthier 1990; Gauthier and Breittmayer 1990), has shown that the osmoprotectant glycine betaine helps *E. coli* both to survive well in sea water and to transfer plasmid RP4 more effectively, thus suggesting a link between survival and plasmid transfer potential. A lot more work of this type needs to be done if we are to predict both the survival and the gene transfer potential of GEMs in nature.

Frequencies

What does the term 'transfer frequency' mean? Many authors use transconjugants per recipient or per donor, where the count is defined at 0 h, to define the success of transfer. In practice, does the use of viable counts, taken at zero time, give different results (interpretations) to using 6-, 8- or 24-hour counts? If you choose 0 h and if mating is prolonged or delayed then cell growth can occur and those changes in densities, ratios and population sizes will influence the frequency calculations.

For these reasons we have always counted bacteria before and after mating. Our experience has been that results expressed in transconjugants per recipient obtained after mating have most faithfully reflected the trends observed from all the counts obtained before and after mating. Simonsen *et al.* (1990) have suggested an end-point method for estimating plasmid transfer frequencies in liquid, based on the mass action model of Levin *et al.* (1979). The end-point method gives a transfer frequency (ml/cell/litre), which is independent of initial density, donor-to-recipient ratio, mating time and substrate concentration. Although this model might give a genuinely useful measure of transfer frequency, the units are obscure and the frequencies are between 10^{-18} and 10^{-10}, compared to the range of 10^{-9} to 10^{0} transconjugants per donor or recipient from the conventional calculations. Thus, comparison with the large body of existing data would be difficult.

In some cases, such as with *Streptomyces* spp., for which viable counts are not very meaningful because of the mycelial nature of the organism, there can be no alternative to reporting counts without calculation of transfer frequencies. Clearly, more investigations are needed to decide on what are relevant measurements to make before we can calculate transfer frequencies in nature effectively.

Cell death

Chapter 8 discusses aspects of mortality and survival. Our present knowledge about transformation both *in situ* and in microcosms does not allow us to know how significant a genetic resource 'free' or bound extracellular DNA might be to competent cells (Chapter 6). It may be that successful populations deposit more DNA from cell death and secretion, which would allow growing populations

to 'sample' this gene bank as though it were part of an enlarged genome. Thus, cell death could be envisaged not always as a lack of success, but as a necessary process in the successful evolution of a community.

Monitoring populations

Selective isolation

Microbial geneticists have a long history of using antibiotic resistance genes as convenient, easy-to-use markers. These resistances, along with those for the heavy metals, have proved useful markers for the selective identification of the donor, recipient and transconjugant strains isolated from transfer experiments done in soil microcosms and *in situ* experiments in water. There would appear to be little risk in using such genes and hospital and municipal effluents have been discharging multiple antibiotic-resistant bacteria for many years into our waterways, with little sign of these bacteria causing harm. Even the least polluted habitats, like upland tarns in the English Lake District and pristine streams in the Andes, have been shown to contain many multiple antibiotic-resistant bacteria. (Jones *et al.* 1986; Bello *et al.* 1987). So antibiotic-resistant bacteria are part of the normal aquatic, and probably also the soil, microflora. Thus, it would seem unnecessarily restrictive for regulatory authorities to continue with their apparent current belief that antibiotic-resistance genes should not be used in the recombinant sequences used in GEMs.

Strain monitoring

Strain monitoring is currently one of the most intensively investigated areas of assessment and it is to be anticipated that, within the next 5 years, monitoring populations (by some form of hybridization and PCR-based procedures) will be done routinely with much greater efficiency than is possible now. This process will allow an accurate measurement of population size, distribution, growth, migration and perhaps even permit the measurement of transfer events. However, monitoring does not make dealing with a release any easier. Unlike cars or transgenic plants, but like rabbits in Australia and other rapidly reproducing, mobile organisms, microbes cannot be recalled if they are found to be faulty and have some unexpected and unwanted ecological effect. In this instance we will not need to monitor the GEM because the adverse ecological effect, if it ever occurs, will provide all the indication needed!

    There is a need for survey work to produce the basic genetic and ecological database necessary to develop an understanding of how the environment works. Although this basic work may appear mundane and uninteresting to funding authorities, because it is in competition with the more exciting areas of biotechnology, it remains worthy of support. Until such a database is compiled we will not develop the degree of security that allows the safe exploitation of these other products of science.

    To give an example of the type of work needed: the authors have a unique range of exogenously isolated plasmids, which could form the basis for an analysis of natural plasmids present in epilithic bacteria. No survey has yet been done on natural plasmids obtained in this way and details of their transfer range and abilities, genetic organization, stability functions, host range, phenotypes, etc. would give us an appreciation of their ecological importance.

Gene distribution

Some consideration should be given to the genetic source of all the sequences involved in the production of the GEM. For example, a safe GEM might be produced from the use of an appropriate transposable element. It would be sensible to use a transposable element found widely in the microbial populations in which the GEM is most likely to be used. It would consequently make little ecological sense to utilize a rare transposon like Tn5. If this approach is to be adopted we need to know a lot more about the types of plasmids, transposons, IS sequences etc. that are common in habitats like soil and water. Again this stresses the need for basic data, which can only be achieved by incorporating survey work into research programmes.

General questions

In the light of the above discussion, and to understand the relevance of the data in Table 12.1, additional questions about naturally occurring organisms in soil and water need to be answered. Some of these questions are:

1. Conjugation
   (a) What proportion of plasmids are self-transmissible?
   (b) What proportion of plasmids can mobilize, conduct or retromobilize genes?
2. Transformation
   (a) What proportion of natural bacteria can take part in the transformational process?
   (b) Are those that take up DNA (i.e. are transformable) the only types involved in DNA release?
3. Transduction
   (a) What proportion of phages is capable of transduction?
   (b) What are the real transfer ranges of phages?
4. General
   (a) Are natural strains recombination proficient ($recA^+$)?
   (b) Are all bacteria equally capable of donating and receiving DNA from different donor cells in the population?
   (c) Are all donors and recipients equivalent in their transfer efficiencies?
   (d) Are evolutionary boundaries at the species or at the genus level or higher?
   (e) Is it likely that some environments or conditions are better for transfer than others. Can we define these?
   (f) How important is gene exchange to microbial evolution? Does gene transfer really influence success, survival, growth and colonization *in situ*?
   (g) What is the mutation rate *in situ*?
   (h) Are different types of transposon and IS sequences more widespread in some environments than others?

Conclusion

The currently accepted regulatory view seems to be that all genes, whether recombinant or not, are highly likely to transfer between bacteria. If this view is maintained then it appears that much of the work into designing and incorporating safety features into GEMs is unnecessary. Although these strategies (Chapter 7) reduce horizontal gene transfer in laboratory experiments, they still only produce relatively small reductions in the likelihood of transfer in nature. By what factor and how effective this is needs to be determined through experiments in microcosms and *in situ*.

Generally, there is a need to avoid irrational concerns that ignore the significant body of case histories that indicates a lack of adverse effects on the environment (Dwyer and Timmis 1990). However, we need to be able to recognize and test those types of GEMs that we predict are likely to be a risk. This means we have to build up a larger series of case histories to establish the correctness of our approach and nullify the concerns. At present the sorts of questions posed in this chapter are not being addressed from an ecological view point. These ecological studies must be done to provide the basic data to allow the full application of the very detailed molecular biological approaches that have received so much investment in recent years. If financial support is not directed into this area of investigation then the proper rewards for the effort being expended will not be fully realized.

Acknowledgements

Part of the work reported in this chapter was supported by Natural Environment Research Council and by the European Economic Community Biotechnology in Action Program BAP 0379).

References

Bale, M.J., Day, M.J., and Fry, J.C. (1988). Novel method for studying plasmid transfer in undisturbed river epilithon. *Applied and Environmental Microbiology*, **54**, 2756–8.

Bello, H., Norambuenca, R., Mandaca, M.A., Montoya, R., and Zemelman, R. (1987). Inactivation of some third generation cephalosparins by β-lactamases of non-fermentative gram-negative bacteria isolated from freshwater of a remote environment. *Letters in Applied Microbiology*, **5**, 123–6.

Bender, C.L., and Cooksey, D.A. (1986). Indigenous plasmids in *Pseudomonas syringe* p.v. *tomato*: conjugative transfer and role in copper resistance. *Journal of Bacteriology*, **165**, 534–41.

Breittmayer, V.A., and Gauthier, M.J. (1990). Influence of glycine betaine on the transfer of plasmid RP4 between *Escherichia coli* strains in marine sediments. *Letters in Applied Microbiology*, **10**, 65–8.

Colwell, R.R. (1987). From counts to clones. *Journal of Applied Bacteriology*, Symposium Supplement **16**, 1S–6S.

Domsch, K.H. (1984). Effects of pesticides and heavy metals on biological processes in soil. *Plant and Soil*, **76**, 367–78.

Domsch, K.H., Jagnow, G., and Anderson, T.H. (1983). An ecological concept for the assessment of side-effects of agrochemicals in soil microorganisms. *Residue Reviews*, **86**, 65–105.

Dwyer, D.F., and Timmis, K.N. (1990). Engineering microbes for function and safety in the environment. In *SCOPE 44 Introduction of genetically modified organisms into the environment*, (Ed. H.A. Mooney and G. Bernardi), pp. 79–98. John Wiley and Sons, Chichester.

Fry, J.C., and Day, M.J. (1990). Plasmid transfer in the epilithon. In *Bacterial genetics in natural environments*, (Ed. J.C. Fry and M.J. Day), pp. 172–81. Chapman and Hall, London.

Gauthier, M.J., and Briettmayer, V.A. (1990). Gene transfer in marine environments. In *Bacterial genetics in natural environments*, (Ed. J.C. Fry and M.J. Day), pp. 100–10. Chapman and Hall, London.

Gealt, M.A., Chai, M.D., Alpert, K.B., and Boyer, J.C. (1985). Transfer of plasmids pBR322 and pBR325 in wastewater from laboratory strains of *Escherichia coli* to bacteria indigenous to the waste system. *Applied and Environmental Microbiology*, **49**, 836–41.

Goodgal, S.H. (1982). DNA uptake in *Haemophilus* by transformation. Annual Review of Genetics, **16**, 169–92.

Hartman, P.E., Hartman, Z., and Serlman, D. (1960). Complementation mapping by abortive transduction of histidine requiring Salmonella mutants. *Journal of General Microbiology*, **22**, 354–60.

Jones, J.G., Gardener, S., Simon, B.M., and Pickup, R.W. (1986). Antibiotic resistant bacteria in Windermere and two remote upland tarns in the English Lake District. *Journal of Applied Bacteriology*, **60**, 443–53.

Levin, B.R., Stewart, F.M., and Rice, V.A. (1979). The kinetics of conjugative plasmid transmission: fit of a simple mass action model. *Plasmid*, **2**, 247–60.

Mergeay, M., Springael, D., and Top, E. (1990). The potential for gene exchange between rhizosphere bacteria. In *Bacterial genetics in natural environments*, (Ed. J.C. Fry and M.J. Day), pp. 172–81. Chapman and Hall, London.

Morita, R.Y. (1982). Starvation-survival of heterotrophs in the marine environment. *Advances in Microbial Ecology*, **6**, 171–98.

Morita, R.Y. (1985). Starvation and miniturization of heterotrophs with special emphasis on maintenance of the starved viable state. In *Bacteria in their natural environments*, (Ed. M. Fletcher and G.D. Floodgate), pp. 111–30. Academic Press, London.

Morrison, W.D., Miller, R.V., and Sayler, G.S. (1978). Frequency of F116 mediated transduction of *Pseudomonas aeruginosa* in a freshwater environment.

Applied and Environmental Microbiology, **36**, 724–30.

Roberts, R.J. (1985). Restriction and modification enzymes and their recognition sequences. *Nucleic Acids Research*, **13** (Suppl.), r165–r200.

Rochelle, P.A., Day, M., and Fry, J.C. (1988). Occurrence, transfer and mobilisation in epilithic strains of *Acinetobacter* of mercury-resistance plasmids capable of transformation. *Journal of General Microbiology*, **134**, 2933–41.

Rochelle, P.A., Fry, J.C., and Day, M.J. (1989), Factors effecting conjugal transfer of plasmids encoding mercury resistance from pure cultures and mixed natural suspensions of epilithic bacteria. *Journal of General Microbiology*, **135**, 409–24.

Saunders, V.A., and Saunders, J.R. (1987). *Microbial genetics applied to biotechnology: Principles and techniques of gene transfer and manipulation.* Croom Helm, London.

Saye, D.J., Ogunseitan, O., Sayler, G.S., and Miller, R.V. (1987). Potential for transduction of plasmids in a natural freshwater environment: Effect of plasmids donor concentration and a natural microbial community on transduction in *Pseudomonas*

aeruginosa. Applied and Environmental Microbiology, **53**, 987–95.

Simonsen, L., Gordon, D.M., Stewart, F.M., and Levin, B.R. (1990). Estimating the rate of plasmid transfer: an end point method. *Journal of General Microbiology*, **136**, 2319–25.

Thiry, G., Mergeay, M., and Faelen, M. (1984). Back mobilisation of Tra^- Mob^+ plasmids mediated by various IncM, IncN and IncP1 plasmids. *Archives of the International Journal of Physiology and Biochemistry*, **92**, 64–5.

Wilkins, B.M. (1990). Factors influencing the dissemination of DNA by bacterial conjugation. In *Bacterial genetics in natural environments*, (Ed. J.C. Fry and M.J. Day), pp. 22–30. Chapman and Hall, London.

Willetts, N., and Skurray, R. (1980). The conjugation system of F-like plasmids. *Annual Review of Genetics*, **14**, 41–76.

Williams, S.T. (1985). Oligotrophy in soil: Fact or Fiction. In *Bacteria in their natural environments*, (Ed. M. Fletcher and G.D. Floodgate), pp. 81–110. Academic Press, London.

Index